Mathematical Modeling in Machine Transport Dynamics

Andrey Vladimirovich Eliseev
Irkutsk State Transport University, Russia

Nikolay Konstantinovich Kuznetsov
Irkutsk National Research Technical University, Russia

IGI Global
Scientific Publishing
Publishing Tomorrow's Research Today

Vice President of Editorial — Melissa Wagner
Director of Acquisitions — Mikaela Felty
Director of Book Development — Jocelynn Hessler
Production Manager — Mike Brehm
Cover Design — Jose Rosado

Published in the United States of America by
IGI Global Scientific Publishing
701 East Chocolate Avenue
Hershey, PA, 17033, USA
Tel: 717-533-8845 | Fax: 717-533-7115
Website: https://www.igi-global.com E-mail: cust@igi-global.com

Library of Congress Cataloging-in-Publication Data

Names: Eliseev, A. V. (Andreĭ Vladimirovich) author http://id.loc.gov/authorities/names/n2017053040 http://id.loc.gov/rwo/agents/n2017053040 | Kuznet͡sov, N. K. (Nikolaĭ Konstantinovich) author http://id.loc.gov/rwo/agents/n91043996

Title: Mathematical modeling in machine transport dynamics / by: Andrey Eliseev, Nikolay Kuznetsov.

Description: Hershey PA : IGI Global Scientific Publishing, [2026] | Includes bibliographical references. | Summary: "The monograph is devoted to the development of the scientific and methodological basis of modern mechanical engineering, formed by the scientific areas of machine science, the dynamics and strength of machines, applied mechanics, the theory of mechanisms and machines, vibration theory, applied system analysis and the theory of automatic control"-- Provided by publisher.

Identifiers: LCCN 2025000067 (print) | LCCN 2025000068 (ebook) | ISBN 9798337304472 hardcover | ISBN 9798337304489 paperback | ISBN 9798337304496 ebook

Subjects: LCSH: Machinery, Dynamics of--Mathematical models | Vibration--Mathematical models http://id.loc.gov/authorities/subjects/sh2010117786

Classification: LCC TJ173 .E45 2025 (print) | LCC TJ173 (ebook) | DDC 621.8/11--dc23/eng/20250320

LC record available at https://lccn.loc.gov/2025000067

LC ebook record available at https://lccn.loc.gov/2025000068

British Cataloguing in Publication Data
A Cataloguing in Publication record for this book is available from the British Library.

Table of Contents

Foreword

In the modern era of technological progress, the role of mathematical modeling in ensuring the efficient operation of transport machinery is becoming increasingly significant. The complexity of modern transport systems, combined with the need for precise dynamic analysis, necessitates the development of advanced modeling techniques that can accurately predict and optimize their performance.

The monograph «Mathematical Modeling in Machine Transport Dynamics» represents a comprehensive exploration of this critical field, offering a unique blend of theoretical foundations and practical applications. It serves as an invaluable resource for professionals seeking to understand and implement advanced modeling techniques in the analysis and design of transport machinery.

At the heart of this work lies the innovative application of structural mathematical modeling techniques, which are essential for analyzing the complex dynamics of transport systems. By employing advanced mathematical methods, including Laplace transformations and dynamic system analysis, the authors provide a robust framework for understanding and predicting the behavior of transport machinery under various operating conditions.

The monograph delves deeply into the intricacies of modeling transport machinery dynamics, examining the complex interactions between different system components. This comprehensive approach makes it an essential reference for engineers and researchers involved in the design, development, and optimization of transport systems across various industries.

The authors present a thorough examination of methods for evaluating the dynamic behavior of transport machinery, exploring techniques to analyze and modify their performance characteristics. They introduce advanced methodologies for determining motion parameters and system responses through the implementation of transfer functions and frequency analysis techniques. The utilization of multiple modeling approaches facilitates the discovery of novel dynamic phenomena and optimization opportunities.

This monograph stands as a pioneering work in the field of transport machinery dynamics, offering readers a comprehensive understanding of mathematical modeling techniques and their practical applications. It equips professionals with the knowledge and tools necessary to enhance the performance, safety, and efficiency of transport systems in an increasingly demanding technological environment.

By bridging the gap between theoretical foundations and practical implementation, this work serves as both a reliable guide and a source of inspiration for researchers, engineers, and practitioners involved in the development of modern transport machinery. It represents a significant contribution to the advancement of mathematical modeling techniques in the field of transport dynamics.

Pavel Abramovich Lontsikh

Doctor of Technical Sciences, Head of the Department of Theoretical Mechanics, Irkutsk National Research Technical University, Irkutsk, Russia

Preface

INTRODUCTION

Mechanical engineering and applied mechanics play a key role in technological development, influencing progress across industries. The study of machine dynamics has become an interdisciplinary field involving theoretical mechanics, mathematics, physics, computer science, and cybernetics. The complexity and multifunctionality of modern technical systems require a deep understanding of their behavior under dynamic loads. Machines operate in diverse conditions, from mining to high-tech production, where dynamic processes affect reliability and safety. Oscillatory phenomena can lead to part wear and emergencies, making dynamics research strategically important. This field is crucial for robotic systems and automated production lines requiring precise control and reliability. Research in machine dynamics is essential for modern science and practice, especially under high-speed and extreme conditions. Over recent decades, mechanical engineering has transformed due to automation and robotics. Robotic systems perform complex tasks with minimal human involvement in assembly, welding, and medical operations. Automatic technological complexes streamline production, enhancing safety and productivity in the chemical industry. Automation is also prevalent in construction, using technologies like 3D printers and robotic assemblers to improve efficiency. Even small industries adopt automation for packaging, quality control, and logistics, enhancing competitiveness and meeting safety standards.

Modern mechanical engineering is characterized by increasing automation and robotics, transforming industries and enhancing production efficiency and safety. Automation minimizes human error in managing complex technical systems, but requires fault-tolerant control systems with predictable behavior. Key to reliable automated systems is their adaptability to changing conditions. Modern control systems use AI and machine learning to adjust operations dynamically. For example, mining robots monitor vibrations and temperatures in real time to prevent failures. Safety depends on automated systems' ability to prevent dangerous situations, such as shutting down mechanisms under excessive loads. Redundancy and multi-level controls ensure safe operation even with component failures. Mechanical engineering methodology integrates multiple disciplines for designing and optimizing machines. Theoretical mechanics, mechanism theory, and oscillation theory analyze motion, forces, and vibrations. Control theory enables real-time regulation of dynamic processes. Systems analysis considers machines as integrated entities, accounting for mechanical, electronic, and software interactions. This comprehensive approach ensures the development of efficient and reliable technical solutions across various industries.

The contribution of Russian scientists to the development of the theory of machine dynamics has deep roots and continues to play a key role in the formation of modern scientific foundations in this field. The diversity of machines and mechanisms predetermines the diversity of dynamics problems; Many problems, the specifics of their research and solutions were reflected in the works of Russian scientists I. I. Artobolevsky, K. V. Frolov, R. F. Ganiev, V. A. Zinoviev, V. A. Shchepetilnikov, V. L. Veits, M. Z. Kolovsky, I. I. Vulfson, M. D. Genkin, N. I. Levitsky, S. V. Eliseev, S. F. Yatsun and others. The research conducted by outstanding representatives of the Russian school of mechanics and mechanical engineering became the foundation for many modern approaches to solving problems of dynamics.

Russian scientists developed fundamental methods for analyzing and synthesizing kinematic and dynamic characteristics of mechanical systems, particularly crucial for complex mechanisms like robots and automated manipulators. Their research advanced stability analysis and vibration control methods, especially important for extreme-condition machinery in mining. They pioneered vibration analysis for large structures such as cranes, pipelines, and buildings, forming the basis of modern vibration protection techniques that enhance industrial safety. Current Russian research focuses on dynamic analogies and mathematical modeling, creating universal models for managing technical processes. In modern technology, machine vibrations significantly impact reliability and safety. Originating from dynamic interactions and external forces, vibrations can cause material fatigue, reduced accuracy, and operational issues. This is critical in CNC machines, construction equipment, and transport, where vibrations may lead to quality defects, increased wear, and safety risks. Effective vibration protection remains essential for reliable machine operation.

Vibration protection is a set of measures aimed at reducing the vibration level to acceptable values. This is achieved through the use of special devices such as vibration isolators, dampers and vibration absorbers. These devices allow either completely isolating the vibration source from the rest of the system or absorbing the vibration energy, converting it into heat. For example, in construction equipment such as excavators or bulldozers, vibration isolators are installed in operator cabins to reduce the transmission of vibrations from operating mechanisms. Special dampers are used in aircraft engines that reduce the vibration level of turbine blades, preventing their destruction. However, vibration protection is not only a technical task, but also a scientific problem that requires a deep analysis of dynamic processes. For this purpose, mathematical models are developed that allow describing the behavior of the machine under vibration conditions and offering optimal solutions for their elimination. For example, structural mathematical modeling methods allow representing the machine as an oscillatory system with lumped parameters, which simplifies the analysis of its dynamic properties. Based on such models, it is possible to conduct numerical experiments, study the influence of various factors on the vibration level and test the effectiveness of various protection measures.

Vibration isolation, being a part of vibration protection, is a process aimed at preventing the transmission of vibrations from one element of the system to another. This is achieved through the use of elastic elements such as springs, rubber inserts or air cushions. For example, in rail transport, vibration isolation is used to reduce the transmission of vibrations from the wheels to the car body, which improves passenger comfort and reduces wear of the structure. In industrial installations such as pumps or compressors, vibration isolation is used to reduce the transmission of vibrations to the foundation, which prevents its destruction.

The tasks of vibration protection and vibration isolation are significant for modern mechanical engineering, since they allow not only to improve the performance characteristics of machines, but also to ensure their durability and safety. These tasks require an integrated approach, including both theoretical research and practical implementation of the developed solutions. The proposed monograph develops an original scientific approach based on the use of principles of dynamic analogies and methods of structural mathematical modeling. These principles allow creating a universal toolkit for analyzing and controlling the dynamic states of technical objects, considering them as oscillatory systems that perform small oscillations relative to the equilibrium position or steady-state process. The main idea of the author's approach is that any technical object operating under dynamic loading can be represented as a physical model describing its dynamic properties through a mechanical oscillatory system with lumped parameters. Such a model allows simplifying the analysis of complex dynamic processes and using well-developed methods of automatic control theory for their study. Structural mathematical modeling proposed by the authors is a method based on representing a mechanical oscillatory system as a structural diagram that is equivalent in dynamic terms to an automatic control system. This approach allows transferring the methods of analysis and synthesis developed for control systems to the problems of machine dynamics. For example, mechanical elements such as masses, springs, and dampers can be represented as transfer functions, which makes it possible to use the apparatus of circuit theory and automatic control to describe dynamic processes. This equivalence allows not only to simplify the mathematical description, but also to apply standard methods of analysis such as frequency characteristics, root locus plots, and optimization methods.

The principles of dynamic analogies play a key role in this approach, as they allow us to establish correspondences between various physical processes and use universal methods for their analysis. For example, oscillations of a mechanical system can be described similarly to electrical oscillations in a circuit with inductance, capacitance and resistance. This allows us to use methods of electrical circuit analysis to study mechanical oscillations, which significantly simplifies the process of modeling and analysis. In addition, the use of dynamic analogies allows us to unify approaches to solving problems for various types of technical objects, be they machines, mechanisms or equipment. The author's position in this monograph is disclosed in detail through a description of the methodology for constructing mathematical models and their equivalent transformations. The authors offer a systematic approach to structuring the dynamic properties of technical objects, which allows us to identify the key parameters that affect their behavior. For example, to analyze vibration processes in a machine, you can select the main units, such as the engine, transmission and working element, and present them as separate elements of the oscillatory system. Then, using structural modeling methods, you can study the interaction of these elements and assess their impact on the overall dynamics of the system. This approach allows not only to analyze existing systems, but also to develop new solutions to improve their efficiency.

One of the key advantages of the author's approach is the possibility of using universal methods to solve a wide range of machine dynamics problems. For example, structural modeling methods can be used to analyze vibration protection, control dynamic states and predict the behavior of technical objects under various operating conditions. This makes this approach especially valuable for modern problems related to the creation of complex automated systems, such as robots, automated lines and production complexes. Thus, the author's approach, based on the principles of dynamic analogies and structural modeling, is a powerful tool for solving machine dynamics problems and opens up new horizons for their research and development.

The history of the development of structural mathematical modeling methods in the context of machine dynamics and controlled systems begins in the middle of the 20th century, when engineers and scientists faced the need to analyze complex technical objects operating under intense dynamic loads. Initially, such problems were solved using classical methods of theoretical mechanics and oscillation theory, but with the increasing complexity of machine and equipment designs, it became obvious that traditional approaches were not effective enough to analyze multi-level dynamic interactions. It was during this period that the active development of new methods began, which could combine mathematical rigor with practical applicability. The first steps in the development of structural mathematical modeling were made at the Leningrad Polytechnic Institute (now Peter the Great St. Petersburg Polytechnic University) under the supervision of Professor, Doctor of Technical Sciences, Honored Scientist and Engineer of Russia Mikhail Zakharovich Kolovsky, an outstanding researcher of the mechanics of complex modern machines. In the late 1970s, his work laid the theoretical foundations for the analysis of dynamic systems with lumped parameters, which made it possible to translate complex mechanical oscillatory processes into structural schemes similar to control systems. These studies were aimed at finding universal methods that could be used to analyze both mechanical and electromechanical systems, which is especially important for machines operating under high loads and vibrations. One of the key achievements of this period was the creation of methods for converting differential equations of motion into transfer functions, which made it possible to use the apparatus of automatic control theory to analyze mechanical vibrations.

An important stage in the development of structural modeling methods was the collaboration between the Leningrad and Irkutsk Polytechnic Institutes, where the research begun in Leningrad was continued and deepened. In Irkutsk, under the supervision of S. V. Eliseev and his colleagues, practical methods for constructing equivalent structural diagrams for mechanical systems were developed, which made it possible to significantly simplify the analysis of dynamic processes. These methods found application in vibration protection and vibration isolation problems, where it was necessary not only to describe the dynamic properties of the system, but also to propose specific engineering solutions for reducing the vibration level. Particularly significant was the introduction of these methods into the practice of designing machines for the mining and construction industries, where vibrations are one of the main causes of premature wear of equipment. In the 1980s and 1990s, research in the field of structural mathematical modeling was further developed at the Irkutsk Computing Center of the Siberian Branch of the Russian Academy of Sciences (now the V. M. Matrosov Institute of System Dynamics and Control Theory of the Siberian Branch of the Russian Academy of Sciences). A team of scientists was created here, which focused on the application of the analytical apparatus of the theory of circuits and automatic control to solve problems of machine dynamics. One of the leaders in this area was Professor N.K. Kuznetsov, whose works were devoted to increasing the reliability of technical objects by optimizing their dynamic properties. His research made a significant contribution to the development of methods for analyzing and controlling the dynamic states of machines, which made it possible to create new approaches to the design of vibration protection devices and control systems. An important addition to this research was the work of

Candidate of Technical Sciences A.V. Eliseev, who focused on the problems of the dynamics of mechanical oscillatory systems with non-retaining links. His methods made it possible to solve complex problems related to the analysis of impact processes and contact interactions, which is especially important for machines operating in extreme conditions. These studies have found application in the design of transport systems, construction equipment and equipment for material processing. Later, the development of structural mathematical modeling methods continued at the Irkutsk State Transport University and the Irkutsk National Research Technical University, where new applications of the analytical apparatus of circuit theory and automatic control were developed for solving machine dynamics problems. These studies made it possible to create universal methods for analyzing and controlling dynamic processes, which have found wide application in various fields of industry.

Thus, the history of the development of structural mathematical modeling methods is an example of successful cooperation between scientists from different regions of Russia, who were able to combine their efforts to create a powerful toolkit applicable to analyzing and controlling the dynamic states of machines. These methods continue to develop and find new areas of application, which makes them one of the key areas of modern mechanical engineering.

The methods of structural mathematical modeling developed within the framework of the described approach have found wide practical application in various areas of transport dynamics. For example, in the automotive industry, these methods are used to analyze suspension dynamics, optimize aerodynamic characteristics and reduce noise levels in the cabin. In the railway industry, the methods are used to analyze the interaction of wheels with rails, reduce vibrations in passenger cars and increase the stability of trains at high speeds. The results of studies carried out using these methods not only made it possible to solve a number of complex technical problems, but were also protected by Russian patents. This confirms their novelty and significance for scientific and technical progress in the field of transport dynamics.

This monograph is a natural development of the ideas formed under the guidance of Professor S.V. Eliseev within the framework of the scientific school of active vibration protection and dynamic synthesis. She summarizes the results of many years of research carried out in close collaboration with the leading scientists of the group: doctors of Technical Sciences, professors A.P. Khomenko, Yu.N. Reznik, A.I. Artyunin, N.K. Kuznetsov, V.I. Sobolev, P.A. Lontsikh, S.V. Belokobylsky, as well as candidates of technical sciences, associate professors I.S. Sitov, A.V. Dimov, R.S. Bolshakov, V.B. Kashuba, Yu.V. Ermoshenko, E.V. Kaimov, R.Yu. Upyr, A.S. Logunov, D.N. Nasnikov, A.O. Moskovskikh, I.V. Fomina, A.N. Trofimov, M.V. Drach, E.A. Parshuta. The work continues to develop the key concepts of the school — methods of structural modeling, the theory of quasi-springs and virtual levers — in relation to the dynamics of modern machines with controlled elastic systems.

ORGANIZATION OF THE BOOK

The content of the monograph covers the class of problems of dynamics of mechanical engineering objects operating under conditions of intense vibration loading. The basis of the approach is the formation of mathematical models of technical objects that can be displayed using mechanical oscillatory systems with several degrees of freedom with concentrated parameters.

The first chapter introduces the basics of modern machine science, focusing on dynamics and vibrations in technical system design. It provides a framework for understanding complex technical object dynamics, analysis, synthesis, and vibration protection methods. The concept is based on computational schemes, dynamic states, vibration protection tasks, element interaction forms, and a structural approach. Computational schemes serve as a foundation for modeling machine and mechanism dynamics. The chapter emphasizes scheme adequacy to real objects, degree of freedom selection, distributed parameter consideration, and parameter reduction methods. Methods for assessing and controlling system states are analyzed, including parameter identification from experimental data, vibration frequency and form determination, and stability analysis. Traditional and modern vibration protection approaches are presented, with attention to elastic, dissipative elements, and lever connections in protection systems. The concepts of "generalized dynamic connections" and "generalized spring" extend standard element and connection concepts. The transition from classical methods to structural approaches using automatic control and circuit theory is highlighted. The importance of computational modeling (MatLab, ANSYS) and interdisciplinary approaches (mechanics, oscillation theory, control theory) is emphasized.

The second chapter focuses on the methodology of analyzing the dynamics of mechanical oscillatory systems through structural transformations and the concept of compacts. Key aspects: Structural schemes and compacts, Model construction methods, Lever interactions, Systems with solid links, Practical significance. It is shown that the representation of mechanical systems in the form of structural schemes of automatic control systems makes it possible, through the introduction of compacts, to aggregate typical elements into blocks with dynamic rigidity depending on frequency, which leads to a simplification of the initial models by reducing the number of degrees of freedom. Model construction methods include the development of algorithms for the transformation of structural circuits, the analysis of the influence of additional connections and lever mechanisms on the matrix coefficients of the equations of motion, as well as the study of modes of dynamic damping and resonance through transfer functions. Lever interactions are described through the mathematical representation of kinetic and potential energies, as well as on the basis of determining the reduced stiffness depending on the gear ratio of the lever and the transformation of lever systems into chain structures with compacts. Systems with solid links are modeled by analyzing motion conversion devices and accounting for articulated joints through "zeroing" coordinates. The chapter offers tools for analyzing systems with levers, hinges, and non-standard connections. The emphasis is placed on the systematization of transformations of computational schemes and the expansion of the conceptual framework (compacts, quasi-springs) to solve applied problems of machine dynamics.

The third chapter is devoted to the methodology of constructing and analyzing mathematical models of mechanical systems with various types of connections. The main aspects include: algorithms for converting structural circuits for various applications of forces, considered using the example of a system with 3 degrees of freedom; analysis of the influence of additional connections and lever mechanisms (levers of the first kind, gears) on matrices of coefficients of equations of motion; research of dynamic damping and resonance modes through transfer functions. The chapter provides a comprehensive approach to modeling mechanical systems, taking into account various types of interactions and their impact on the dynamic behavior of the system. The material is supported by specific examples, formulas, and graphical representations for a better understanding of the phenomena under study.

The fourth chapter explores innovative advancements in structural mathematical modeling methods, emphasizing their application to the dynamic analysis of mechanical systems. The chapter focuses on developing computational algorithms for determining natural frequencies and vibration modes, particularly in systems with distributed parameters. It introduces modern numerical techniques, such as the finite element method and boundary element method, to address complex dynamic problems. A significant portion of the chapter is dedicated to analyzing how external factors, such as elastic bonds, damping, and external forces, influence system behavior, supported by detailed graphical representations and numerical data. Additionally, the chapter examines transient processes and steady-state modes under various disturbances, offering a systematic approach to understanding the interplay between system parameters and dynamic responses. This material is particularly valuable for engineers and researchers seeking to apply advanced modeling techniques to real-world mechanical systems.

The fifth chapter delves into the intricacies of dynamic states in mechanical oscillatory systems with linkage mechanisms, focusing on the transformation and simplification of structural models. It investigates the elimination of coordinates as a formal technique to simplify systems with multiple degrees of freedom, enabling the representation of partial systems as more manageable structures, such as those with two degrees of freedom. The chapter highlights the role of quasi-springs and lever mechanisms in shaping dynamic properties, with detailed discussions on transfer functions, feedback connections, and the formation of dynamic invariants. Graphical representations, such as structural diagrams and design schemes, are extensively used to illustrate the transformation of motion and the interaction between system elements. Furthermore, the chapter addresses practical applications, including vibration protection and isolation, making it a valuable resource for specialists working on the optimization of mechanical systems with complex linkage configurations.

In the sixth chapter, a methodological basis is developed for constructing specific mathematical models of dyads, autonomous structural formations consisting of two mass — inertial elements. The peculiarity of dyads is that their elements can elastically interact and move in isolation. The possibilities of estimating the dynamic properties of dyads on mathematical models in two coordinate systems are investigated. A method for determining the motion parameters of dyads based on the introduction of a transfer function and a frequency energy function is proposed. Using different coordinate systems makes it possible to identify new dynamic effects that arise from new connections between elements. These relationships form the inertial and inertial-elastic type dependencies. The concept of dyads and the features of their interaction can become the basis for new methods of analysis and dynamic synthesis in the theory of mechanical vibrations and nanomechanics.

The seventh chapter examines the methods of structural mathematical modeling used to solve problems of assessment, formation and correction of dynamic states of mechanical oscillatory systems. Attention is paid to systems that allow for the implementation of one-way interactions of elements. The purpose of the study is to determine the motion parameters of the system, at which the contact between the elements of a solid body is not disrupted. The equations of static and dynamic equilibrium of the system, as well as the contact conditions of the elements are presented. The Dalembert principle is used to determine the dynamic component of the reaction. Particular attention is paid to estimating the range of variation in the dynamic component of the contact reaction and estimating the parameters that prevent contact between the elements of a composite solid during a period of steady motion.

The eighth chapter explores the methods of controlling the dynamic states of mechanical oscillatory systems (MCS) through the concepts of a generalized lever and the energetic formation of vibrational fields. Key aspects: control of dynamic states through a generalized lever; energy approach to the formation of vibrational fields; construction of smoothing and interpolating vibrational fields; theoretical foundations. A system apparatus is proposed for the assessment, formation and correction of dynamic states and vibration fields of technical objects. The methods are applicable in the tasks of vibration protection, vibration isolation and vibration technologies (transportation, material processing). The emphasis is placed on variational principles and energy criteria as tools for controlling complex vibrational processes.

CONCLUSION

The presented monograph reflects the details of the technology of constructing mathematical models, their equivalent transformations, which are essential for practical applications, providing the advantages of using the automatic control theory apparatus in assessing the dynamic properties of mechanical oscillatory systems. The authors carried out a laborious and meaningful work on the structuring of extensive and versatile material, which made it possible to thoroughly reveal the possibilities and features of the method of structural mathematical modeling as a promising direction in the development of the dynamics of transport systems

The monograph is devoted to the study of mathematical models of transport objects. It will be useful to specialists in the field of modern mechanical engineering, dynamics and strength of machines, theoretical and applied mechanics.

A.V. Eliseev

Irkutsk State Transport University, Irkutsk National Research Technical University, Irkutsk, Russia

N.K. Kuznetsov

Irkutsk National Research Technical University, Irkutsk, Russia

Acknowledgment

First and foremost, we pay profound tribute to the memory of our esteemed scientific supervisor, Professor Sergey Viktorovich Eliseev (1940–2021), Doctor of Technical Sciences, Professor, Honored Scientist of the Russian Federation. This work was initiated under his foundational guidance. Prof. Eliseev was a towering figure in mechanics and mechanical engineering. A graduate of the Omsk Institute of Railway Transport Engineers (1963), he earned his Doctorate in 1973 from the Institute of Mechanics of the Ukrainian Academy of Sciences with his seminal thesis on «The Theory of Active Vibration Protection Systems and Its Application». His distinguished career included extensive international research and significant contributions at institutions including the Irkutsk Computing Center (now Matrosov Institute of System Dynamics and Control Theory, Siberian Branch of the Russian Academy of Sciences), Irkutsk Polytechnic Institute (now INRTU, iPolytech), Sibexpocentre, and Irkutsk State Transport University (2002–2021). A full member of numerous prestigious academies, his prolific legacy includes over 800 scientific works, 40 monographs, and more than 100 patents. In his final decade, he pioneered methods of structural mathematical modeling for mechanical oscillatory systems, developing innovative concepts like quasi-springs and virtual levers, and uniquely integrating automatic control theory to solve complex dynamics problems under intensive loading. His intellectual leadership and pioneering spirit formed the bedrock upon which this project began.

A.V. Eliseev

Irkutsk State Transport University, Irkutsk National Research Technical University, Irkutsk, Russia

N.K. Kuznetsov

Irkutsk National Research Technical University, Irkutsk, Russia

Introduction

Modern mechanical engineering and engineering face complex challenges related to ensuring the reliability, durability and safety of technical facilities under intense dynamic loads. Vibrations, dynamic interactions of elements, oscillatory processes — these phenomena not only determine the operational characteristics of machines, but also form new requirements for methods of their analysis, design and control. The presented monograph is devoted to the fundamental and applied aspects of the dynamics of mechanical systems, combining the theory of oscillations, mathematical modeling methods and innovative approaches to dynamic state control.

Mathematical modeling in solving problems of the dynamics of technological machines in recent years has been developing intensively, which is reflected in the works of domestic and foreign authors. The increase in the requirements for the level of reliability of machines and the safety of their operation is associated with trends in increasing the productivity of technical objects and expanding the possibilities of their effective operation in conditions of intensive dynamic loading.

Many technical facilities for transport purposes operate in conditions of vibration influences of various nature, this is especially typical for technological processes of transport and vibration transportation. A wide range of dynamics tasks is formed when solving problems with vibration protection of equipment, instruments and equipment from vibration. The operation of many process machines, in which the working elements interact with the medium, is a source of a wide range of vibration disturbances. In general, the effect of vibration on the operation of technical objects is significant, which requires the development of a scientific and methodological basis that allows the development of appropriate methods and means for assessing dynamic states and searching for structural and technical solutions.

The theoretical foundations of solving problems of machine dynamics are determined by the developed apparatus of theoretical mechanics, theory of oscillations and theory of automatic control. The monograph proposes a structural mathematical modeling method based on dynamic analogies between mechanical systems and equivalent automatic control circuits. Its key advantage is the ability to analyze complex interactions of elements (including lever gears and motion conversion devices) through transfer functions and dynamic invariants, which is unattainable in classical approaches. The developed method is applied to the design of active vibration protection systems for vehicles with lever dampers, the analysis of the dynamic properties of quasi-springs in structural formations such as dyads, the formation of optimal vibration fields of technological machines based on variational principles, allowing to identify new effects of dynamic vibration damping and parametric adaptation of systems.

The noted generality of dynamic states in this monograph is interpreted as the dynamics of mechanical oscillatory systems with concentrated parameters and several degrees of freedom. The developed approach is the development of technical ideas laid down in the works of Kolovsky M.Z., Genkin M.D., Weitz V.L., which suggests the possibility of representing technical objects operating under conditions of dynamic (vibration) loading in the form of mechanical oscillatory systems with concentrated parameters.

In turn, the mechanical oscillatory system, due to the feedback effects, can be represented in the form of a structural diagram, dynamically equivalent to an automatic control system, based on the principles of dynamic analogies. As for the methods of structural mathematical modeling based on the apparatus of the theory of automatic control, it is characterized by the possibility of one-to-one reduction of structural schemes obtained on the basis of the rules of circuit theory to structural schemes built on the basis of the rules and technologies of the theory of automatic control.

The method developed in the monograph has certain advantages, since it allows solving problems of machine dynamics based on the use of transfer functions of systems that are universal for analyzing complex interactions, including the dynamic rigidity of quasi-springs and reactions in unilateral ties. This is due to the determination of reduced masses and stiffness by equivalent transformations of mathematical models, taking into account the coherence of movements, etc. Comparative analysis and review of structural and technical forms and features of elements of technical objects shows characteristic features of the structure of systems, facilitating transitions to the construction of structural models. This process is quite simple, but involves the transition from previously obtained time-domain differential equations to their Laplace transform-based counterparts.

The work is addressed to researchers in the field of machine dynamics, design engineers of transport and technological systems, as well as specialists in mathematical modeling.

Chapter 1
Analysis and Modeling of Vibrations and Dynamic States in Technical Systems:
Forms of Interaction Between Elements

ABSTRACT

The issues of vibration interactions, dynamic states and forms of interaction of elements in technical objects with an emphasis on modern mechanical engineering are considered. Basic methods such as the use of systems of equations related to the dynamic state are discussed, as well as a special focus on interdisciplinary interactions applied in robotics and mechatronics. It has been shown that modern technical systems require detailed descriptions, which leads to the use of idealized calculation schemes that help in the analysis and prediction of machine behavior under the influence of dynamic influences. Thus, the need for a systematic approach to the analysis and design of vibration properties of technical objects is justified, which requires the use of interdisciplinary methods, the latest mathematical and computational technologies.

INTRODUCTION

Modern machine science is a complex scientific and technical direction, closely connected in its interdisciplinary interactions with the theory of mechanisms and machines, the dynamics and strength of machines, instruments and equipment. In

DOI: 10.4018/979-8-3373-0447-2.ch001

recent years, methods of rational design and construction of machines, drives and their components have received significant development. This was reflected in the use of the methodological basis of theoretical mechanics, the theory of oscillations, which provide opportunities for the development of methods of system analysis and mathematical modeling in solving problems related to the corresponding calculations of machines, mechanisms and parts for reliability, endurance, vibration stability, etc.

Vibrations in the operation of technological and transport machines are quite multifaceted and their influence can lead to negative consequences, which stimulates the development of theoretical foundations in the methods of calculating machines, mechanisms and their parts, as well as methods and means of protection against vibrations and control of the dynamic state of machines as a whole.

The variety of solved problems predetermined the development of theoretical approaches and methods of mathematical modeling, among which methods of computational modeling using MatLab, MathCad, ANSYS, Maple and other application packages occupy a prominent place.

Modern machine science is an actively developing scientific field, in which more and more attention is paid to detailing ideas about the dynamic properties of machines from nodes and parts obtained on the basis of mathematical modeling. The basis of such approaches is the preliminary development and refinement of physical models, which most often take the form of mechanical oscillatory systems of one complexity or another. Such approaches are based on the idea that the studied models have a well-defined adequacy in relation to real processes, and typical elements of mechanical oscillatory systems are correlated with real prototypes of machine parts and mechanisms.

At the stages of preliminary research and subsequent design and calculations of technical objects - machines, equipment and their components, quite often there is a need to draw up idealized and simplified schemes, which could reflect the features of the physical nature of the elements, connections and structure essential for the tasks being solved, in general. For mathematical analysis and subsequent calculations, it is necessary to take into account the main factors that determine the dynamic state of a certain mechanical system, which generally contains elements and nodes of devices of different physical nature. The calculation scheme precedes the subsequent stage associated with the creation of a mathematical model.

In recent years, the dynamics of machines has been actively developing in a number of areas that are associated with the widespread use of methods of the theory of automatic control. This is reflected in the solution of theoretical and practical issues in robotics, mechatronics, vibration diagnostics, the development of methods and means to ensure the reliability and safety of the operation of complex technical objects. Considerable attention is paid to the tasks of protecting machines, equipment and apparatus from vibration external influences, which stimulates the development of scientific and applied research in the field of system analysis and the development of methods of dynamic synthesis in systems with a complex structure and a variety of dynamic interactions between elements of different physical nature.

CALCULATION DIAGRAMS IN TECHNICAL OBJECTS DYNAMICS PROBLEMS

In solving the problems of machine dynamics, methods of circuit theory are used, based on taking into account the peculiarities of constructing chain structures characteristic of many technical objects of a mechanical, electrical and electromechanical nature. Many problems of machine dynamics are solved in an interdisciplinary space formed by various approaches and methods of theoretical mechanics, theory of mechanisms and machines, theory of oscillations and applied system analysis. The development of methodological foundations for solving the problems of research, design and calculation of modern machines was reflected in the works (Dubrovsky & Podvolotskaya, 2008; Eliseev, Reznik, Khomenko, & Zasyadko, 2008; Khomenko, Eliseev, & Ermoshenko, 2012), focused on the apparatus of system analysis, vibration theory and control theory. Possible applications of theoretical approaches are associated with the dynamics of robotic systems, the tasks of the dynamics of various mechanisms and machines, the creation of methodological methods for calculating and ensuring the reliability of the functioning of technical objects in vibration conditions (Bolotin, 1978) .

The design scheme is used when designing a new object, when it is necessary to theoretically determine its characteristics in advance (for an oscillatory system - amplitude-frequency or phase-frequency characteristics, impedances, etc.); the calculation scheme is also needed to assess its condition, capabilities and compliance with the specified operating conditions.

The development of computer technology has created the conditions for controlling the dynamic state using mechatronics or active control devices that combine the capabilities of fast processing of information from sensor systems and their implementation by servos.

In this regard, it would be possible to note an increase in interest in the formation of mathematical models of processes for managing the state of technical objects, taking into account detailed ideas about the properties of systems and the conditions of their functioning, the influence of features of structural and technical forms and requirements for dynamic quality. Traditional approaches to solving these problems are based on the implementation of a rather complex system of interconnected stages of studying and constructing design schemes of objects that reflect their characteristic properties and features; constructing mathematical models of various forms: from systems of differential equations to structural schemes and chains of analogue systems; development of methods and means for solving various problems in assessing dynamic properties, appropriate selection of parameters, etc.

Traditional approaches based on the use of mathematical models in the form of systems of ordinary linear differential equations with constant coefficients have the potential necessary for engineering practice to adapt to the features of various technical objects and allow solving many problems. At the same time, in recent years, attention has noticeably increased to methods for studying the dynamic state of objects of protection against vibrations and shocks within the framework of ideas about the generalized tasks of vibration protection and vibration isolation. Mathematical models of protection objects in such directions are formed on the basis of the analytical apparatus of the theory of automatic control and circuit theory, which makes it possible to consider dynamic processes of various physical nature from a single position. Structural interpretations of protection objects in the theory of automatic control and the theory of mechanical circuits have differences between themselves, however, they can be a system of mating elements when fulfilling the determined conditions characteristic of mechanical oscillatory systems. Many tasks of machine dynamics are related to consideration of peculiarities of passage of external influences through some system, which predetermines the possibility of development of generalized approaches in which external influences of various nature are identified with signals (Eliseev, 2009).

Determination of frequencies and forms of free movements, analysis of dynamic stability and determination of forced vibrations for the designed and real objects, as a rule, begins with the choice of a design scheme, which is necessary to solve the problem with full consideration of all the properties of the real object. When solving dynamics problems, it is necessary to schematize the physical phenomena and properties of the system elements: for example, the forces of resistance to movement are usually taken proportional to speed or independent of speed (dry friction force). Forces arising in elastic elements at low vibrations are considered linearly dependent on displacements. Along with the schematization of physicalphenomena and properties of individual elements of oscillatory systems, the formation of the design scheme is largely due to the choice of the number of degrees of freedom.

PECULIARITIES OF APPROACHES IN JUSTIFICATION OF CALCULATION SCHEMES

When considering the features of the original physical system, determining the number of degrees of freedom and their corresponding generalized coordinates can be a rather difficult task, since real systems often have an infinite number of degrees of freedom. Therefore, several calculation schemes can be proposed for the same system, depending on the initial conditions, the required accuracy, the nature of the forces involved and the research tasks.

The reduction in the number of degrees of freedom is permissible only in cases where frequencies are considered that differ significantly in magnitude from the frequencies of the disturbing external influence or from the frequencies with which the system fluctuates at these initial deviations. It is necessary to take into account the features of parts or elements of the system.

So, for example, if the natural frequencies of the spring as an element with distributed parameters are significantly lower than the main frequency of the periodic external force, then the movement of the object resting on the spring can be expressed by one generalized coordinate. In this case, the mass of the spring should be small compared to the mass of the object. An object or load suspended on a weightless spring is usually taken as a design scheme.

If the mass m_{pr} of the spring is commensurate with the mass m of the load, and the frequency of change of the external force is close to the spring $\sqrt{k/m}\, k - rate)$, then the mass of the spring must be taken into account. Taking the movements of the spring coils proportional to the distance from the suspension point, a more accurate expression (Dimentberg & Kolesnikov, 1980) can be obtained to calculate the natural frequency of the system:

$$\omega = \sqrt{\frac{k}{m + \frac{1}{3} m_{np}}} \tag{1}$$

Methods of compiling equivalent calculation schemes with the given parameters of elements in the dynamics of machines are used quite often.

If the frequency of change of the external force is commensurate with the frequencies of free oscillations of the extension-compression of the spring, then the design scheme with one degree of freedom may be unacceptable; in this case, the degrees of freedom due to the tension-compression oscillations of the spring as an elastic system with distributed parameters are taken into account. The number of such degrees of freedom depends on the ratio of the frequencies of perturbation and the natural vibrations of the spring. With significant fluctuations during compression, the spring may lose stability - bend. It is known that buckling in such a case occurs when the frequency of change of the external force is twice, equal to or a multiple of the frequency of ω *tons* of free bending elastic vibrations of the spring (parametric excitation of vibrations). If the frequencies satisfy the above relationship, then in the calculation scheme it is necessary to introduce additional degrees of freedom that take into account the bending vibrations of the spring as an elastic system with distributed parameters.

When choosing a design scheme for solving the problems of forced vibrations of an object or load mounted on an elastic cantilever beam, it is necessary to take into account some features. The simplest design scheme can be a system with one degree of freedom in the form of a point mass suspended on a weightless elastic beam. The circuit corresponds to the lowest (fundamental) free frequency, which in this case will be overestimated. It is possible to clarify the fundamental natural frequency by attaching a part of the beam mass to the load mass and taking into account the moment of inertia of the load relative to the axis passing through the neutral line of the beam. If it is necessary to take into account the bending vibrations of a beam with higher natural frequencies, then the calculation can be based on the equations of transverse vibrations of an elastic beam. For a long beam in the equations, you can not take into account the cutting forces and moments of inertia for river sections of the beam:

$$EJ\frac{\partial^4 y(x,t)}{\partial x^4} + \rho F\frac{\partial^2 y(x,t)}{\partial t^2} = 0 \tag{2}$$

Where E-is the modulus of elasticity; the moment of inertia and the cross-sectional area of the beam, $J.F$-respectively; material ρ-density; y-deflection.

For a short beam, shear strain and rotational inertia of the cross sections may be considered. Equations of transverse vibrations of the beam in this case are taken in the form:

$$EJ\frac{\partial^4(y,t)}{\partial x^4} + \frac{\rho F\,\partial^2(x,t)}{\partial t^2} - \left(\rho J + \frac{EJ\rho}{kG}\right)\frac{\partial^4 y(x,t)}{\partial x^2 \partial t^2} + \frac{\rho^2 J\,\partial^4 y(x,t)}{\partial t^4} = 0 \tag{3}$$

where is the shear G-modulus; k- numerical coefficient depending on the cross-section shape (Dimentberg & Kolesnikov, 1980). The influence of the weight fixed on the beam can be taken into account in the boundary condition as a force of inertia, or as a force and moment of inertia. Examples of interaction of elastic beams with mass inertia elements, as well as methods of accounting for distributed elastic properties, taking into account the design features of springs are given in (Karamyshkin, 1988; Harris & Allan, 2002).

The choice of those generalized coordinates that should be taken into account (or which can be neglected) also depends on the nature of the forces involved.

CONSIDERATION OF VIBRATION EFFECTS FEATURES

When analyzing vibrations of more complex mechanical systems, the corresponding features are taken into account. So, when considering the shafts of turbomachines, the classical theory of a flexible shaft, based on the consideration of "small" bending displacements and linear relations between forces and displacements, gives sufficient accuracy In this case, critical speeds can be determined under the assumption of an undamped system. However, if the stability of motion in the supercritical region of rotation is analyzed, then internal and external friction must be taken into account. Internal friction, that is, friction between rotating elements, in the supercritical region generates instability, and external friction, that is, friction between rotating and stationary parts, contributes to increased stability, pushing the boundary of the instability region towards high rotation frequencies. In this case, the law of internal friction has an effect; which is tested most often experimentally.

Consideration of the connectedness of oscillations is most often taken into account in the analysis of many systems, which contributes to the refinement of the design scheme in comparison with the scheme in which oscillations of parts are considered separately or independently.

Depending on the design, the oscillating system can be simple, open or branched, as well as closed, annular. Such a system can have many natural frequencies, therefore, to determine the amplitudes of torsional vibrations, it is necessary to know the amplitudes of force effects consisting of many harmonics. In the presence of special couplings in the shaft system, non-linear properties are manifested, which should be reflected in the design scheme. Damping significantly reduces amplitudes in resonant and near-resonant regions of excitation frequencies.

The complexity of the studied system, in particular, in the study of machine structures, is very often determined not so much by the number of degrees of freedom as by the extent to which individual elements can be interpreted as standard elements: rods, beams, plates, etc. Determination of the design scheme for an active object is called identification. The design scheme of the operating object (system) is also necessary, for example, to clarify the parameters and evaluate the behavior in a given range of excited frequencies, determine resonant states, etc. Second, the operating object (system) may represent a subsystem to be included in a system that is only being designed, and upon joining to which associated oscillations may occur, which must be predetermined. Identification tasks are solved on the basis of the analysis of the dynamic behavior of the system observed either in the conditions of a special experiment or in normal operation conditions (Dimentberg & Kolesnikov, 1980).

IDENTIFICATION OF SYSTEMS UNDER CONSIDERATION

Most of the tasks and methods of identification are related to the study of systems for which the structure is considered known in advance; you only need to find the parameter values or certain functional dependencies of the adopted model. For mechanical systems, it is most often necessary to determine from the experiment the frequencies of free oscillations and the damping coefficient, which for linear systems can be considered constant within one form of free oscillations; for nonlinear systems, it can generally be a function of generalized velocities and coordinates. If it is possible to conduct a special experiment, the system can be subjected to certain types of influences (Veitz, Kachura, & Martynenko, 1971).

As a result of the impact, damped vibrations are excited, by which it is possible to judge the frequencies of free vibrations and damping. If the system is sufficiently pliable at the place of impact, then the method of striking ("soft" or "hard" impact) plays a role, depending on which, in addition to the pitch, certain high harmonics are excited.

Excitation of stationary harmonic oscillations with gradual change of frequency. It is used in dynamic tests of individual parts of machines (disks, shells, etc.). This method allows you to determine the amplitude-frequency and phase-frequency characteristics of the object, the characteristic of vibration shapes, as well as damping at individual frequencies.

The question of determining the unknown structure of a system is least developed in identification theory. Consideration of the nature of the structure depending on the spectrum of natural frequencies can be based on the position that the solution of the frequency equation of any discrete system reduced to the main coordinates is reduced to determining the values that C_{ii} turn the product $(c_{11} - \lambda^2)(c_{22} - \lambda^2)\ldots(c_{nn} - \lambda^2)$ (Bolotin, 1978).

The structure of the system is essentially related to the nature of the distribution of numbers c_{ii}, which are the eigenvalues of the oscillatory system. It is known that for a beam system, these numbers are either all (for beams of constant section) or, starting from some not very high number (for beams of variable section), are proportional to the squares of the numbers of the natural series. Multiple frequencies are not assumed in the beam. The same applies to a simple (unbranched) torsional system. These considerations may indicate a beam-type structure or a simple torsional structure. The numbers c_{ii} in the system can be completely arbitrary. In particular, some of them may coincide (multiples of frequencies). Such phenomena occur for plates, shells or arbitrary spatial systems of bodies and springs. In general, the spectrum of frequencies determines the structure ambiguously, therefore, additional data should be used when deciding on the structure.

Attention to calculation schemes is understandable, since their adequacy determines the accuracy of subsequent calculations, especially those related to the determination of forces arising at the points of connection of individual elements of the system with each other or at the points of contact with support surfaces. Such forces are called system reactions and can consist of static and dynamic components. Static components depend on conditions of pre-loading of the machine, its mechanisms and units, which depends on gravity forces of individual elements, pre-deformation of springs, mutual arrangement of elements in space, availability of leverage and other factors. Dynamic reactions occur during the movement of the system and are formed under the influence of inertial forces and can take large values that are many times higher than the reactions from static loads. In general, determining dynamic reactions is a rather difficult task. Firstly, this is due to the specific features of the mechanisms under consideration, determined by the type of the latter, the presence of a certain set of kinematic pairs, and the design of the links. Secondly, the magnitude of the reaction depends significantly on the type of external influences, the presence of connectivity between disturbances and the type of influences, in terms of whether they will be forceful or kinematic. Of great importance is the application of disturbance. In addition, the magnitude of reactions is greatly influenced by non-linearities that appear during significant deformations of system elements and movements along selected coordinates.

POSSIBLE FORMS OF VIBRATION REDUCTION AND DYNAMIC STATE MANAGEMENT

The design features of technical objects predetermine the variety of dynamics tasks, in which attention is paid to the protection and vibration isolation of objects from vibrations, vibration diagnostics, and the development of methods and means for ensuring reliable operation. Such views can be considered within the framework of generalized approaches to the possibilities of changing and managing the dynamic state of technical systems. Many technical objects have design schemes that are mechanical oscillatory systems with many degrees of freedom, which are characterized by the presence of such fragments as solids making movements in a plane or space, as well as systems of elastic-inertial and elastic-dissipative bonds distributed in a certain way in a plane or in space. With respect to such linkage systems, one might agree with the definition that they possess a spatial "metric." Such a representation of a mechanical oscillatory system is associated with the separation for a separate consideration of leverage, which can manifest itself in the form of relationships between system parameters or take the form of leverage mechanisms (Vargunin et al., 2006). The "metric" of mechanical oscillatory systems in the assessment of dynamic properties leads to the need to take into account a number of factors of geometric and dynamic nature that have not previously been taken into account. We also note that the proposed approaches are based, in particular, on a detailed consideration of the joints of solids involved in dynamic interaction. In the works(Buryan, Sorokin, & Galuza, 2011), in particular, it is shown that dynamic connections in mechanical oscillatory systems can be diverse and take complex forms of oscillatory structures, including mechanical chains, flat mechanisms (lever, screw, gear), which predetermined, to a certain extent, the introduction of the concept of "generalized dynamic connection" (Eliseev, Reznik, Khomenko, & Zasyadko, 2008). A number of detailed developments in this direction are presented in the works (Buryan, Sorokin, & Galuza, 2011; Eliseev, Khomenko, & Upyr, 2009). The development of the idea of generalized dynamic connections naturally leads to the introduction into the structures of mechanical oscillatory systems of controlled or active forces realized by servo drives. In (Ivanov, 2007; Eliseev, Belokobylsky, & Upyr, 2009), the features of mechanical oscillatory systems with controlled dynamic connections received detailed lighting, which stimulated theoretical developments and technical applications for the creation of active vibration protection systems, transport suspensions, pneumatic suspension systems and mechatronic spring suspension systems.

TASKS OF EVALUATION AND FORMATION OF DYNAMIC STATES

The natural development of approaches in the study of mechanical oscillatory systems focused on taking into account dynamic features associated with the introduction of generalized dynamic connections, including those based on the concept of "generalized spring" (Eliseev, Belokobylsky, & Upyr, 2009), was the proposal to expand the standard set of elements of mechanical oscillatory systems. If in the traditional form such a set of elements or elementary links consists of elastic (spring) and dissipative (damper - viscous friction) elements, then such elementary links as a differentiating link of the second order, as well as integrating links of the first and second orders (Eliseev, Reznik, & Khomenko, 2011) are introduced in the extended set. This allows us to significantly expand the range of dynamic properties of mechanical oscillatory systems and to search for and develop means of changing the state of systems in the desired direction. The generalized approach to the study of mechanical oscillatory systems is based on the use of the apparatus of the theory of automatic control and is reflected in (Eliseev, Moskovskikh, Bolshakov, & Savchenko, 2012; Khomenko, 2000) related to the development of the structural theory of vibration protection systems.

In essence, the initial design scheme in the form of some mechanical oscillatory system is compared (and further investigated) by a dynamically equivalent automatic control system. The latter is not unexpected, since the development of suspension and springing systems, including state sensors, converters, regulators, amplifiers and pneumo-, hydro- and electroautomatics devices, form a class of specialized automatic control systems or mechatronic systems (Eliseev, Reznik, Khomenko, & Zasyadko, 2008).

Design schemes of vehicles are systems with one, two or more degrees of freedom, among which a solid body (beam type) located on elastic supports is often used as a base model. As a rule, the motion of the base (kinematic disturbance) acts as external influences. The dynamics of such systems is quite well studied in the applications to the tasks of ensuring the dynamic quality of suspensions for road transport, which will be discussed later, as well as for railway rolling stock (Galiev, Nekhaev, & Nikolaev, 2010). At the same time, many issues have not yet received proper development and assessment of new opportunities in changing the dynamic properties of systems, which could be assumed when additional connections are introduced into systems, the use of non-traditional elementary links, dynamic vibration dampers, etc. Little studied is the question of taking into account the features introduced into the system by joints, hinges, levers and movement conversion mechanisms. Note that the properties of the systems significantly depend on the type and features of the introduced connections. The latter can lead to changes in the type of cross interactions between partial subsystems, ensure the emergence of forms of self-organization of movement of individual elements of the system, the emergence of non-traditional forms of dynamic cancellation, which, in general, opens up the possibility of rational selection of parameters that ensure the necessary dynamic qualities of systems.

METHODS AND MEANS OF VIBRATION PROTECTION AND VIBRATION ISOLATION TECHNICAL FACILITIES

As objects of vibration protection and vibration isolation, not only machines and equipment are considered, but also vehicles, as well as other objects operating under conditions of dynamic interactions with support surfaces (Khokhlov, 2002; Vershinsky, Danilov, & Khusidov, 1991), where common design schemes of transport objects are considered. Such systems may have depending on the detail of consideration (flat or spatial; longitudinal, transverse or vertical oscillations) of more than two tens of degrees of freedom and even in linearized form is described by a complex system of differential equations. For practical purposes, depending on the specifics of the tasks of transport dynamics,simpler models are used.

In the tasks of vibration protection of technical objects, traditional approaches are often used, in which industry specifics are reflected. At the same time, we can definitely note the formation of a direction focused on generalizing the tasks of vibration protection and vibration isolation based on the methods and means of the theory of automatic control and their corresponding structural research methods. Even on the example of a four-axle car, it can be seen that transport systems and their elements belong to complex technical objects, consisting, in turn, of units and assemblies.

Calculation schemes used to solve local problems of rolling stock dynamics are presented in Table 1; possible ways of using lever connections are given in Table 2 (Vershinsky, Danilov, & Khusidov, 1991).

The real needs of solving design and technical problems associated with the development of individual units of the system or the assessment of dynamic properties in certain types of motion, as a rule, are realized on simplified models, which in their basic forms take the form of mechanical oscillatory systems with one, two or more degrees of freedom.

As the main elements of vibration protection systems, elastic elements are used, as a rule, these are twisted springs or leaf springs, as well as devices for dissipating vibration energy in the form of hydraulic dampers, friction dampers and, in some cases, pneumatic shock absorbers.

Naturally, inertia-mass links of the oscillatory contour in the form of solid bodies having the shape of a material point act as system-forming elements; Beam type solid a solid body that is fixed at a point or has the ability to make more complex movements.

A complication with respect to the designated forms of representation of inertial mass elements is the connection of solids realized through hinges or kinematic pairs of III, IV or V classes, therefore, leverage plays a special role in calculation schemes. The latter are used in various chains, however, their fixation in the geometric space of spring suspension, in most cases, is not “geometrically” fixed. Thus, a mechanical oscillatory system, as a calculated scheme, in most problems of transport dynamics is an oscillatory structure of one or another complexity, in which a set of typical elements is used, consisting of elastic elements, dampers, mass inertia links connected by hinges and lever mechanisms (in a generalized sense) that help form the spatial structure of interaction of elements of the vibration protection system.

Figure 1. Design schemes in dynamic models of rolling stock

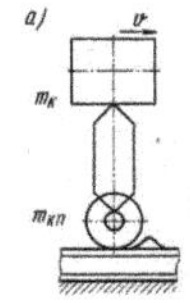

Schemes for calculating the force transmitted from the track to the wheelset during vertical oscillations

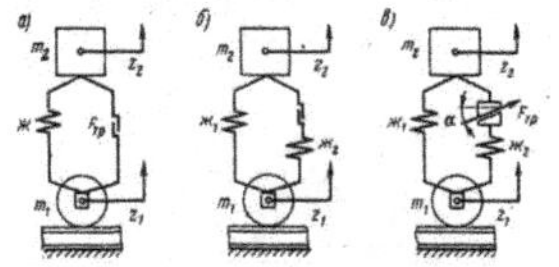

Diagrams and characteristics of elastic-friction links

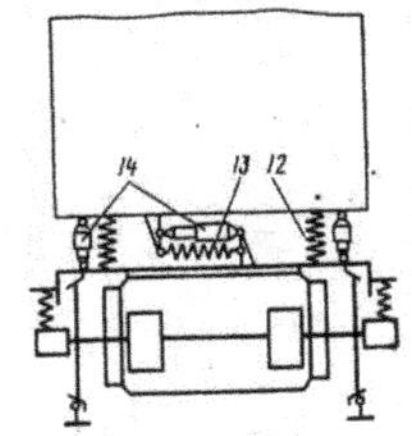

Design diagram of the truck and body instrument vibrations

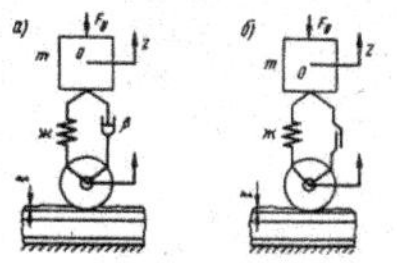

Model of locomotive vertical oscillations as a system with one degree of freedom

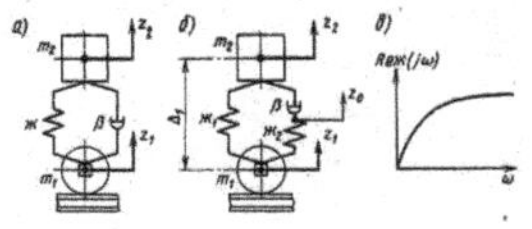

Diagrams and characteristics of combined elastic-viscous bonds

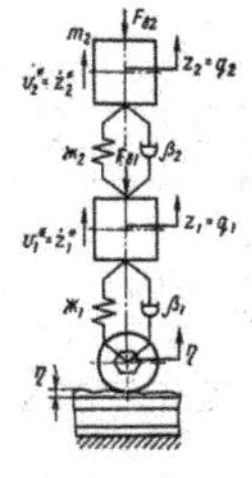

Linear single-axis model of a locomotive as a system with two degrees of freedom

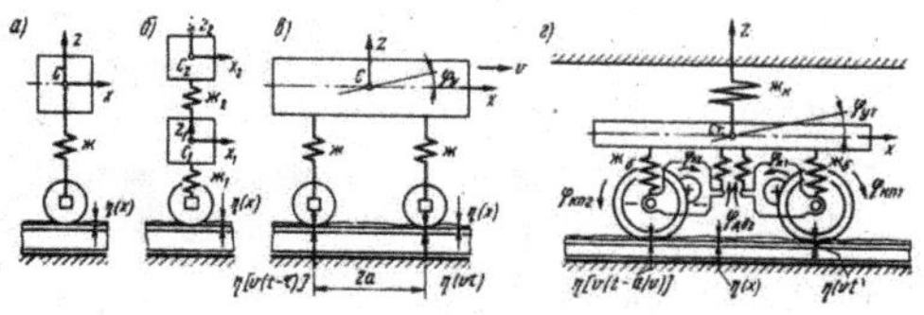

Flat locomotive vibration models

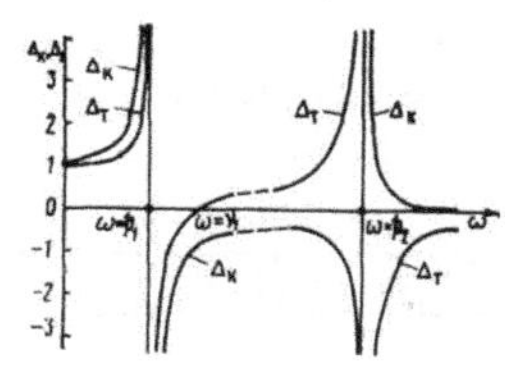

Amplitude frequency response curve

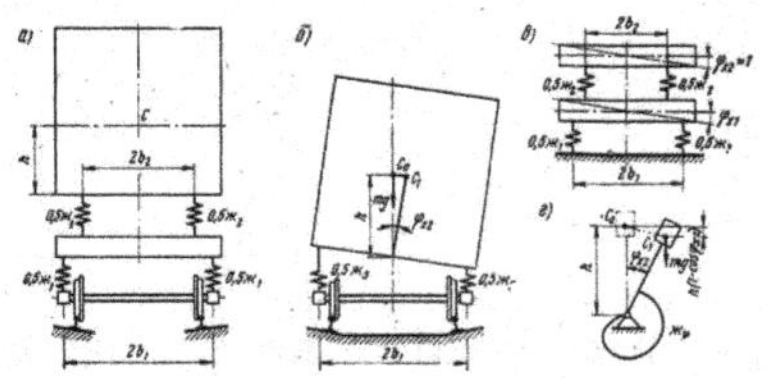

Diagram of two-stage spring suspension

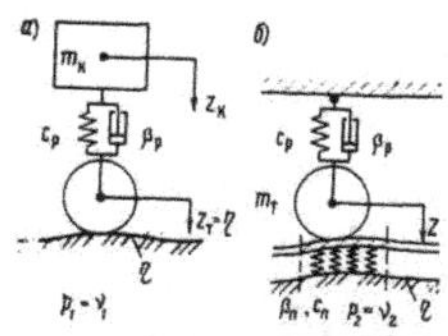

Design schemes for studying body vibrations

Figure 2. Models for the study of rolling stock vibrations taking into account lever-type links

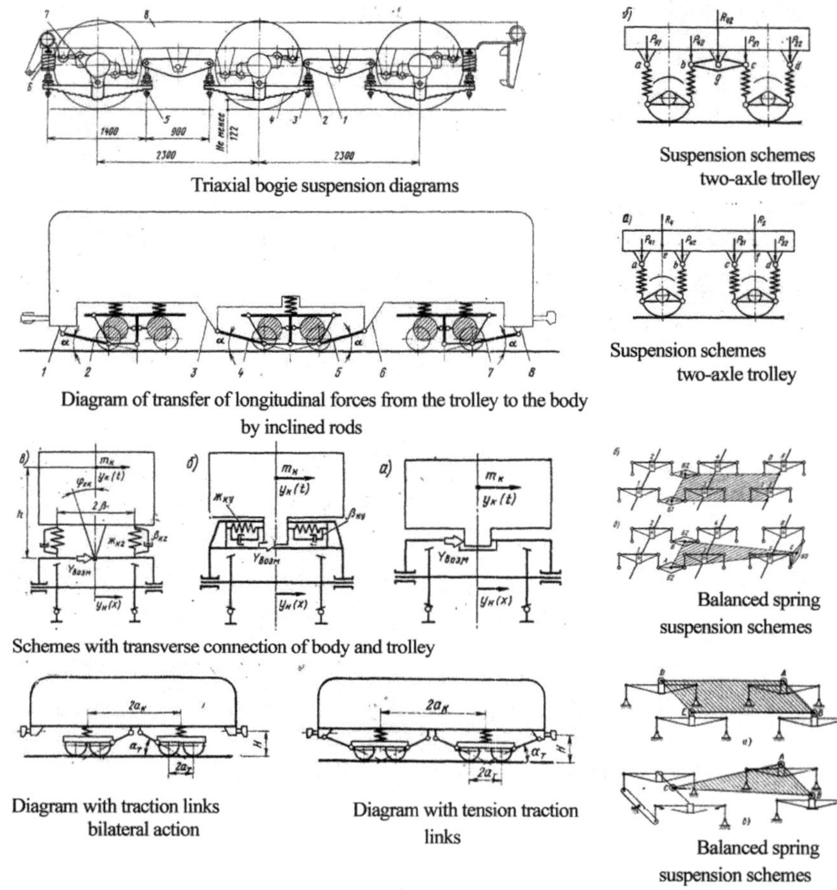

The generalized idea of lever interactions is based on the fact that the attachment points of springs, dampers, friction dampers are spaced apart in the spatial scheme, which suggests the possibility of features in the dynamics of the system as a whole. It should be noted that a fundamental consideration of the existing situation testifies to the great commonality of the tasks of vibration protection and vibration isolation. In this direction, there have recently been certain shifts that are associated with the use of approaches to the dynamics of controlled systems and the theory of active vibration protection systems, which initiates the expansion of the set of traditional means for controlling the dynamic state of objects by introducing servo drives or power actuators. This is reflected in the means of protection of vehicles, in particular in the form of controlled pneumatic spring suspension systems, mechatronic spring suspension systems and active vibration damping systems.

Leverage, the use of which is quite common in the design schemes of transport dynamics, is taken into account during the design and technical study of vibration protection systems. However, lever mechanisms, as a rule, are not considered as elements of a dynamic nature (Khokhlov, 2002).

Figures 3 and 4 (Lavrus, 2006) show the structural-lever forms of lever-articulated vibration isolators.

Figure 3. Design solutions for lever-hinge vibration isolators

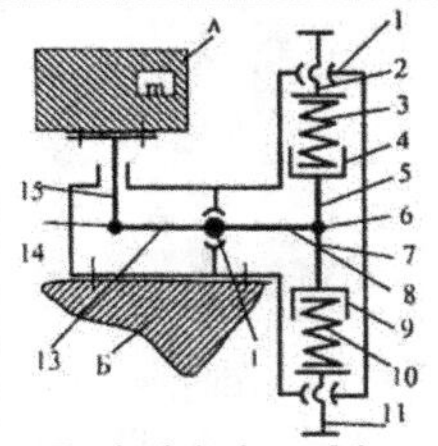

Typical design of the spring lever-articulated vibration isolator

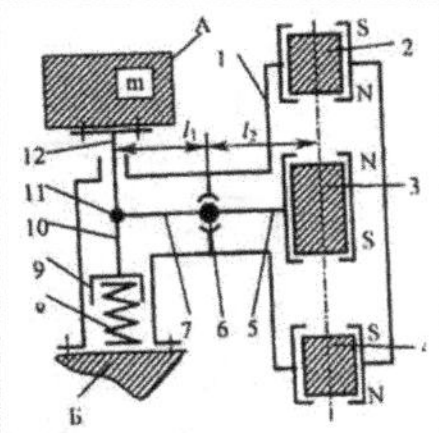

Spring lever-hinge vibration isolator with corrector of stiffness of elastic elements based on permanent magnets

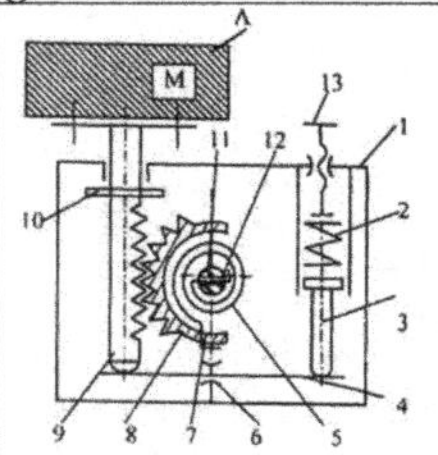

Spring-loaded vibration isolator with lever-type power dissipation mechanism

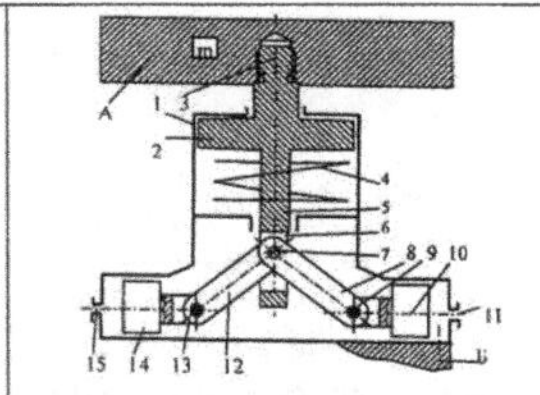

Lever-articulated vibration isolator with coil spring unit

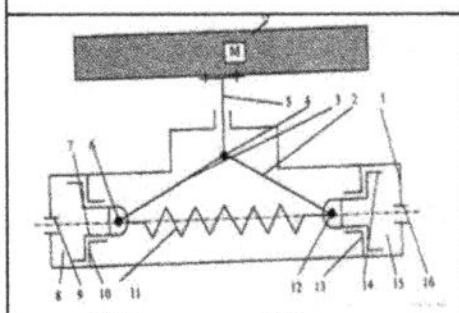

Diagram of lever-articulated vibration isolator in combination with gas damper

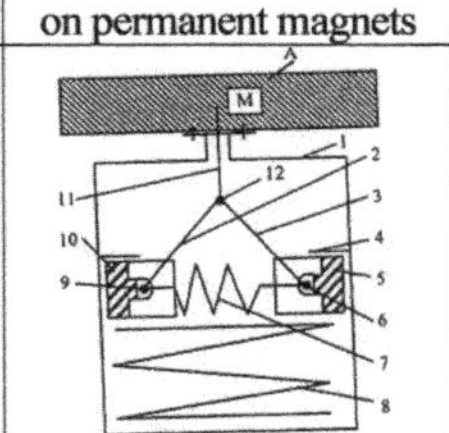

Diagram of lever-articulated vibration isolator in combination with dry friction damper

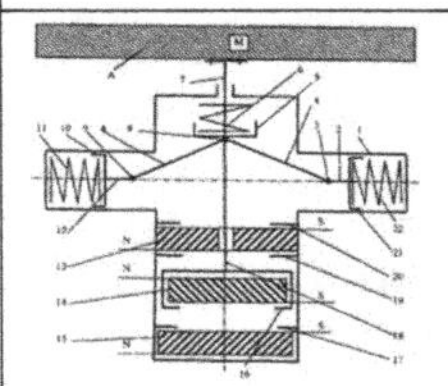

Diagram of lever-hinge vibration isolator in combination with magnetic damper

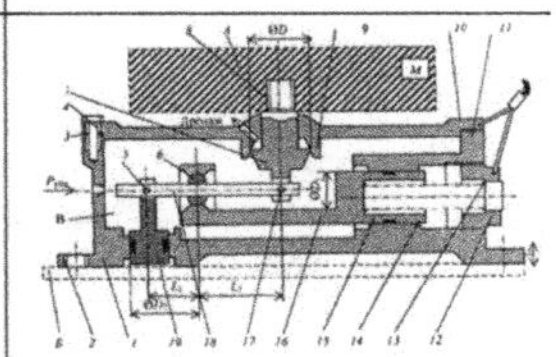

Vibration isolator with changeable axis of lever-piston mechanism

(Lavrus, 2006)

Figure 4. Design solutions for lever-hinge vibration isolators

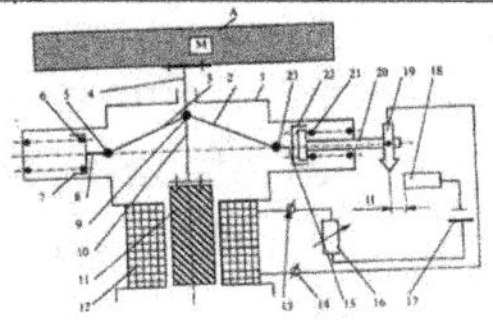 Diagram of lever-articulated vibration isolator in combination with electromagnetic damper	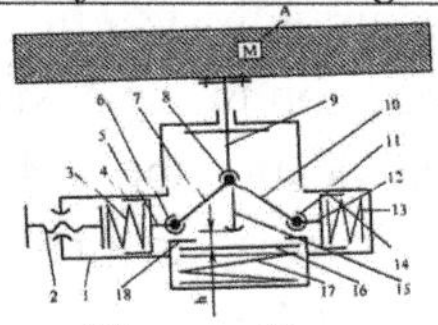 Diagram of lever vibration isolator with built-in anti-shock device	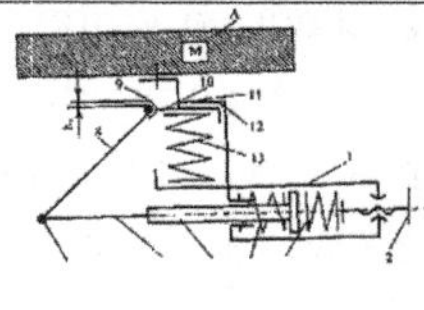 Diagram of lever vibration isolator with switchable rigidity of elastic element	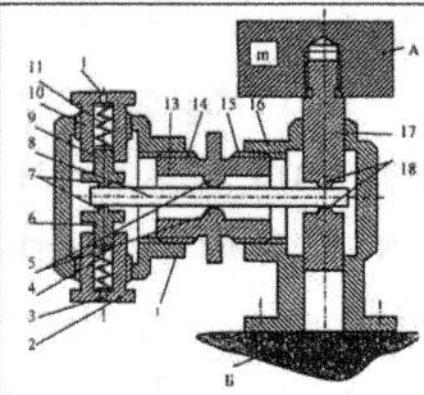 Lever-articulated vibration isolator with mechanical adjustment of stiffness of elastic elements
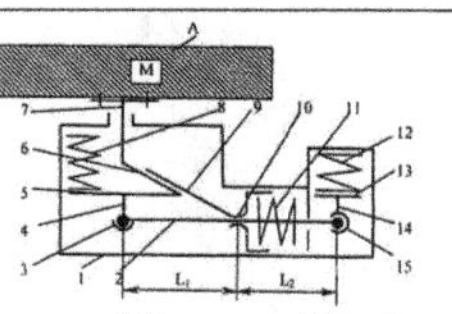 Diagram of lever-wedge vibration isolator with variable ("floating") axis of rotation of the lever	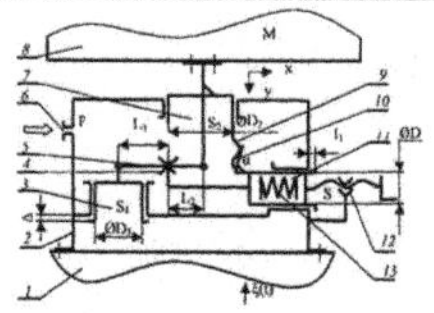 Design diagram of a two-arm lever-piston damper with a tunable axis of rotation in the form of a plunger orthogonally located relative to the pistons with a wedge loading mechanism	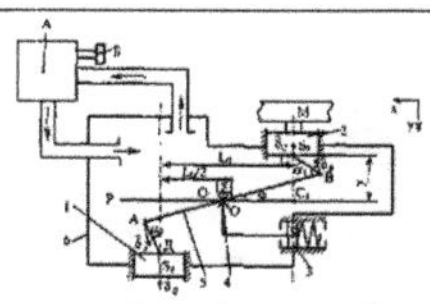 Design diagram of pneumatic vibration isolator with lever-piston control mechanism	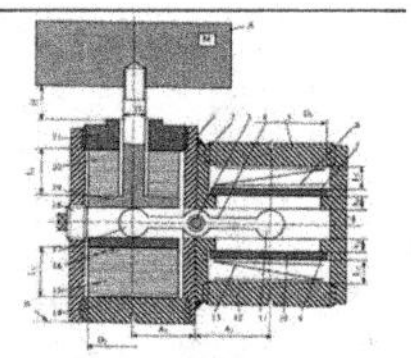 Design diagram of a vibration isolator based on a dry friction damper with an anti-shock lever-hinge mechanism

(Lavrus, 2006)

ON THE DIRECTION OF DEVELOPMENT OF THE THEORY OF VIBRATION PROTECTION SYSTEMS

The previous sections discussed the relationship of calculation schemes and their mapping in mathematical models related to the dynamics problems of various technical objects. The basis of generalized approaches to various problems of dynamics of machines, equipment of mechanisms and their elements is the orientation to compile mathematical models in the form of differential equations for mechanical oscillatory systems with concentrated or distributed parameters. As typical connecting elements of design schemes, elastic devices in the form of springs are used, as well as dissipative elements called dampers, shock absorbers, vibration dampers, etc. The possibilities of abstract representations of typical elements, and these include mass-inertial links, make it possible to implement the principles known in theoretical mechanics for obtaining mathematical models. The universality of the principles makes it possible to compile complex mathematical models, consistently presenting the possibilities of transition from a general model to models of its individual nodes, and then details (Eliseev, Reznik, & Khomenko, 2011; Eliseev, Khomenko, & Upyr, 2009).

The use of modern computing tools makes it possible to automate the process of compiling mathematical models in various forms of differential equations combined into systems. The subsequent solution of equations taking into account the initial conditions, non-linearities, properties and features of external perturbations is widely represented in the modern dynamics of machines.

Computational modeling occupies a significant place in the hierarchy of methods and means for assessing the dynamic properties of technical objects at various stages of their life cycle: from pre-design studies to operation. However, the complexity of technical objects, especially vehicles, often leads to complex mathematical models and the need to take into account the action of a large number of different factors, which causes the loss or "shading" of important and fundamental properties of objects. It could be noted that often in such cases the idea of the integral essence of the processes taking place is lost. In this regard, in the development of calculation schemes and the construction of mathematical models, great attention is paid to the issues of simplifying the models of linearization of equations, etc.

The possibilities of transition from computational schemes explaining the details of information transformation, as well as kinematic and power parameters characterizing the dynamic state of an object predetermine the development of generalized ideas about input and output signals interacting with a dynamic system, which creates conditions for combining the analytical apparatus of communication and control theory, and also expands the methodological basis of the theory of automatic control and modern circuit theory (Eliseev, Moskovskikh, Bolshakov, & Savchenko, 2012; Eliseev, Khomenko, & Upyr, 2009; Eliseev, Trofimov, Bolshakov, & Savchenko, 2012).

Preferences for the use of computational schemes observable in terms of the number of degrees of freedom when expanding and detailing knowledge about the properties of complex objects may be due to the fact that not all aspects of dynamic interactions in systems are studied. Note that in recent years, many methods have been developed and are used based on the use of concepts such as dynamic stiffness, dynamic compliance, complex resistance and compliance, electromechanical analogies that bring a number of new ideas to theoretical mechanics. Transfer functions based on Laplace and Fourier integral transformations play an important role in the evaluation of dynamic systems, which makes it possible to pay attention to the structure of the system, as well as to the possibility of taking into account various connections between partial systems. The dependence of the system structure on the choice of generalized coordinates allows generalizing a number of dynamics problems by introducing them into the class of problems of the theory of automatic control.

Vibration protection and vibration isolation of technical objects is, in a certain sense, a landmark problem, the solution of which could have a positive impact on interdisciplinary ties formed by the development of mechatronics, robotics, vibration diagnostics, etc.

COUPLINGS IN MECHANICAL OSCILLATORY SYSTEM

Connections can occur when moving between mass inertia elements and support surfaces or the environment, in addition, connections arise when moving one mass inertia element relative to another. The most common element is the spring (in its idealized representation as an element with linear properties). The elastic element forms a coupling reaction in the mechanical oscillatory system, the value of which linearly depends on the amount of deformation of the spring and its coefficient of stiffness (or elasticity). The spring in oscillatory systems provides, in a physical sense, the return of the system to its original state, which is considered as a form of implementation in systems of the feedback principle. In particular, in the simplest embodiments, the spring provides so-called negative feedback. The features of such approaches are reflected in sufficient detail in the works (Eliseev, Reznik, Khomenko, & Zasyadko, 2008; Eliseev, Khomenko, & Upyr, 2009). Mechanical oscillatory systems can be attributed to the most common form of design schemes of machines and mechanisms, equipment and apparatus subjected to periodic internal and external loads, in many tasks of modern mechanical engineering, which include, in particular, the tasks of vibration protection and vibration isolation of technical objects. Evaluation of dynamic interactions between elements of mechanical oscillatory systems makes it possible to determine on the basis of mathematical modeling methods to determine the capabilities of real objects to ensure reliable and safe operation.

Connections and their reactions can be attributed to the traditional sections of theoretical mechanics and such applications as the theory of mechanisms and machines, machine science. Many scientific and engineering areas of research were developed, which determined the analytical apparatus of the theory of oscillations, structural mechanics, material resistance, dynamics of machines, structures and structures. In recent years, finite element methods have received great development, based on a developed system of software for studying the deformed state of various objects. Although the works of their predecessors created and sufficiently developed the methodological foundations for solving many problems of the dynamics of machines and mechanisms, due to the specifics of movements, mechanical oscillatory systems, in particular, used for vibration protection and vibration isolation, turned out to be insufficiently studied. First of all, this relates to the issues of assessing static and dynamic reactions in the contacts of elements of oscillatory systems, taking into account the structure of systems and its effect on the distribution of the system and its effect on the distribution of the system of force interactions. Note also that mechanical oscillatory systems differ from mechanisms in the set of their typical elements and the specifics of the connection of the latter. In general, reduction makes it possible to determine a fairly representative range of tasks related to taking into account the features of mechanical oscillatory systems, if we keep in mind the possibilities of developing some generalized representations. One of such approaches can be determined using mathematical models of mechanical oscillatory systems in the form of structural schemes of dynamically equivalent automatic control systems with the subsequent application of operator research methods. It should be noted that the advantage of this approach is limited by the scope of linear theory and analysis of small movements, which significantly narrows the class of problems under consideration, however, such approaches have a certain prognostic potential and offer the possibility of finding and developing non-traditional methods and means of solving machine dynamics problems.

Static and Dynamic Reactions

The definition of bond reactions in mechanical systems can be attributed to quite advanced issues, if we mean various mechanisms consisting of solid bodies (links) connected to each other by kinematic pairs. In mechanism and machine theory, the definition of dynamic reactions is based on kinetostatic methods. At the same time, generalized approaches based on variational principles of mechanics and methods of analytical mechanics are used for complex mechanisms (Lurie, 1968).

At the same time, the issues of determining communication reactions can be classified as problematic in detailing ideas about the ratio of static and dynamic components of reactions. Such issues are of particular importance when considering the type of links. The main difficulty is associated with taking into account the irrepressible nature of the connections (Eliseev, Selvinsky, & Eliseev, 2015). Opening of links, even with preliminary compression with the formation of gaps and subsequent collisions. The identified problems are characteristic of transport dynamics.

In general, for the development of scientific directions in the dynamics of machines, taking into account the problems of robotics, vibration diagnostics and vibration technologies, dynamic interactions with the possibility of contact disruption are of particular interest. The theoretical foundations of such interactions are presented in (Lurie, 1968). The definition of static and dynamic reactions is most developed in methodological and methodological aspects, mainly focused on kinematic pairs of I, IV and III classes with force and kinematic closures with links in the form of solids. At the same time, many calculation schemes, as shown above, are mechanical oscillatory systems. Such systems usually have several degrees of freedom and can have a complex structure, which makes it difficult to determine both static and dynamic reactions. In this regard, difficulties are associated with the need to use complex mathematical models. Therefore, attention to methods for determining reactions seems quite appropriate and determines the current direction of scientific research.

With all the variety of known analytical approaches, which was reflected in the work(Eliseev & Artyunin, 2016), structural research methods (Eliseev, Reznik, & Khomenko, 2011; Khomenko, Eliseev, & Ermoshenko, 2012; Eliseev, Volkov, & Kukharenko, 1990) have certain prospects, within which mathematical models in the form of systems of linear differential equations are transformed after Laplace transformations into dynamically equivalent structural schemes of automatic control systems. The use of operator methods creates the basis for appropriate structural transformations and simplifications with the subsequent determination of the transfer functions of the system as a whole and its individual blocks. Existing developments (Eliseev, Reznik, & Khomenko, 2011; Kiryukhin, Tikhonov, Chistyakov, & Yablonsky, 2012) make it possible to develop original technologies for constructing mathematical models based on generalization of ideas about both elementary units and their structural formations (compacts), consisting of standard elementary units connected to each other. Note that structural formations (or compacts, as well as generalized springs (Eliseev, Kovyrshin, & Bolshakov, 2012)) have the property of entering into transformation with other elements or structures according to the rules of serial and parallel connection of structural circuit links(Khomenko, Eliseev, & Ermoshenko, 2012).

STRUCTURAL METHODS FOR BUILDING MATHEMATICAL MODELS OF MECHANICAL OSCILLATORY SYSTEMS

The possibilities of connecting the most common links in the form of springs and damping devices are reflected in works on the theory of mechanical chains (Druzhinsky, 1977; Genkin & Ryaboy, 1988; Samarsky & Mikhailov, 2005), the theory of vehicle suspensions, which is associated with the consideration of equivalent springs and dampers. Initial ideas about the possibilities of combining the properties of typical elements of mechanical oscillatory systems can be found in the works on the theory of mechanisms and machines, where the mentioned approaches were used to determine the reduced masses and reduced stiffnesses. The given characteristics have been used not only in solving the problems of dynamics of mechanical systems, but also in statics in connection with the tasks of reducing power factors (Eliseev, Parshuta, & Bolshakov, 2013) . The widespread task of reducing complex calculation schemes in the dynamics of units, in particular, power transmissions, is reflected in (Bakalov, 2006).

In a more developed form, the formation of structures from typical elements is considered in circuit theory and its applications in relation to electrical systems. The possibilities of simplifying the initial mechanical systems and their mathematical models, especially linear ones, have become the basis of a number of approaches related to the development of algorithms for building technologies for automatic compilation of mathematical models and the implementation of ideas for automating scientific research and designing systems for specific purposes (Eliseev, Zasyadko, Karpukhin, Kukharenko, & Metzeger, 1988; Kron, 1972).

In the dynamics of mechanical oscillatory systems, which are design schemes for many machine dynamics problems, structural research methods related to the use of the analytical apparatus of circuit theory and automatic control theory have certain prospects. The main idea of the method is to determine an expanded set of typical elementary links of mechanical systems as elements that have the property at the input signal in the form of displacement - to have an output signal in the form of force. Typical elements are devices with transfer functions of amplifying differentiating and integrating links of the first and second orders, from which structural diagrams of dynamically equivalent automatic control systems can be formed. The homogeneity of typical links in the "input" and "output" laid in the initial positions creates the basis for the construction of reduced structures consisting of blocks, which, in turn, can also be considered as peculiar typical elements, but more complex. Such elements may be referred to as typical second level elements, etc. In essence, taking into account the features of mechanical oscillatory systems, in which there is a dynamic interaction of mass-inertial and elastic-dissipative forces, naturally predetermines the possibility of structuring the system into blocks, taking into account external disturbances and research tasks. To the greatest extent, such approaches are convenient for systems with a finite number of degrees of freedom; to a lesser extent, block construction technologies are still developed for systems with distributed parameters, although works are known in this direction (Lurie, 1959; Khomenko & Eliseev, 2015). Analytical approaches and applied software complexes in the development of finite element methods have become widespread, however, many issues with the assessment of the dynamic properties of even linear mechanical oscillatory systems are still far from a final solution.

Figure 5, *a* - *f* show examples of calculation schemes used in solving problems of vibration protection and vibration isolation of objects. The diagrams reflect the variety of possible contacts (Figure 5, *a*) between the elements of the system and the form of introducing additional connections into the structure (Figure 5, b and *c* in the form of articulated linkage mechanisms), details of the dynamic interactions of such systems are given in (Vargunin et al., 2006; Eliseev, Khomenko, & Upyr, 2009; Ivanov, 2007; Lavrus, 2006).

Figure 5. Examples of the representation of elastic elements in mechanical oscillatory systems are a) the simplest system with ideally spring k; b) the system comprises a dyad m, l with two kinematic pairs; c) the system comprises two double articulated linkage mechanisms; d) the system has a complex vibration protection system of a solid body m, J and four elastic elements; e) the system has, in addition to the spring k, a chain k_1, m_1, k_2, m_2, k_3; f) generalized calculation scheme with a protection object of mass M and a generalized elastic element with reduced stiffness $k_{\text{пр}}$

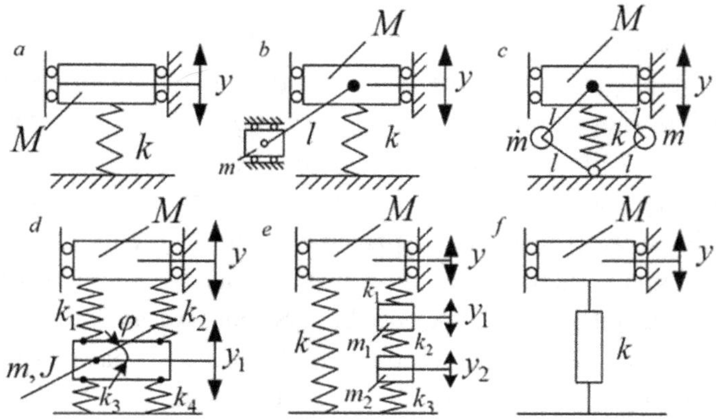

In complex objects of protection, design schemes can include solid bodies and mechanical chains (Figure 5, *d* and *e*), which can be transformed into a generalized form (Figure 5, *e*) using the concepts of reduced stiffness of generalized springs.

Principles of Connection of Typical Elementary Links

When analyzing the dynamic properties of complex objects, it is quite natural to choose design schemes in the form of mechanical oscillatory systems with one and two degrees of freedom, since more complex systems become inaccessible for analytical estimates. In this case, the mass of the object of protection, elastic and dissipative properties of the elements form the reduced properties of the system. When studying the dynamic properties of vibration protection systems related to taking into account the features of structures and systems of interaction of their fragments, such concepts as the reduced mass of an object or its component part, as well as the concepts of reduced stiffness, reduced damping, etc. are often used.

Thus, an elastic device, taking into account its real structural and technical design, can be considered as a fundamental concept based on an integrated idea of an elastic reaction seeking to return the system to its original position. Note that such an action can be associated with the presence of negative feedback in the elastic-oscillatory system (Eliseev, Volkov, & Kukharenko, 1990; Ryabov, Chernyshov, Eliseev, & Upyr, 2010) . At the same time, the operation of the elastic element is accompanied by overcoming the resistance of external and internal forces, which leads to dissipation of energy per cycle, reduces the amplitude of free vibrations and manifests itself, in a specific way, in forced vibrations. Due to the specifics of the research tasks, certain properties of springs can be considered at the appropriate level of detail and represent a parallel connection of several elements (elastic, dissipative and mass inertia). Each of the elements, from this point of view, is indivisible into simpler links by a representative of an expanded set of typical elements of vibration protection systems (amplifying, differentiating of the first and second order), as is done in (Eliseev, Reznik, Khomenko, & Zasyadko, 2008). Each of the mentioned elements at the input has an offset, and at the output (if we use the apparatus of the theory of automatic control) - a power signal or force. In the further development of such approaches, the rules of parallel addition and sequential connection of springs are used (Fig. 2). In the simplest forms, these connections take the form of kinematic pairs of various types, most often of class V, although as the structural type of elastic elements becomes more complicated, the contacts between the elements of the system can be determined by kinematic pairs.

Some Properties of Generalized Springs

The internal logic of the development of an approach based on the possibility of generalized ideas about the properties of springs predetermines not only the use of concepts about reduced stiffness (or elasticity), reduced resistance, reduced mass, but also the introduction of the concept of a generalized spring, which has already been mentioned (Eliseev, Kovyrshin, & Bolshakov, 2012; Genkin & Ryaboy, 1988), as some device of a fairly general form, acting as a link between the object and the base, between fragments of the system, etc.

Table 1. Typical links and their connections (switching)

Link name	Parallel connection	Consecutive connection
I. Proportional link (spring)	$k = k_1 + k_2$	$k = \dfrac{k_1 \cdot k_2}{k_1 + k_2}$
II. 1st order differentiating link	$b = b_1 p + b_2 p = p(b_1 + b_2)$	$b = \dfrac{b_1 p \cdot b_2 p}{b_1 p + b_2 p} = \dfrac{b_1 \cdot b_2}{b_1 + b_2} \cdot p$
III. 2nd order differentiating unit	$L = L_1 p^2 + L_2 p^2 = p^2 (L_1 + L_2)$	$L = \dfrac{L_1 p^2 \cdot L_2 p^2}{L_1 p^2 + L_2 p^2} = \dfrac{L_1 \cdot L_2}{(L_1 + L_2)} \cdot p^2$
IV. 1st order integrating link	$A = \dfrac{A_1}{p} + \dfrac{A_2}{p} = \dfrac{A_1 + A_2}{p}$	$A = \dfrac{\frac{A_1}{p} \cdot \frac{A_2}{p}}{\frac{A_1}{p} + \frac{A_2}{p}} = \dfrac{A_1 \cdot A_2}{(A_1 + A_2)} \cdot \dfrac{1}{p}$
V. 2nd order integrating unit	$A = \dfrac{A_1}{p^2} + \dfrac{A_2}{p^2} = \dfrac{A_1 + A_2}{p^2}$	$A = \dfrac{\frac{A_1}{p^2} \cdot \frac{A_2}{p^2}}{\frac{A_1}{p^2} + \frac{A_2}{p^2}} = \dfrac{A_1 \cdot A_2}{(A_1 + A_2)} \cdot \dfrac{1}{p^2}$

Figure 6. Mechanical system conversions

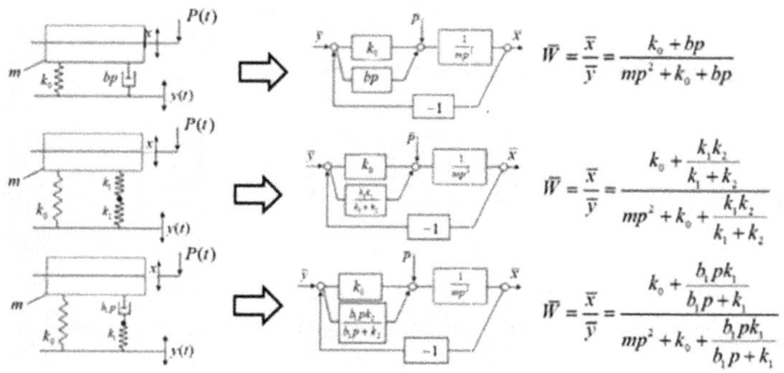

A formalized representation of a generalized spring can give a transfer function in the form of a fractional rational expression

$$W_{об.пр}(p) = \frac{a_0 + a_1 p + ... + a_n p^n}{b_0 + b_1 p + ... + b_m p^m} \tag{5}$$

where coefficients $a_i(i = \overline{0,m})$and $bj(j = \overline{0,m})$are determined by specific parameters of structural and technical design of the generalized spring. The convenience of the proposed approach lies in the possibility of a formalized assessment of the entire variety of solution space associated with the choice of the transfer function of a generalized spring, which can take the form of combinations in the connection of the above mentioned elementary links, represent various mechanisms or oscillatory structures (Table 1, Figure 6). Details of such representations are presented, for example, in works (Khomenko, Eliseev, & Ermoshenko, 2012; Eliseev, Khomenko, & Upyr, 2009; Eliseev, 2009).

Generalized Ideas About Problems of Technical Objects Considered as Mechanical Oscillatory Systems

Developing ideas about the basic properties of dynamic systems, we note that the introduction of additional connections is realized through the attachment of some elements to others. At the same time, it is important to take into account the features and design forms of connections, since they often determine the dynamic capabilities of systems. For example, it can be noted that the introduction of a dynamic damper as an additional device corresponds to the introduction of so-called additional bonds into the mechanical oscillatory system. The design features of additional links can be seen in the fact that additional links must have attachment points, that is, take the form of dual elements, and provide the ability to create from them in a dynamic sense, some structures or blocks, which is determined by joints.

In some problems of dynamics of articulated bodies or mechanical systems with joints of elements, it is advisable to set the configuration or structure of the system with parameters. $q_1 \dots q_n \dots q_{n+m}$.They are called redundant generalized coordinates.

Figure 7. Examples of design implementations of vibration protection systems with elastic-inertial connections

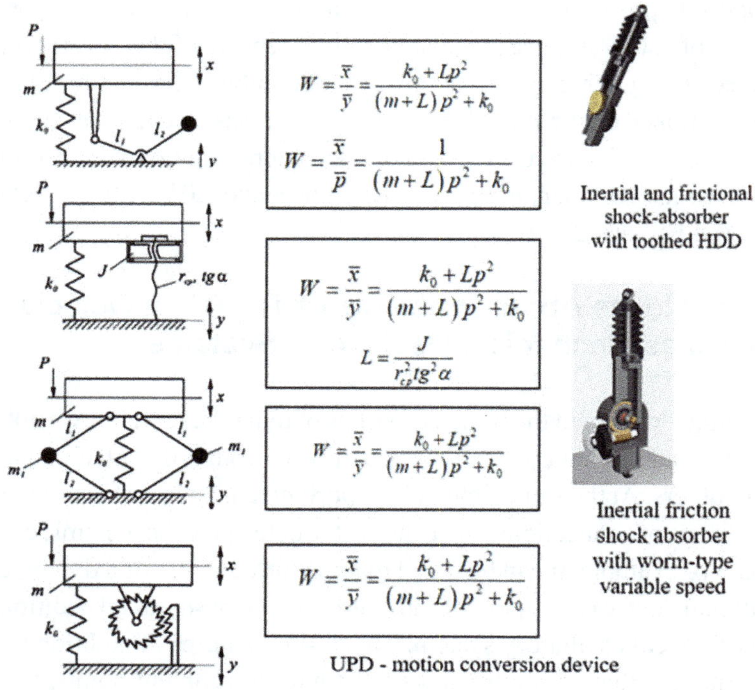

$q_1 \dots q_{n+m}$ There are m relations between the $n+m$ parameters, which can contain time:

$$F_k(q_1 \dots q_{n+m};\ t) = 0\ (k\ =\ \overline{1,m}). \tag{6}$$

These relations, representing generalized coupling equations, must be solvable with respect to m of $n+m$ quantities q_s, and the defect of the Jacobian matrix must be zero. In general, the system configuration is given by n independent parameters $q_1 \dots q_n$; however, when determining the velocities of its points, you can use not directly generalized velocities $q_1 \dots q_n$, but some of their linear forms with coordinates depending on generalized coordinates.

$$\omega_s\ =\ a_{s1}\dot{q}_1 + \dots + a_{sn}\dot{q}_n\ =\ 0\ (s\ =\ \overline{1,n}). \tag{7}$$

The concept of "links" given above corresponds to ideas about interactions between elements of mechanical systems, as described in analytical mechanics. In the physical sense of communication in systems, if mechanisms and devices are considered in which the structural and technical implementation has features, they are defined as kinematic pairs of different classes.

Structural models of mechanical oscillatory systems are an adequate representation of mathematical models in the form of linear differential equations with constant coefficients. Although the linearity of the properties imposes limitations on the possibilities of approaches, their development is quite justified when considering small movements relative to the positions of stable static or dynamic equilibrium. Moreover, a significant part of the information about the dynamic state of the system can be obtained on the basis of knowledge of the elastic property. The system is considered under the assumption of a small resistance force. With all the development of the theoretical basis for solving the problems of determining the forces in the elements of the connection of the links of mechanisms, the issues of determining dynamic reactions in mechanical oscillatory systems are less studied. Reactions also occur in static states when determining the equilibrium position, in mechanical oscillatory systems with several degrees of freedom. The relevance of research and study of the features of the formation of dynamic reactions during dynamic interactions of system elements and elements with surfaces is of particular importance for situations in which connections in systems become non-holding (Eliseev, Selvinsky, & Eliseev, 2015).

The development of structural approaches based on the use of the apparatus of the theory of automatic control has the capabilities of a prognostic plan, within which proposals for expanding the set of typical elements and the form of their interaction with each other are possible, which can be implemented in various applications.

REFERENCES

Bakalov, A. N. (2006). *Theory of chains*. Nauka.

Bolotin, V. V. (Ed.). (1978). *Vibrations in technology: A reference book of 6 volumes* (Vol. 1). Mechanical Engineering.

Buryan, Y. A., Sorokin, B. N., & Galuza, Y. F. (2011). Active hydromechanical system for damping car vibrations. *Omsk Scientific Bulletin*, *103*, 122–126.

Dimentberg, F. M., & Kolesnikov, K. S. (Eds.). (1980). *Vibrations in technology: A reference book in 6 volumes* (Vol. 3). Mechanical Engineering.

Druzhinsky, I. A. (1977). Mechanical chains. *Mechanical Engineering (New York, N.Y.)*.

Dubrovsky, V. A., & Podvolotskaya, N. I. (2008). Problems of dynamics and strength of actuators and machines. *Problems of Mechanical Engineering and Automation*, *3*, 35–42.

Eliseev, A. V., Selvinsky, V. V., & Eliseev, S. V. (2015). *Dynamics of vibration interactions of elements of technological systems, taking into account non-holding bonds* [Monograph]. Nauka.

. Eliseev, S. V. (2009). *Mechatronics of vibration protection systems*. Elements of the theory. IrGUPS. (Deposited at VINITI 27.11.09. No. 738-V 2009)

Eliseev, S. V. (2009). New approaches in the theory of oscillations. Tasks of controlling the dynamic state of oscillatory systems based on the introduction of additional connections. In Wiener readings: Materials of the IV All-Russian Scientific-Practical Conference (pp. 46-60). IrSTU.

Eliseev, S. V., & Artyunin, A. I. (2016). *Applied theory of oscillations in problems of dynamics of linear mechanical systems*. Nauka.

Eliseev, S. V., Belokobylsky, S. V., & Upyr, R. Y. (2009). Generalized spring in the tasks of machinery and equipment. Zbirnik Naukovy Pratsy (Galuzev Machine-Working, Budivnitsvo). *Poltava National Technical University Imeni Yuri Kondratyuk*, *1*(3(25)), 79–89.

. Eliseev, S. V., & Ermoshenko, Y. V. (2012). *Articulations of links in the dynamics of mechanical oscillatory systems*. IrGUPS.

Eliseev, S. V., Khomenko, A. P., & Upyr, R. Y. (2009). Mechatronics of vibration protection systems with lever links. Modern Technologies. Systems Analysis. *Modeling*, (3), 104–119.

Eliseev, S. V., Kovyrshin, S. V., & Bolshakov, R. S. (2012). Features of constructing compacts of elastic elements in mechanical oscillatory systems. Interactions with system elements and connection forms. Modern Technologies. *Systems Analysis. Modeling*, (4(36)), 61–70.

Eliseev, S. V., Moskovskikh, A. O., Bolshakov, R. S., & Savchenko, A. A. (2012). Possibilities of integration of methods of circuit theory and theory of automatic control in machine dynamics problems. *Science and Education: Electronic Scientific and Technical Publication*, 6. http://technomag.edu.ru/doc/378699.html

Eliseev, S. V., Parshuta, E. A., & Bolshakov, R. S. (2013). Transfer functions of mechanical oscillatory systems. Ability to assess reduced stiffness. *International Journal of Applied and Basic Research*, (1C), 11–18.

Eliseev, S. V., Reznik, Y. I., Khomenko, A. P., & Zasyadko, A. A. (2008). *Dynamic synthesis in generalized problems of vibration protection and vibration isolation of technical objects*. ISU.

Eliseev, S. V., Reznik, Y. N., & Khomenko, A. P. (2011). *Mechatronic approaches in the dynamics of mechanical oscillatory systems* [Monograph]. Nauka.

Eliseev, S. V., Trofimov, A. N., Bolshakov, R. S., & Savchenko, A. A. (2012). The concept of feedback in the dynamics of mechanical systems and dynamic damping of vibrations. *Science and Education: Electronic Scientific and Technical Publication*, (5). http://technomag.edu.ru/doc/378353.html

Eliseev, S. V., Volkov, L. N., & Kukharenko, V. P. (1990). Dynamics of mechanical systems with additional connections. *Science*.

Eliseev, S. V., Zasyadko, A. A., Karpukhin, E. L., Kukharenko, V. P., & Metzeger, P. K. (1988). *Software-algorithmic complex of optimization design of complex vibration protection systems. In Sixth All-Union Conference on Control in Mechanical Systems* (p. 61). Institute of Applied Mechanics of the Academy of Sciences of the Ukrainian SSR.

Galiev, I. I., Nekhaev, V. A., & Nikolaev, V. A. (2010). *Methods and means of vibration protection of railway crews*. SEI "Educational and Methodological Center for Education in Railway Transport."

Genkin, M. D., & Ryaboy, V. M. (1988). *Elastic-inertial vibration isolation systems*. Ultimate capabilities, optimal structures. Nauka.

Harris, C. M., & Allan, G. (2002). *Shock and vibration handbook.* McGraw-Hill.

Ivanov, B. G. (2007). Development of methods for calculating the dynamics and strength of transport equipment units with lever-hinge links [Doctoral dissertation, Samara State Transport University]. Samara.

Karamyshkin, V. V. (1988). Dynamic vibration dampers. *Mechanical Engineering (New York, N.Y.).*

Khokhlov, A. A. (2002). *Dynamics of complex mechanical systems.* MIIT.

Khomenko, A. P. (2000). *Dynamics and control in the tasks of vibration protection and vibration isolation of mobile objects.* ISU.

Khomenko, A. P., & Eliseev, S. V. (2015). On some properties of dynamic damping of vibrations in mechanical systems. Modern Technologies. Systems Analysis. *Modeling, 2*(46), 8–19.

. Khomenko, A. P., Eliseev, S. V., & Ermoshenko, Y. V. (2012). *System analysis and mathematical modeling in mechatronics of vibration protection systems.* IrGUPS.

Kiryukhin, A. V., Tikhonov, V. A., Chistyakov, A. G., & Yablonsky, V. V. (2012). Active vibration protection - purpose, principles, condition. Active vibration protection and noise insulation of pipelines and experimental research. *Problems of Mechanical Engineering and Automation,* (4), 102–110.

Kron, G. (1972). *Study of complex systems in parts - diakoptics (Trans. from English).* Nauka.

Lavrus, V. V. (2006). Improvement of pneumatic lever-hinge systems of railway transport [Doctoral dissertation, Orel State Technical University]. Orel.

Lurie, A. I. (1959). *Operational calculus and its applications to problems of mechanics.* Nauka.

Lurie, A. I. (1968). *Analytical mechanics.* Nauka.

Ryabov, I. M., Chernyshov, K. V., Eliseev, S. V., & Upyr, R. Y. (2010). Experimental studies of inertia-friction dampers in vehicle suspension systems (part 1). *Bulletin of Irkutsk State Technical University, 2*(42), 146–154.

Samarsky, A. A., & Mikhailov, A. P. (2005). *Mathematical modeling: Ideas.* Methods. Examples. Fizmatlit.

Vargunin, V. N., Gusarov, V. N., Ivanov, B. G., Levchenko, A. S., & Mulyukin, O. P. (Eds.). (2006). *Design and calculation of lever-articulated means and units.* SamGAPS.

Veitz, V. L., Kachura, A. E., & Martynenko, A. M. (1971). Dynamic calculations of machine drives. *Mechanical Engineering (New York, N.Y.)*.

Vershinsky, S. V., Danilov, V. I., & Khusidov, V. D. (1991). Wagon dynamics. *Transport*.

ADDITIONAL READING

Den-Gartog, D. P. (1960). *Mechanical vibrations*. Fizmatgiz.

Eliseev, A. V., Kuznetsov, N. K., & Mironov, A. S. (2024). Energy characteristics in the evaluation of elastic and lever connections in the dyad of a mechanical oscillatory system. *Proceedings of the MAI*, 135.

Eliseev, A. V., Kuznetsov, N. K., & Mironov, A. S. (2024). Elastic bonds in the formation of the structure and dynamics of mechanical systems under conditions of vibrational loads of a forceful nature. *Proceedings of MAI*, 137.

Eliseev, A. V., Sitov, I. S., & Kuznetsov, N. K. (2024). Formation of dynamic effects in mechanical oscillatory systems within the framework of the generalized lever concept, taking into account additional kinematic relationships. The system. *Metals Technology*, *1*(61), 10–20.

Gatti, P. L. (2021). *Advanced mechanical vibrations: Physics, mathematics, and applications. CRC Press*. Taylor & Francis.

Karnovsky, I. A., & Lebed, E. (2016). *Theory of vibration protection*. Springer International Publishing Switzerland. DOI: 10.1007/978-3-319-28020-2

Lyan, I. P., Panovko, G. Ya., & Shokhin, A. E. (2019). Comparative analysis of the energy efficiency of the use of vibration technological machines in resonant and post-resonant modes of operation. *Ore Beneficiation*, *6*, 42.

Rocard, Y. (1949). *General dynamics of vibrations*. Masson.

Tsze, F. S., Morze, I. E., & Khinkl, R. T. (1966). *Mechanical vibrations*. Mashinostroenie.

Chapter 2
Peculiarities of Structural Mathematical Modeling Method

ABSTRACT

Methods of evaluating the dynamic state of mechanical oscillatory systems and the possibility of changing their dynamic characteristics are investigated. Particular attention is paid to the search for new approaches to the development of protective systems against vibrations, based on the use of structural interpretations of mathematical models of systems. A method for determining static and dynamic reactions based on transformations of the initial structural model of an oscillatory system is proposed. The idea of the method is that an elastic element in an oscillatory system is considered as a feedback circuit for an object making periodic movements relative to static equilibrium. Methodological bases of reaction determination method using Dalembert principle are considered. The idea of constructing compact systems including standard links, lever mechanisms and other devices forming a quasi-element or a system of quasi-elements interacting according to the rules of structural transformations is proposed.

INTRODUCTION

Determination of static and dynamic characteristics in interactions of elements and parts of machines and equipment is an important step in solving the tasks of ensuring reliability of operation and safety of operation. In this regard, the issues of assessing the reduced parameters of systems with mass-inertial, elastic and dissipative elements are of great importance. Particular attention is paid to the possibilities of identifying various dynamic effects that may arise when establish-

DOI: 10.4018/979-8-3373-0447-2.ch002

ing non-traditional forms of communication between individual types of system movements and their elements, as well as during interactions of disturbing factors. The structural approaches developed in the proposed studies may be useful due to their visibility and the ability to obtain integral estimates associated with the use of mathematical models in the form of structural schemes and transfer functions.

COMPACTS IN STRUCTURAL DIAGRAMS OF MECHANICAL OSCILLATORY SYSTEMS

As shown in the previous sections, structural approaches in the dynamics of mechanical oscillatory systems are based on ideas about the possibility of comparing the physical initial model of a dynamically equivalent structural scheme of an automatic control system. The identity of the description of processes is predetermined by the fact that a mathematical model in the form of a linear system of differential equations is common. Structural transformations of models allow you to obtain the transfer functions of systems. This is used in design details to solve many machine dynamics problems. As for the ideas about the elementary base of mechanical oscillatory systems, its expansion, first of all, due to the introduction of an elementary link with a second-order differentiating link transfer function, significantly expanded the possibilities of changing the dynamic properties of systems. Equally important in this regard is the use of leverage, which is reflected in the development of ideas for building compacts. The essence of such representations lies in the fact that blocks or structures of elementary links obtained on the basis of the rules for transforming structural schemes can also be further included in subsequent transformations according to the same rules, forming in a sense megastructures in which typical elementary links are replaced by compacts. Such transformations can be considered as quasi-springs, which allows, when building mathematical models of complex mechanical oscillatory systems in solving a number of specific problems (for example, vibration protection) related to the allocation of an object with a controlled dynamic state, to obtain. At the same time, the structures of complex systems can be considered as the results of transforming compacts or quasi-structures according to the rules of transformation and connection of typical elementary units.

Features of Building Compacts

We consider a mechanical oscillatory system with three degrees of freedom, making small vibrations in the absence of friction forces (Figure 1*a*, *b*).

Figure 1. Design (a) and structural (b) diagrams of a mechanical oscillatory system with 3 degrees of freedom

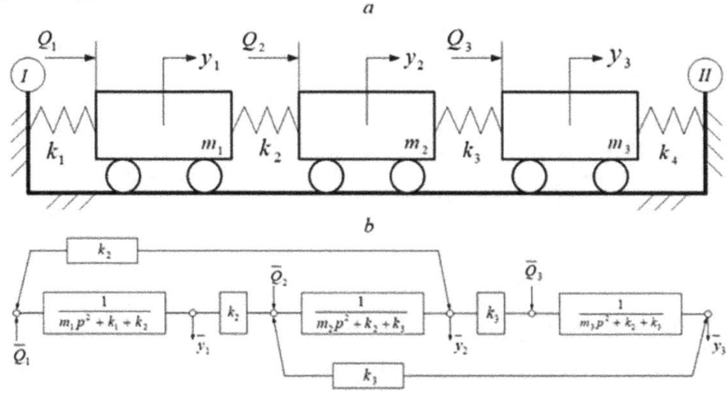

In general, external forces can be represented by three harmonic effects Q_1, Q_2, Q_3 applied to elements with masses m_1, m_2, m_3. The system has two bearing surfaces *I* and *II* (Figure 1*a*). We use the coordinate system y_1, y_2, y_3 associated with the fixed basis. Stiffness of the corresponding elastic elements is indicated by k_1-k_4 (Figure 1*a*). Motion is considered relative to the position of static equilibrium. The structural diagram of the system (Figure 1*b*) is built with the allocation of partial systems formed in the usual way with sequential "suspension" of movement at all coordinates except for the one under consideration (Eliseev A.V., Selvinsky, & Eliseev S.V, 2015; Bolshakov & Eliseev, 2013). However, such a technique is not the only one. In particular, the block diagram in Figure 1*b* can be transformed with the exception of some variables, which reduces the total number of degrees of freedom of movement. At the same time, the quasi-springs or compacts mentioned above appear in the structural diagram of the system (Eliseev A.V., Selvinsky, & Eliseev S.V, 2015).

The objective of the following studies is to develop methodological foundations for the transformation of structural schemes, taking into account the peculiarities of the location of external forces in various options for choosing the object of protection.

Building Mathematical Models

If mass inertia element m_2 is taken as the object of protection against vibration forces ($Q_2 \neq 0$, $Q_1 = Q_3 = 0$), then the block diagram in Figure 1*b* can be reduced to the view, as shown in Figure 2. In this case, the determination of external forces in relation to the object of protection is essential.

Figure 2. Block diagram with protection object m_2

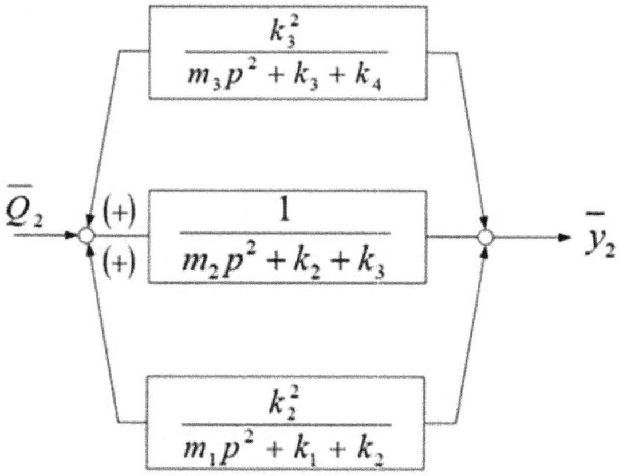

In the direct circuit of the structural scheme, a transfer function of a partial system of the usual form can be found:

$$W_2(p) = \frac{1}{m_2p^2 + k_2 + k_3} \tag{1}$$

In turn, in positive feedback circuits, the transfer functions can be denoted:

$$W_1(p) = \frac{k_2^2}{m_1p^2 + k_1 + k_2}, \tag{2}$$

$$W_3(p) = \frac{k_3^2}{m_3p^2 + k_2 + k_3}. \tag{3}$$

Expressions (2) and (3), in their physical essence, correspond to the stiffness coefficient of some elastic inertial compact or quasi-spring. The stiffness of such a spring depends on p ($p = j\,\omega$ is a complex variable used in frequency analysis). In circuit theory and its applications, the concept of dynamic stiffness is also used, which takes into account the dependence of the elastic properties of the system on the frequency of external influence. In this case, fragments of the elastic system (2), (3) are considered as compacts or quasi-springs, therefore, the structural diagram in Figure 2 can be converted to a view, as shown in Figure 3.

Figure 3. Converted structural diagram reduced to a system with one degree of freedom

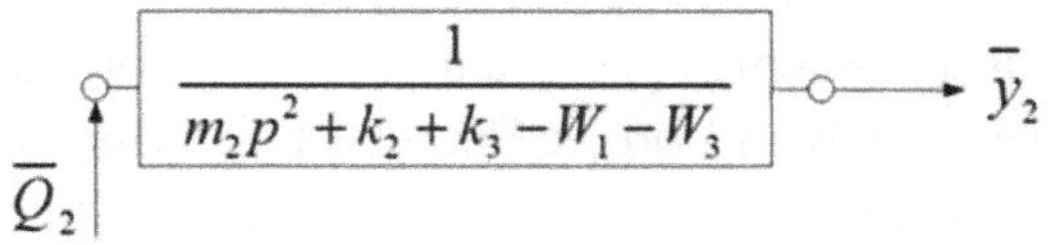

The transfer function of the transformed structure is defined by:

$$W_2(p) = \frac{\bar{y}_2}{\bar{Q}_2} = \frac{1}{m_2p^2 + k_2 + k_3 - W_1 - W_3} \tag{4}$$

In expanded form (4) has the form:

$$W_2(p) = \frac{\bar{y}_2}{\bar{Q}_2} = \frac{(m_1p^2 + k_1 + k_2) \cdot (m_3p^2 + k_3 + k_4)}{A_0} \tag{5}$$

Where

$$\begin{aligned} A_0 = {} & (m_1p^2 + k_1 + k_2) \cdot (m_2p^2 + k_2 + k_3) \cdot (m_3p^2 + k_3 + k_4) \\ & -k_2^2 \cdot (m_3p^2 + k_3 + k_4) - k_3^2 \cdot (m_1p^2 + k_1 + k_2). \end{aligned} \tag{6}$$

Expression (6) corresponds to the characteristic frequency equation of the system. Thus, taking into account the peculiarities of the force impact and the selection of the m_2 protection object, the initial structural diagram can be (conditionally) simplified to a system with one degree of freedom (Figure 3), however, two new links with transfer functions $W_2\ (p)$ and $W_3\ (p)$ appear in the system structure, defined by the expressions (2) and (3), respectively. These structures have the same dimension as k_2 and k_3, that is, they are quasi-springs, but the rigidity of these springs depends on the frequency of external influence and on the mass inertia parameters m_2 and m_3.

Note the following feature that in the structural diagram (Figure 2) the feedback is positive, which leads to a negative sign of the reduced rigidity. Note that in the structural theory of vibration protection systems, the elastic element is interpreted as a negative feedback in relation to the object of protection, that is, an element with the transfer function of the second-order integrating link - $\frac{1}{mp^2}$(Eliseev A.V., Selvinsky, & Eliseev S.V, 2015). Let's convert the structural diagram in Figure 2 to a view, as shown in Figure 4a - d. Thus, elastic elements (meaning quasi-springs or compacts) can be considered as typical elements of mechanical oscillatory systems, in the sense of applying the usual rules of structural transformations to them. From Figure 4a, b it follows, in particular, that k_1 and k_2 can be transferred from the forward circuit to the reverse circuit with the corresponding sign. In Figure 4b, due to such rules, it is shown that quasi-springs form a negative feedback. In Figure 4, the mechanical vibration system is presented in the form of a design diagram of a vibration protection system with a protection object m_2, which is supported by two elastic structures. The latter can be considered as springs taking into account their inertia ($m_{1,}$ $m_{3)}$. In this regard, the design diagram of the vibration protection system is indicative. In Figure 4d, the object of protection rests on two quasi-springs or on two elastic-inertial compacts operating in parallel.

Figure 4. Block diagram (Figure 1) with positive feedbacks (a); block diagram for negative connections (b); converted calculation scheme (in); computation scheme with compacts in coordinates $\bar{y}_2$(Laplace images) (d)

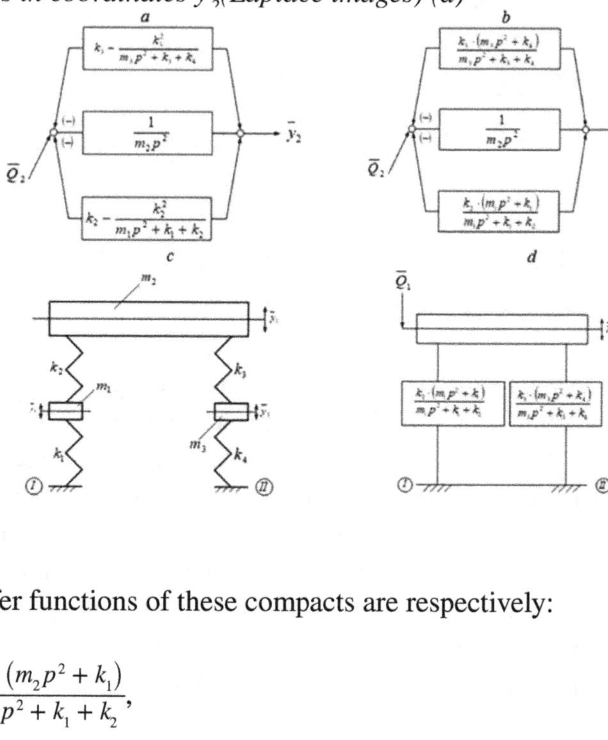

The transfer functions of these compacts are respectively:

$$W'_1(p) = \frac{k_2 \cdot (m_2 p^2 + k_1)}{m_1 p^2 + k_1 + k_2}, \tag{7}$$

$$W'_3(p) = \frac{k_3 \cdot (m_3 p^2 + k_4)}{m_3 p^2 + k_3 + k_4}. \tag{8}$$

The construction of the transfer functions (7), (8) can be carried out according to the scheme of Figure 4, *in* based on the rules for converting circuits, assuming that the object m_2 is based on mechanical circuits. The issues of determining dynamic reactions in contacts with bearing surfaces *I* and *II*, as well as with the object of protection m_2 have their own characteristics and are considered in the work (Khomenko, Eliseev, & Bolshakov, 2014).

If $m_1 = 0$ and $m_3 = 0$, then compacts, or quasi-springs, are transformed into a spring constituting a series connection of elastic elements k_1, k_2, k_3, k_4. Note that this corresponds to the condition $p = 0$, that is, the case of static equilibrium. At very high frequencies of external influence, quasi-springs transform into ordinary elastic elements with stiffnesses k_2 and k_3. Characteristic in this approach is the fact that the stiffness of quasi-springs is "zeroed" at frequencies:

$$\omega_1^2 = \frac{k_1}{m_1}, \tag{9}$$

$$\omega_3^2 = \frac{k_4}{m_3}. \tag{10}$$

If these frequencies are equal, then the amplitude of movement along the coordinate y_2 increases to ∞, which corresponds to the resonance mode. If "zeroing" does not occur simultaneously, then resonant phenomena do not occur. At the same time, at frequencies:

$$\omega_{10}^2 = \frac{k_1 + k_2}{m_1}, \tag{11}$$

$$\omega_{30}^2 = \frac{k_1 + k_3}{m_2}. \tag{12}$$

the quasi-spring stiffness becomes infinitely large, which corresponds to dynamic blanking modes. In this case, the mass m_2 will be stationary. When ω_{10} and ω_{30} coincide, a dynamic blanking mode may occur in the system, which determines the features in the amplitude-frequency characteristics (multiple zeros of the transfer function of the system).

Consideration of Peculiarities in the Location of External Forces

Consider the case when $Q_1 \neq 0$, and $Q_2 = Q_3 = 0$, then we select the object of protection in the form of an element with a mass of m_1. We will use a mathematical model that can be reconstructed from the structural diagram in Figure 1*b*, since in the proposed approaches the structural diagram (or model) is a graphical interpretation of the original differential equations of motion:

$$m_1\ddot{y}_1 + y_1(k_1 + k_2) - k_2 y_2 = Q_1 \; ; \tag{13}$$

$$m_2\ddot{y}_2 + y_2(k_2 + k_2) - k_2 y_1 - k_3 y_3 = 0 \; ; \tag{14}$$

$$m_3\ddot{y}_3 + y_3(k_3 + k_4) - k_3 y_2 = 0. \tag{15}$$

Let's write that in Laplas variable images:

$$\bar{y}_1 = \frac{\bar{Q}_1 + k_2\bar{y}_2}{m_1p^2 + k_1 + k_2}, \tag{16}$$

$$\bar{y}_3 = \frac{k_3\bar{y}_2}{m_3p^2 + k_3 + k_4}. \tag{17}$$

Using (14), (16), (17), we find:

$$\bar{y}_2 \cdot \left(m_2p^2 + k_2 + k_3\right) - \frac{k_2^2 \cdot y_2}{m_1p^2 + k_1 + k_2} - \frac{k_3^2 \cdot \bar{y}_2}{m_3p^2 + k_3 + k_4} = \frac{k_2 \cdot Q_1}{m_1p^2 + k_1 + k_2}. \tag{18}$$

Equation (18) is transformed into a structural diagram, as shown in Figure 5*a* - *c*.

Figure 5. Design (a), structural (b) and simplified (c) diagrams of the mechanical oscillatory system (Figure 1,a) with force Q_1 acting on the element with mass m_1

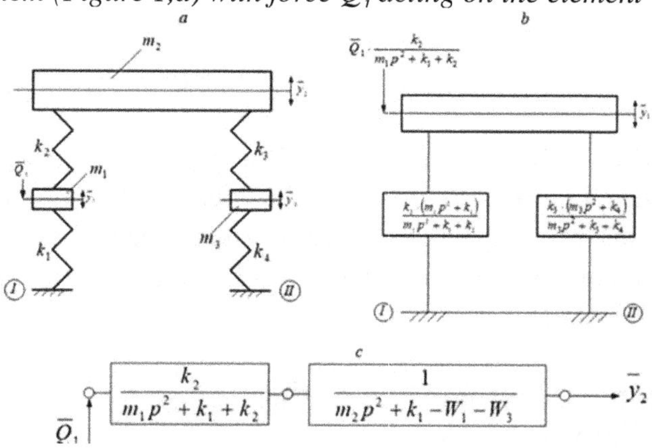

If an external impact is applied to element m_1, and the object of protection is element m_2, then an equivalent transfer of force Q_1 to mass m_2 is possible. In this case, the equivalent value of the external disturbance takes the form:

$$Q_{2eq} = \frac{k_2}{m_1p^2 + k_1 + k_2}. \tag{19}$$

Let's check by determining the transfer functions for disturbance by $_{y1}$. Then:

$$W_1(p) = \frac{\bar{y}_1}{\bar{Q}_1} = \frac{a_{22} \cdot a_{33} - a_{23}^2}{A_0}; \tag{20}$$

$$W_2(p) = \frac{\bar{y}_2}{\bar{Q}_1} = \frac{-a_{12} \cdot a_{33}}{A_0}; \tag{21}$$

$$W_3(p) = \frac{\bar{y}_3}{\bar{Q}_1} = \frac{a_{21} \cdot a_{32}}{A_0}. \tag{22}$$

In turn,

$$W'_1(p) = \frac{\bar{y}_1}{\bar{Q}_{2eq}} = \frac{a_{11} \cdot a_{32} - a_{33} \cdot a_{12}}{A_0}; \tag{23}$$

$$W'_2(p) = \frac{\bar{y}_2}{\bar{Q}_{2eq}} = \frac{a_{11} \cdot a_{33}}{A_0}; \tag{24}$$

$$W'_3(p) = \frac{\bar{y}_2}{\bar{Q}_{2eq}} = \frac{-a_{11} \cdot a_{32}}{A_0}. \tag{25}$$

Thus, it can be proved that the use of formulas (23) (25) leads to the same results as the use of (20) - (22). However, structural transformations of circuits (starting with the structural diagram in Figure 1b) can be quite laborious. Note that the transfer leads to a change in the magnitude of the force by a factor, which, in its physical essence, is the transmission coefficients of the amplitude of vibrations in a partial system with parameters $m_{1,}$ $k_{1,}$ $k_{2.}$ In this case, the kinematic disturbance is transmitted through the elastic element $k_{2,}$ which makes it possible to obtain a transfer function (or transfer coefficient):

$$W(p) = \frac{k_2}{m_1 p^2 + k_1 + k_2}. \tag{26}$$

Issues of equivalent transformations in problems of dynamics of mechanical oscillatory systems were also considered in (Eliseev, 2018; (Eliseev & Artyunin, 2016).

Effect of Additional Coupling Between Partial Systems

Consider the case when the movement along the coordinates y_1, y_2, y_3 is such that there is an additional spring k_{13} connecting the mass inertia elements m_1 and m_3. In this case, the system of equations will take the form:

$$\bar{y}_1 \cdot (m_1 p^2 + k_1 + k_2 + k_{13}) - y_2 k_2 - y_3 k_{13} = Q_1 \; ; \tag{27}$$

$$\bar{y}_2 \cdot (m_2 p^2 + k_2 + k_3) - y_1 k_2 - y_3 k_3 = 0 \; ; \tag{28}$$

$$\bar{y}_3 \cdot (m_3 p^2 + k_3 + k_4 + k_{13}) - y_2 k_3 - y_1 k_{13} = 0. \tag{29}$$

Let's compile Table 1 of the coefficients of the system of equations (27) - (29), which describe the movement in the coordinate system y_1, y_2, y_3.

Table 1. Coefficients of the system of equations in coordinates y_1 - y_3.

a_{11}	a_{12}	a_{13}
$m_1p^2 + k_1 + k_2 + k_{13}$	$-k_2$	$-k_{13}$
a_{21}	a_{22}	a_{23}
$-k_2$	$m_2p^2 + k_2 + k_3$	$-k_3$
a_{31}	a_{32}	a_{33}
$-k_{13}$	$-k_3$	
External forces		
Q_1	Q_2	Q_3
Q	0	0

Note: Q_1, Q_2, Q_3 - generalized forces by coordinates y_1, y_2, y_3, respectively; Note here that $Q_1 = Q$, $Q_2 = Q_3 = 0$.

In general, the transfer functions of the system under external influence $Q_1 = Q$ can be found by Cramer formulas:

$$\frac{\bar{y}_1}{\bar{Q}_1} = \frac{a_{22} \cdot a_{33} - a_{23}^2}{A'_0} \; ; \tag{30}$$

$$\frac{\bar{y}_2}{\bar{Q}_1} = \frac{a_{23} \cdot a_{31} - a_{12} \cdot a_{33}}{A'_0} \; ; \tag{31}$$

$$\frac{\bar{y}_3}{\bar{Q}_1} = \frac{a_{21} \cdot a_{32} - a_{22} \cdot a_{31}}{A'_0} \; ; \tag{32}$$

Where

$$A'_0 = a_{11} a_{22} a_{33} - a_{11} a_{23}^2 - a_{22} a_{13}^2 - a_{33} a_{12}^2 + 2 a_{12} a_{23} a_{31} \tag{33}$$

- is the characteristic equation.

To construct a structural diagram reduced to the mass inertia element m_1, we will use the technique, which is to transform (30) in accordance with the following actions.

1. We find the transfer function:

$$W(p) = \frac{\bar{y}_1}{\bar{Q}_1} = \frac{\left(a_{22}a_{33} - a_{23}^2\right)}{a_{11}\left(a_{22}a_{33} - a_{23}^2\right) - a_{22}a_{13}^2 - a_{33}a_{12}^2 + 2a_{12}a_{23}a_{31}}; \tag{34}$$

2. Divide the numerator and denominator by $\left(a_{22}a_{33} - a_{23}^2\right)$ and we get:

$$W(p) = \frac{\bar{y}_1}{\bar{Q}_1} = \frac{1}{a_{11} - \frac{a_{22}a_{13}^2}{[\,]} - \frac{a_{33}a_{12}^2}{[\,]} + \frac{2a_{12}a_{23}a_{31}}{[\,]}}; \tag{35}$$

where $[\,] = \left(a_{22}a_{33} - a_{23}^2\right)$.

3. We bring (35) to the form:

$$W(p) = \frac{1}{m_1p^2 + k_1 + k_2 + k_{13} - \frac{\left(m_2p^2 + k_2 + k_3\right)\cdot k_{13}}{[\,]} - \frac{\left(m_3p^2 + k_3 + k_4\right)\cdot k_{12}^2}{[\,]} - \frac{2k_2k_3k_{13}}{[\,]}}. \tag{36}$$

Expression (36) corresponds to the structural diagram shown in Figure 6*a*.

Figure 6. Structural diagram of the system on the design diagram as per Figure 1, and at $k_{13} \neq 0$ - (a); structural diagram of the same system at $k_{13} = 0$ - (b) ($Q_2 = 0$, $Q_3 = 0$)

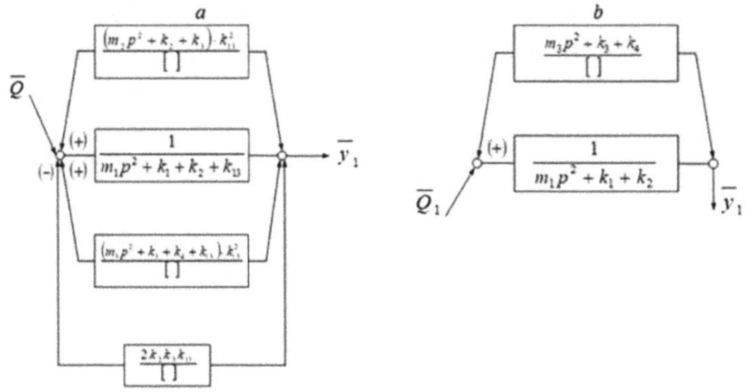

Cross-linking, as follows from Figure 6*a*, significantly complicates the structure of the system. Simplifications in such approaches are possible when transferring k_1 and k_2 in the feedback circuit, taking into account the feedback sign. Detailed studies of design circuits of this type are given in (Eliseev, 2018; (Eliseev & Artyunin, 2016). in connection with the study of the features of complex dynamic vibration dampers.

Consider the structural diagram of the system under the same loading conditions, but with the exception of the coordinate$\bar{y}_2$. In this case, the expression (30) can be divided (numerator and denominator) by the operator $_{a22}$. We get:

$$W(p) = \frac{\bar{y}_1}{\bar{Q}_1} = \frac{a_{33} - \frac{a_{23}^2}{a_{22}}}{(a_{11}a_{33} - a_{13}^2) - \frac{a_{11}a_{23}^2}{a_{22}} - \frac{a_{33}a_{12}^2}{a_{22}} - \frac{2a_{12}a_{23}a_{31}}{a_{22}}}. \quad (37)$$

The block diagram resolved relative to the sequential bundle of blocks in coordinates $_{y1}$ and $_{y3}$ is as shown in Figure 7.

Using Table 1 and expressions (38) construct a structural diagram of the system (Figure 7).

Figure 7. Structural diagram of the system with excluded coordinate y_2 ($Q_2 = 0$)

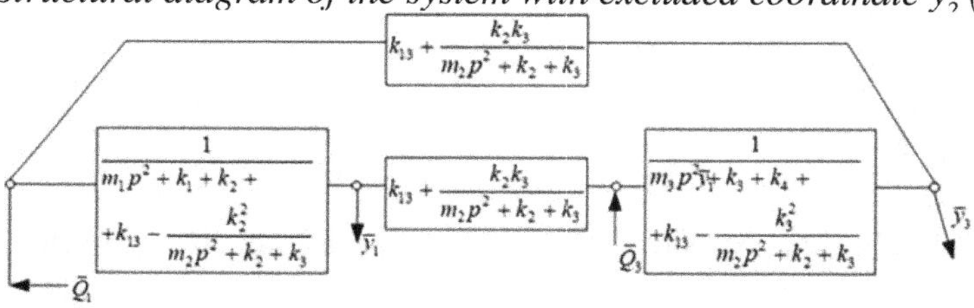

Note that the exclusion of the y_2 coordinate (as well as others) is accompanied by changes in partial systems and interpartial relationships. Figure 8*a*, *b* shows a set of elastic elements. In particular, Figure 8*b* shows that the connection between elements m_1 and m_3 after transformations will be carried out through a compact (or quasi-spring) with dynamic rigidity

$$K_{np} = k_{13} + \frac{k_2 k_3}{m_2 p^2 + k_2 + k_3}. \quad (38)$$

Figure 8. Design diagram of the system with three degrees of freedom; b) - calculation scheme, in which the coordinate y_2 is excluded

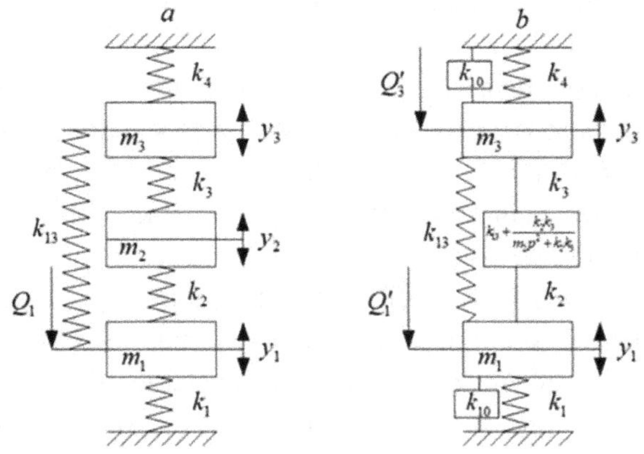

At the same time, changes occur in the external force, which takes into account the exception of the y_2 coordinate. If $Q_2 \neq 0$, then the external force at the y1 coordinate will be determined

$$Q'_1 = Q_1 + \frac{k_2}{m_2 p^2 + k_2 + k_3} Q_2, \tag{39}$$

and by y_3 coordinate, respectively

$$Q'_3 = Q_3 + \frac{k_3}{m_2 p^2 + k_2 + k_3} Q_2, \tag{40}$$

Figure 8*b* shows that the design scheme in comparison with the scheme in Figure 8*a* acquires in the first stage (coordinate y_1) a structural formation of elastic and mass-inertial elements m_3, k_3, k_4. Such a block (or compact) corresponds to ideas about a generalized spring (or quasi-spring) with reduced or dynamic stiffness

$$k_{10}(p) = k_1 + k_{13} + \frac{k_2(m_2 p^2 + k_3)}{m_2 p^2 + k_3}. \tag{41}$$

In the cascade by the coordinate y_3 (Figure 8*b*), we obtain respectively

$$k_{10}(p) = k_4 + k_{13} + \frac{k_3(m_2 p^2 + k_3)}{m_2 p^2 + k_2 + k_3}. \tag{42}$$

In this case, the external forces at the y_2 coordinate (that is, Q_2) are recalculated and have an equivalent effect on the system as a whole, but at the points of application of the action at the coordinates y_1 and y_3, respectively, which is shown by the expressions (41) and (42).

These transformations reveal the possibilities of constructing structural diagrams with the selection of properties necessary for analysis and study of dynamic features. In particular, partial interactions with a decrease in the number of observed coordinates are manifested in the fact that compound links in the form of vibrational elements are formed in interpartial bonds.

Thus, the expressions (41), (42) reflect the properties of quasi-springs, the rigidity of which depends on the frequency of external influences. At low p (or p 0), such a spring is a series connection of elastic elements; p ∞ reduced stiffness increases and tends to some constant. At the same time, the quasi-spring, working in parallel with ordinary (typical) elements, provides special dynamic modes: "neutral grounding" of dynamic stiffness, "resonant" processes, which is observed in real springs when their mass inertia properties are taken into account (parametric resonances - (Bakalov, 2006)). It should also be noted that coordinate exceptions are rather complex procedures in which it is necessary to observe the conditions for symmetry of interpartial bonds.

SOME ISSUES OF INTERACTION DYNAMICS IN MECHANICAL SYSTEMS WITH LINKAGE

Lever mechanisms as part of mechanical oscillatory systems have features that are manifested in changes in dynamic properties in relation to conventional systems and should be taken into account when determining dynamic reactions. Note that the lever mechanisms in the structure of the mechanical system affect the conditions for the formation of relations between the coordinates of the movement of the system elements, as well as the shapes of elastic bonds. In particular, the task of compiling mathematical models of systems with leverage in different coordinates, which is typical for systems that include links in the form of a solid, seems to be poorly understood.

Consider a mechanical system with two degrees of freedom (Figure 9*a*, *b*), which uses levers of the 1st kind (Figure *a*) and a bundle of levers through gear engagement (Figure 9*b*). The introduction of leverage into mechanical chains leads to certain difficulties in using the apparatus of chain theory, which was considered in (Veitz, Kachura, & Martynenko, 1971). In this regard, it is of interest to compare systems with levers in their comparison with known circuit solutions in the form of a chain of two bodies and a solid body on elastic supports.

Figure 9. Design diagrams of mechanical oscillatory systems: a) with a lever of the 1st kind; b) - with motion gearing between links

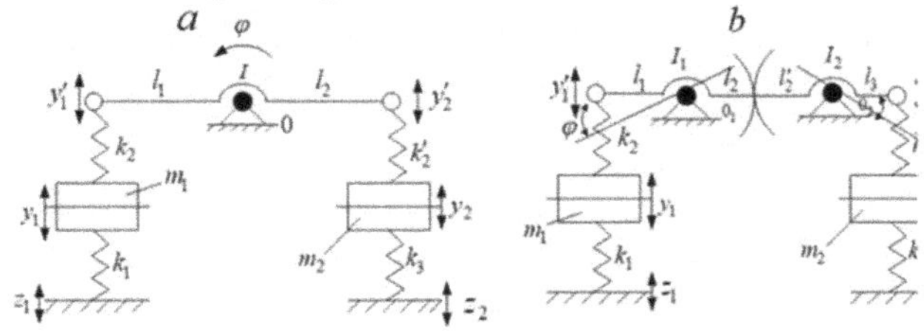

The system under consideration (Figure *9a*) consists of two elements with masses m_1 and m_2, which interact through a lever of the 1st kind. The lever has a center of rotation in the form of a fixed point *O* and arms, respectively: l_1 and l_2. The external disturbance is represented by the base motion $z_1\,(t)$ and $z_2\,(t)$ of the harmonic form; spring strengths are denoted $_{respectively\ k1}$, k_2, k'_2, k_3. To derive the equations of motion, it is assumed that the lever has a moment of inertia *I* relative to the point of rotation: the gear ratio $i_1 = l_2/l_1$ and is characterized in this case by a change in the direction of the speed of movement at the ends of the lever; the properties of the system are assumed to be linear, there is no friction, and the movements of the elements of the system are considered small.

Mathematical Models and Their Features

To build a mathematical model of the system (Figure 9*a*), enter the coordinate system y_1 and y_2 (relative to the fixed base), as well as the angle of rotation of the lever with the moment of inertia *I* around point *0* in the form of a φ. Let's write down the expressions for the kinetic and potential energies of the system:

$$T = \frac{1}{2}m_1\dot{y}_1^2 + \frac{1}{2}m_2\dot{y}_2^2 + \frac{1}{2}I\dot{\phi}^2, \tag{43}$$

$$\Pi = \frac{1}{2}k_1(y_1 - z_1)^2 + \frac{1}{2}k_2(y'_1 - y_1)^2 + \frac{1}{2}k'_2(y'_2 - y_2)^2 + \frac{1}{2}k_3y_2^2, \tag{44}$$

where y'_1, y'_2 are the coordinates of the ends of the: $y'_1 = l_1\varphi$, $y'_2 = l_2\varphi$ in accordance with Figure 9, *a*; $z_2(t) = 0$. Between y'_1 and y'_2 there is a relationship $y'_1/l_1 = y'_2/l_2$, which implies that $y'_2 = iy'_1$ $(i = -l_2/l_1)$. Rewrite (43), (44) with the coordinate relationship condition, then

$$T = \frac{1}{2}m_1\dot{y}_1^2 + \frac{1}{2}m_2\dot{y}_2^2 + \frac{1}{2}I\left(\frac{\dot{y}'_1}{l_1}\right)^2, \tag{45}$$

$$\Pi = \frac{1}{2}k_1(y_1 - z_1)^2 + \frac{1}{2}k_2(y'_1 - y_1)^2 + \frac{1}{2}k'_2(iy'_1 - y_2)^2 + \frac{1}{2}k_3 y_2^2. \quad (46)$$

After a series of transformations, we get a system of equations of motion:

$$m_1\ddot{y}_1 + y_1(k_1 + k_2) - k_2 y'_1 = k_1 z_1, \quad (47)$$

$$m_2\ddot{y}_2 + y_2(k_3 + k'_2) - k'_2 i y'_1 = 0, \quad (48)$$

$$(I/l_1^2)\ddot{y}'_1 + y'_1(k_2 + k'_2 i^2) - k_2 y_1 - k'_2 i y_2 = 0. \quad (49)$$

Coefficients of the system of equations of motion (47) - (49) are given in Table 2.

Table 2. Coefficients of equations of motion (47) - (49)

a_{11}	a_{12}	a_{13}
$m_1p^2 + k_1 + k_2$	0	$-k_2$
a_{21}	a_{22}	a_{23}
0	$m_1 p^2 + k'_2 + k_3$	$-k'_2 i$
a_{31}	a_{32}	a_{33}
$-k_2$	$-k'_2 i$	$(I/l_1^2)p^2 + k_2 + k'_2 i^2$
Q_1	Q_2	Q_3
$k_1 z_1$	0	0

Note: Q_1 - Q_3 are generalized forces corresponding to generalized coordinates y_1, y_2, y 3.

Using (48) at, we find that$y'_1(k_2 + k'_2 i^2) - k_2 y_1 - k'_2 i y_2 = 0$, from where

$$y'_1 = \frac{k_2 y_1 + k'_2 i y_2}{k_2 + k'_2 i^2} = a y_1 + b y_2, \quad (50)$$

where

$$a = {}^{k_2}\Big/_{(k_2 + k'_2 i^2)}; \; b = {}^{k'_2 i}\Big/_{(k_2 + k'_2 i^2)}. \quad (51)$$

We transform equations (47), (48) and get:

$$m_1\ddot{y}_1 + y_1(k_1 + k_2) - k_2(a y_1 + b y_2) = k_1 z_1, \quad (52)$$

$$m_2\ddot{y}_2 + y_2(k'_2 + k_3) - k'_2 i(a y_1 + b y_2) = 0. \quad (53)$$

After reducing to a unified form, we obtain a Table 3 of coefficients of equations (52), (53), in which the y $'_1$ motion coordinate is excluded using the relation (50).

Table 3. System motion equation coefficients in coordinates y_1 and y_2

a_{11}	a_{12}
$m_1p^2 + k_1 + k_2 - k_2a$	$-k_2b$
a_{21}	a_{22}
$-k'_2 ia$	$m_2p^2 + k'_2 + k_3 - k'_2 ib$
Q_1	Q_2
k_1z_1	

Let's check the ratio between the coefficients a_{21} and a_{12}:

$$-k_2b = a_{12} = -\frac{k_2k'_2 i}{k_2 + k'_2 i^2} = \frac{-k_2k'_2 i}{k_2 + k'_2 i^2}, \tag{54}$$

$$-k'_2 ia = a_{21} = -\frac{k_2k'_2 i}{k_2 + k'_2 i^2}. \tag{55}$$

Thus, from (54), (55) it follows that the symmetry of the matrix of coefficients of equations (Table 3) is preserved. The structural diagram of the system at $I = 0$ is as shown in Figure 10.

Figure 10. Block diagram of mechanical system with linkage at I = 0

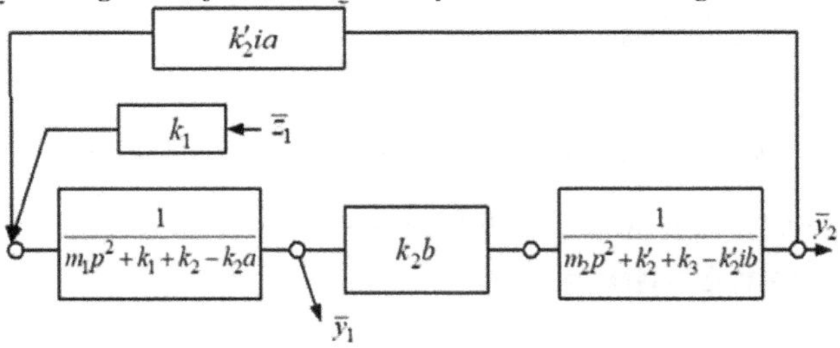

Let's expand the expressions for the reduced stiffness:

$$k_{pr_1} = k_1 + k_2 - k_2a = k_1 + \frac{k_2k'_2 i^2}{k_2 + k'_2 i^2}. \tag{56}$$

In turn:

$$k_{pr_2} = k_3 + k'_2 - k'_2 ib = k_3 + \frac{k_2 k'_2}{k_2 + k'_2 i^2}. \tag{57}$$

If $I \neq 0$, the system structural diagram becomes as shown in Figure 11 taking into account (56), (57).

Figure 11. Structural diagram of the system as per Figure 9, and with inertial lever

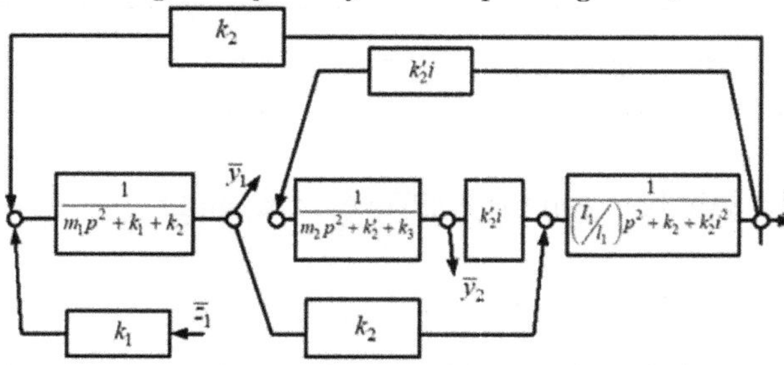

The peculiarity of the system is that there are no connections between partial systems in the coordinates y_1 and y_2. Assuming that $k_1 z_1 = Q_1$, one finds the transfer function of the system:

$$W(p) = \frac{\bar{y}_1}{\bar{Q}} = \frac{[(I/l_1^2)p^2 + k_2 + k'_2 i^2](m_2 p^2 + k'_2 + k_3) - (k'_2 i)^2}{A}, \tag{58}$$

where A is the characteristic equation:

$$\begin{aligned} A = {} & (m_1 p^2 + k_1 + k_2)\left({}^{I}\!/\!_{l_1^2} p^2 + k_2 + k'_2 i^2\right)(m_2 p^2 + k'_2 + k_3) - \\ & -(k'_2 i)^2 (m_1 p^2 + k_1 + k_2) - k_2^2 (m_2 p^2 + k'_2 + k_3). \end{aligned} \tag{59}$$

Reduced System Stiffness

From (58) it follows that the reduced elasticity, that is, the rigidity of the elastic compact in the system in Figure 9 when applying a static force Q_1 to an element with mass m_1, will be determined

$$k_{pr} = \frac{i^2 k_2 k'_2 k_3}{k'_2 k_2 + k_3 k_2 + k'_2 k_3 i^2}. \tag{60}$$

A similar expression using (59), (60) can be obtained directly from the calculation scheme in Figure 10, assuming $I = 0$, $m_2 = 0$ then

$$k_{pr} = \frac{\left(\frac{k'_2 k_3}{k'_2 + k_3}\right) i^2 k_2}{\left(\frac{k'_2 k_3}{k'_2 + k_3}\right) i^2 + k_2} = \frac{i^2 k_2 k'_2 k_3}{k_2 k' + k_2 k_3 + k'_2 k_3 i^2}. \qquad (61)$$

The obtaining algorithm (61) consists in separating two stages from serially connected elastic elements and taking into account the properties of the lever connection. To obtain the expression (60), if you use (58), (59), you must also take $p = 0$ in partial systems $m_2 p^2 + k'_2 + k_3$ and $I/i^2 p^2 + k_2 + k'_2 i^2$, then bring the system to the form $m_1 p^2 + k_1 k_{pr}$. It is shown that taking into account the transformations, the compact (quasi-spring) of elastic elements in the event of a static load, within the framework of the calculation scheme in Figure 22, takes the place of a conventional elastic element.

Figure 12. Design diagram of mechanical oscillatory system with elastic compact (quasi-spring)

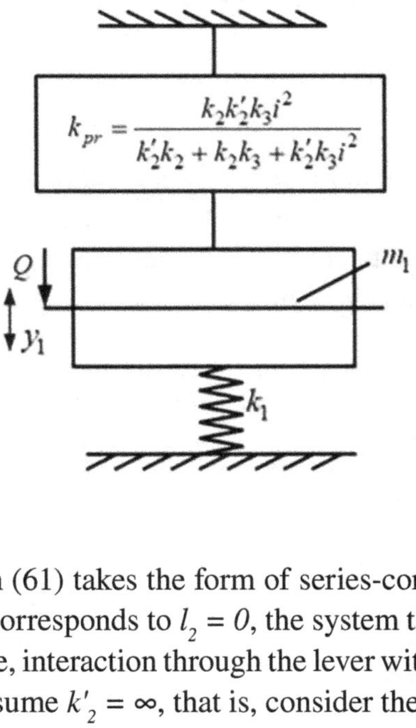

With $i=1$, expression (61) takes the form of series-connected elastic elements. In turn, at $i = 0$, which corresponds to $l_2 = 0$, the system takes a simplified form at which $k_{pr} = 0$. In this case, interaction through the lever with the rest of the elements does not occur. If we assume $k'_2 = \infty$, that is, consider the mass m_2 directly related to the lever, then

$$k_{pr} = \frac{k_2 k_3 i^2}{k_2 + k_3 i^2}, \qquad (62)$$

which coincides with the results, for example, given in (Bolshakov & Eliseev, 2013; Gardner & Bairns, 1949).

Dynamic Interaction Mode With linkage

Taking $I = 0$, you can convert the structure diagram in Figure 10. to the view as shown in Figure 13. In this case, the interaction between partial systems will be carried out through an elastic compact (quasi-spring) with rigidity:

$$W'(p)=\frac{k_2 k'_2 i}{k_2 + k'_2 i^2} \tag{63}$$

Figure 13. Structural diagram of the system with linkage

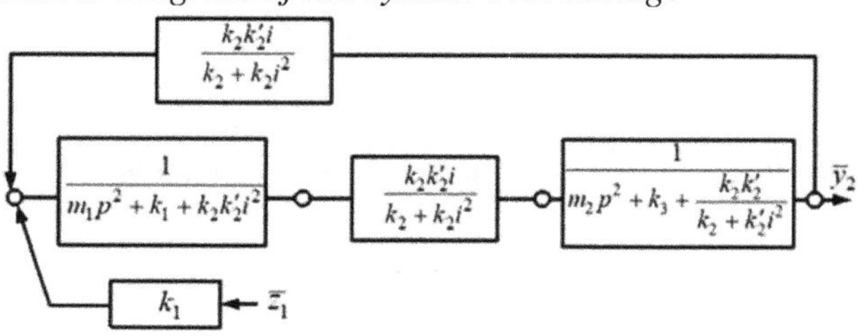

Using the block diagram in Figure 13, a transformation can be made that gives the system in question the appearance of an ordinary chain structure containing compacts of elastic elements connected by lever links. The block diagram with converted elements is shown in Figure 14.

Figure 14. Block diagram of the system reduced to chain view

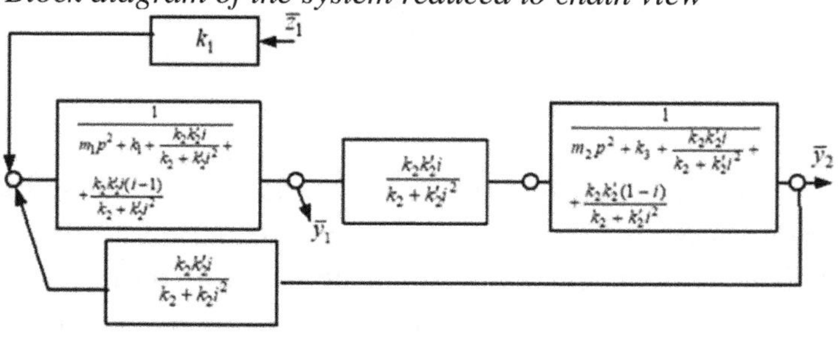

The corresponding calculation scheme at the level of use of individual links, taking into account their physical nature, is shown in Figure 15.

System feature in Figure 15 is that leverage can be introduced into the structure of a compact of elastic elements (or quasi-springs), which has not been previously considered in such a course in the scientific literature. In circuit theory, special techniques are used to account for leverage that do not reflect the commonality of the nature of dynamic bonds. We also note that the connection between partial systems, in the physical sense, is realized through a lever mechanism, which rotational force turns into force factors of interaction between masses m_1 and m_2. Thus, a linkage system can be represented by a chain system with elastic elements that form some compacts (quasi-springs); elastic connections are realized between partial systems (in this case, the mass inertia properties of the lever for static calculations are considered small).

Figure 15. Design diagram of the system reduced to chain form and containing linkage

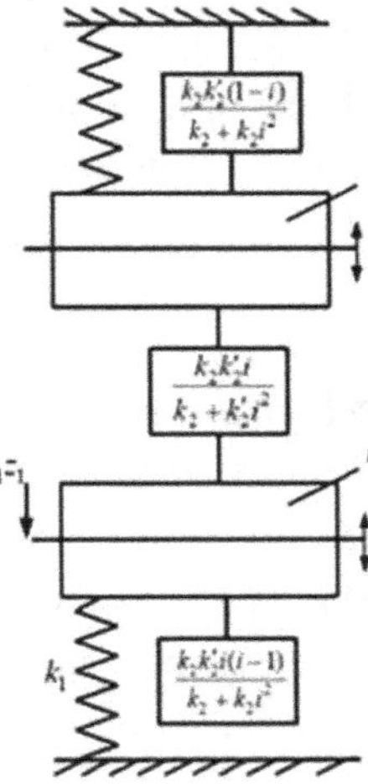

The type of lever device (levers of the first or second kind (Buryan, Sorokin, & Galuza, 2011)) is important for constructing the transfer functions of the system, since the levers of the first kind have a gear ratio with a minus sign. This is important for determining the sign of additional feedback introduced by the levers. The latter is important due to the corresponding changes in the characteristic equation. According to the block diagram in Figure 15, it can be noted that the frequencies of partial systems and the natural oscillation frequency of the system, as well as the dynamic properties, will depend on the type of linkage, which, in particular, is reflected in the work (Eliseev & Artyunin, 2016). To check the correctness of the approach, we determine the static stiffness of the system in Figure 15:

$$k_{pr} = \frac{\frac{[k_3(k_2 + k'_2 i^2) + k_2 k'_2(1 - i)]}{(k_2 + k_2 i^2)} \cdot \frac{k_2 k'_2 i}{k_2 + k'_2 i^2}}{\frac{k_3(k_2 + k'_2 i^2) + k_2 k'_2(1 - i)}{(k_2 + k'_2 i^2)} + \frac{k_2 k'_2 i}{(k_2 + k'_2 i^2)}} + \frac{k_2 k'_2 i}{k_2 + k'_2 i^2} = \frac{k_2 k'_2 k_3 i}{k_2 k'_2 + k_3 k_2 + k_3 k'_2 i^2}, \quad (64)$$

which matches the expression (61).

Taking into account the dynamic properties of a system with an inertial lever requires independent consideration, but is done in the same way. When using a lever of the first kind, the speeds of the lever ends have different directions, therefore, if $i < 0$ is taken, then in accordance with the original equations (47) - (49), the feedback in the structural diagrams will be negative. This implies in the characteristic equation (in Figure 14) the appearance of the plus sign before the last term. It should be taken into account that the "minus" sign of the gear ratio does not change the parameters of partial systems. That is, the lever of the first kind, in this case, provides negative feedback during convolutions of the system. If we use a lever of the second kind, in which $i > 0$, then the feedback in the system will be positive, which changes the sign in the characteristic equation of the system - it becomes negative (minus sign). Taking into account the peculiarities of leverage leads to the fact that the dynamic properties of systems will be different with different types of leverage.

Properties of Complex Linkage Systems

Consider a mechanical system containing a more complex system of linkage, including ear engagement (Figure 9*b*). We write down the expressions for kinetic and potential energies, assuming that the components of the linkage have moments of inertia I_1 and I_2 respectively, and have fixed points of rotation O_1 and O_2 (Figure 9*b*):

$$T = \frac{1}{2} m_1 \dot{y}_1^2 + \frac{1}{2} m_2 \dot{y}_2^2 + \frac{1}{2} I_1 \dot{\phi}^2 + \frac{1}{2} I_2 \dot{\phi}_1^2, \quad (65)$$

$$\Pi = \frac{1}{2} k_1 (y_1 - z_1)^2 + \frac{1}{2} k_2 (y'_1 - y_1)^2 + \frac{1}{2} k'_2 (y'_2 - y_2) + \frac{1}{2} k_3 (y_2 - z_2)^2, \quad (65')$$

where φ and φ_1 are angular coordinates of rotation of gear elements.

We introduce a number of relationships necessary for further calculations:

$$\phi = \frac{y'_1}{l_1};\ \phi_1 = \frac{y'_2}{l_3};\ l_2 \phi = \phi_1 l'_2;\ \frac{y'_1}{l_1} = \frac{y'_2}{l_3} \cdot \frac{l'_2}{l_2};$$
$$y'_2 = \frac{y'_1 l_3 l_2}{l_1 l'_2},\ i = \frac{l_2}{l_1};\ i_1 = \frac{l_3}{l'_2};\ \phi_1 = \phi \frac{l_2}{l'_2}. \quad (66)$$

Taking into account the relations (66), the expressions (64), (65) will take the form:

$$T = \frac{1}{2}m_1\dot{y}_1^2 + \frac{1}{2}m_2\dot{y}_2^2 + \frac{1}{2}I_1\dot{\phi}^2 + \frac{1}{2}I_2\dot{\phi}^2\left(\frac{l_2}{l'_2}\right), \tag{67}$$

$$\Pi = \frac{1}{2}k_1(y_1 - z_1)^2 + \frac{1}{2}k_2(\phi l_1 - y_1)^2 + \frac{1}{2}k'_2(\phi l_1 i i_1 - y_2)^2 + \frac{1}{2}k_3(y_2 - z_2)^2. \tag{68}$$

Since $l_2 = il_1$ and $l'_2 = l_3/i_1$, you can enter the following relationship:

$$l_2/l'_2 = i_1 i\, l_1/l_3 = i_2. \tag{69}$$

The system of equations taking into account (67) - (69) will take the form:

$$m_1\ddot{y}_1 + y_1(k_1 + k_2) - k_2\phi l_1 = k_1 z_1, \tag{70}$$

$$m_2\ddot{y}_2 + y_2(k'_2 + k_3) - k'_2 i i_1 l_1 \phi = k_3 z_2, \tag{71}$$

$$\ddot{\phi}(I_1 + I_2 i_2^2) + \phi\left[k_2 l_1^2 + k'_2(l_1 i i_1)^2\right] - k_2 l_1 y_1 - k'_2 l_1 i i_1 y_2 = 0. \tag{72}$$

Let $I_1 = 0$, $I_2 = 0$, then

$$\varphi = a y_1 + b y_2, \tag{73}$$

where

$$a = \frac{k_2 l_1}{k_2 l_1^2 + k'_2(l_1 i i_1)^2}, \tag{74}$$

$$b = \frac{k'_2 l_1 i i_1}{k_2 l_1^2 + k'_2(l_1 i i_1)^2}. \tag{75}$$

Taking into account (74) - (75), the system of equations (70), (71) is converted to the form:

$$m_1\ddot{y}_1 + (k_1 + k_2)y_1 - \frac{k_2^2 l_1^2}{k_2 l_1^2 + k'_2(l_1 i i_1)^2} - \frac{k_2 l_1 k'_2(l_1 i i_1) y_2}{k_2 l_1^2 + k'_2(l_1 i i_1)^2} = k_1 z_1, \tag{76}$$

$$m_2\ddot{y}_2 + (k'_2 + k_3)y_2 - \frac{(k'_2)^2(l_1 i i_1)^2 y_2}{k_2 l_1^2 + k'_2(l_1 i i_1)^2} - \frac{k_2 l_1 k'_2(l_1 i i_1)}{k_2 l_1^2 + k'_2(l_1 i i_1)^2} = k_3 z_2. \tag{77}$$

Table 4 shows the coefficients of equations (76), (77).

Table 4. Coefficients of equations in coordinates y_1 and y_2

a_{11}	a_{12}
$m_1p^2 + k_1 + \frac{k_2k'_2(l_1ii_1)^2}{k_2l_1^2 + k'_2(l_1ii_1)^2}$	$-\frac{k_2k'_2l_1(l_1ii_1)}{k_2l_1^2 + k'_2(l_1ii_1)^2}$
a_{21}	a_{22}
$-\frac{k_2k'_2l_1(l_1ii_1)}{k_2l_1^2 + k'_2(l_1ii_1)^2}$	$m_2p^2 + k_3 + \frac{k_2k'_2l_1^2}{k_2l_1^2 + k'_2(l_1ii_1)^2}$
Q_1	Q_2
k_1z_1	k_2z_2

Note: Q_1, Q_2 are generalized forces corresponding to generalized coordinates y_1 and y_2. The structural diagram of the system according to Figure 9*b* is shown respectively in Figure 16.

Figure 16. Structural diagram of the system as per Figure 19b

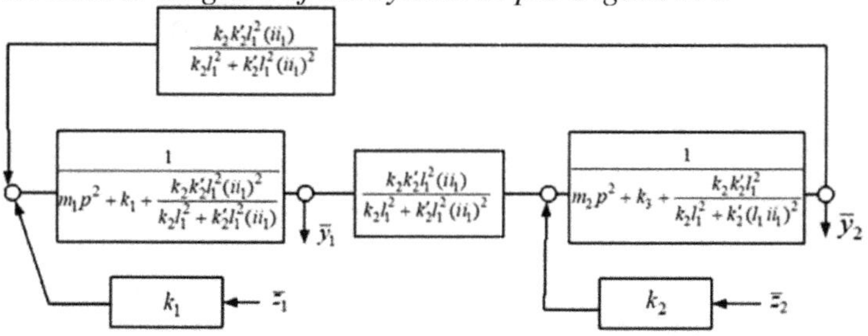

It follows from the structural diagram that the introduction of lever and elastic elements forms a system in which partial blocks have connections determined by the transfer function

$$W'(p) = \frac{k_2k'_2(ii_1)}{k_2 + k'_2(ii_1)^2}. \tag{78}$$

In the structural diagram in Figure 16, a number of equivalent transformations can be made (Figure 17), assuming that

$$ii_1 = i_0. \tag{79}$$

Figure 17. Structural diagram of the system as per Figure 9b, reduced to chain view

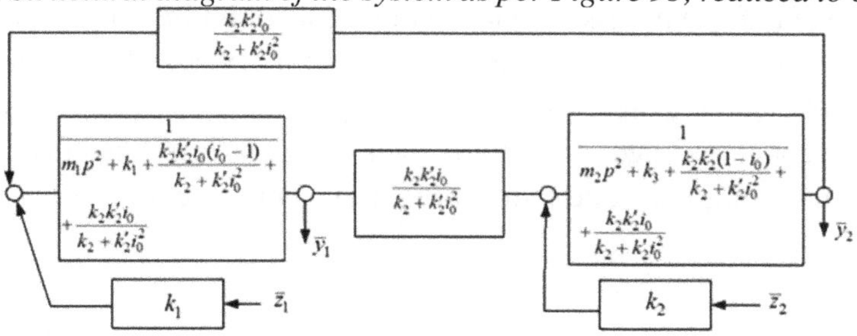

The given transformations for the design scheme in Figure 9*b* are similar to the transformations for the diagram in Figure 9*a*. If you do not take into account the signs of the gear ratio, then for the scheme in Figure 17 there is a system in which positive feedback is introduced between the partial blocks. In this case, the characteristic equation will be as follows:

$$\left(m_1 p^2 + k_1 + \frac{k_2 k'_2 i_0^2}{k_2 + k'_2 i_0^2}\right)\left(m_2 p^2 + k_3 + \frac{k_2 k'_2}{k_2 + k'_2 i_0^2}\right) - \frac{(k_2 k'_2 i_0)^2}{(k_2 + k'_2 i_0^2)^2} = 0. \qquad (80)$$

Let's take a closer look at the structure with the gear ratio $i_0 = ii_1$. Note that $i = l_2/l_1$ is the gear ratio of the lever of the first kind (Figure 9*b*), for which $i = -l_2/l_1$. In turn, $i_1 = l_2/l'_2$ is the transmission ratio of the externally geared gear train allowing rotation of the transmission members in opposite directions; therefore, $i_1 = -l_2/l'_2$. The combination of the two ratios will thus always be positive. The above allows us to conclude that the connection of two levers of the first kind through an external gearing turns the entire block into a lever of the second kind, which in itself is an unconventional representation of the features of lever links.

Thus, leverage in the structures of mechanical systems containing elastic and mass-inertial elements creates spatial (in this case, two-dimensional) interactions. When analyzing static equilibrium, the system can be considered at the levels of separation of structural formations from elastic elements and levers, in turn, interconnected by a gear mechanism. Such structural formations can also be called compacts or quasi-springs, which can have quite complex schemes and consist of various combinations of levers and springs. It is noteworthy that the quasi-spring behaves in static transformations, for example, when determining the reduced stiffness of the mechanical system (stiffness at the point of force application) as well as an ordinary spring in the form of a typical elementary link.

The structural representations considered make it possible to propose a method for determining the reduced stiffnesses based on the use of transfer functions for these purposes. By simple transformations, the calculated transfer functions make it possible to obtain the necessary data on the reduced rigidities of the system and, thereby, take into account the structural and technical features of the system. Such a method for solving dynamics problems opens up the possibility of introducing and taking into account leverage in structural interpretations of mechanical oscillatory systems with the subsequent determination of dynamic reactions in the prescribed manner.

FEATURES OF MECHANICAL OSCILLATORY SYSTEMS, CONTAINING SOLID LINKS

Let us consider the features of mechanical oscillatory systems formed by structural typical elements in the form of devices for converting motion and links as solids containing flat motion. Such devices can be hinge-lever mechanisms, non-self-braking screw and gear mechanisms, etc. At the same time, the emergence of devices for transforming motion can be considered from the standpoint of the general transformation of the structures of mechanical oscillatory systems based on the use of link coupling methods, both an approach for simplifying models and a method for constructing elementary links that are unconventional for the theory of mechanical oscillatory systems.

Figure 18 shows the design diagram of the vibration protection system (ATS), in which there are two units marked with circuits I and II. Block II is based on a solid body with mass and moment of inertia; in addition, elastic ($k_{01,}$ $k_{0,}$ k, $k1$, $k2_{,)}$ elements and the object of protection are involved in the system. It is assumed that the forces of resistance to movement in coordinates are quite small.

Figure 18. Design scheme oscillatory ATS having two interaction loops

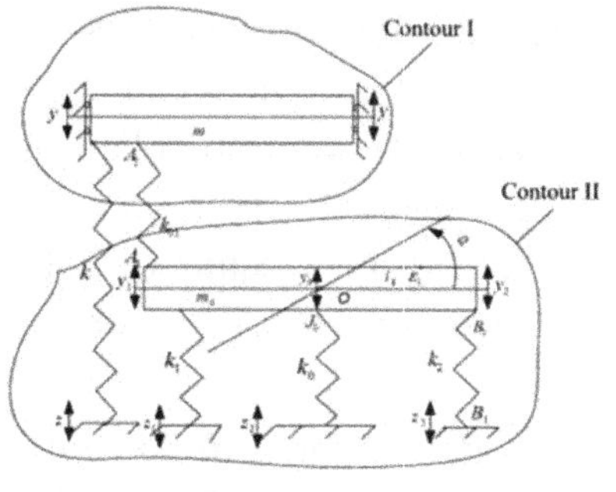

Circuits I and II (Figure 18) interact through an elastic binding element k_{01}. In turn, the solid rests not only on elastic supports k_1 and k_2, but has an elastic connection k_0, the line of action of which passes through the center of gravity of the solid at point O. Let us assume that $A_1O=l_1$, $B_2O=l_2$, and the attached mass m does not cause significant changes in the mass inertia parameters of the system. Developing the provisions on joints of solids as joints taking the form of a rotary hinge, we note that, considering the stiffnesses k_{01} and k_0 large enough, we can convert the design scheme to a view, as shown in Figure 19.

Figure 19. Converted design diagram containing joints

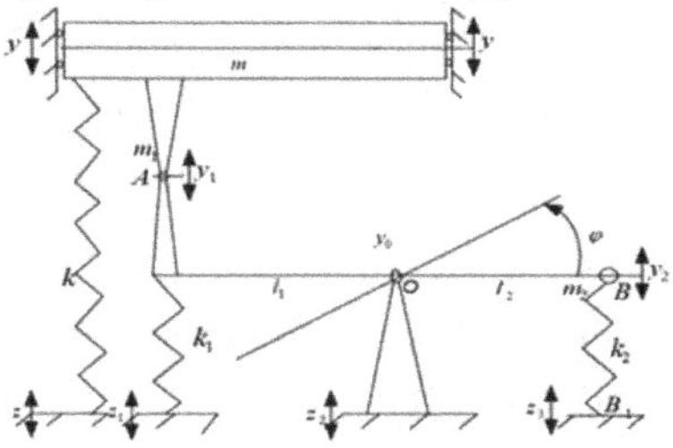

The resulting simplification is shown in Figure 18. calculation diagram in Figure 19. is given in Kiryukhin, Tikhonov, Chistyakov, and Yablonsky (2012), in which it was used to justify the increase in the efficiency of the damping systems of the transport device. However, the question of the peculiarities of the joints has not been studied.

Mathematical Model Construction Methodology

Expressions for kinetic and potential energies for the design scheme of the IBF shown in Figure 19 have the form

$$T = \frac{1}{2}(m + m_1)\dot{y}^2 + \frac{1}{2}m_2\dot{y}_2, \tag{81}$$

$$\Pi = \frac{1}{2}k(y - z)^2 + \frac{1}{2}k_2(y - z_1)^2 + \frac{1}{2}k_2(y_2 - z_3)^2, \tag{82}$$

where $\dot{y}_2$ is the velocity of the element with mass m_2 in absolute motion, determined by the expression

$$\dot{y}_2 = -\frac{l_2}{l_1}\dot{y} + \dot{z}_2. \tag{83}$$

Understanding the gear ratio of the lever in the form$i = \frac{l_2}{l_1}$, we convert (81), (82) to the form

$$T = \frac{1}{2}(m + m_1)\dot{y}^2 + \frac{1}{2}m_2[-\dot{y}i + \dot{z}_2(1 + i)]^2, \tag{84}$$

$$\begin{aligned} \Pi = \frac{1}{2}k(y - z)^2 + \frac{1}{2}k_1(y - z_1)^2 + \frac{1}{2}k_2(y_{2аб} - z_3)^2 = \frac{1}{2}k(y - z)^2 + \\ + \frac{1}{2}k_1(y - z_1)^2 + \frac{1}{2}k_2[-yi + z_2(1 + i) - z_3]^2. \end{aligned} \tag{85}$$

We obtain for the system (Figure 19) the differential equation of motion

$$\begin{aligned} \ddot{y}(m + m_1 + m_2 i^2) + y(k + k_1 + k_2 i^2) = m_2 i(1 + i)\ddot{z}_2 + \\ + z_2 k_2(1 + i) + k_1 z_1 + kz - k_2 i z_3. \end{aligned} \tag{86}$$

To simplify the calculations, we assume that $z_1 = z_2 = z_3$; $k_1 = 0$ and $k_2 = 0$ and find the transfer function of the system

$$W(p) = \frac{\bar{y}(p)}{\bar{z}(p)} = \frac{m_2(1 + i)ip^2 + k}{(m_1 + m + m_2 i^2)p^2 + k}. \tag{87}$$

As an example, Figure 20 shows a family of amplitude-frequency characteristics built on the basis of (87) when the parameter changes i from 0 to 3 in increments of 0.5. The following system parameters are taken as initial values: m=100*kg*; *kg*; k=10000N/m. Figure 20 ω_1–ω_6 shows the dynamic blanking frequencies. Corresponding values of natural oscillation frequencies in comparison with frequencies are given in Table 5.

Figure 20. Transmission function system frequency response family (87)

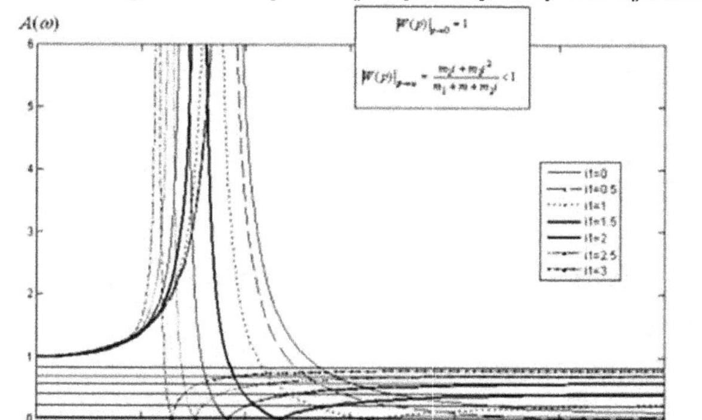

Table 5. Natural vibration frequencies in comparison with dynamic damping frequencies ω_1 - ω_6 for the system in Figure 20

i **(gear ratio)**	**0,5**	**1**	**1,5**	**2**	**2,5**	**3**
Value of dynamic blanking frequencies, 1/sec	ω_{dyn1}	ω_{dyn2}	ω_{dyn3}	ω_{dyn4}	ω_{dyn5}	ω_{dyn6}
	25.81	15.81	11.54	9.12	7.55	6.45
Value of natural frequencies fluctuations	ω_{sob1}	ω_{sob2}	ω_{sob3}	ω_{sob4}	ω_{sob5}	ω_{sob6}
	8.94	8.45	7.78	7.07	6.38	5.77

The peculiarity of the amplitude-frequency characteristics lies in the fact that as *i* increases, the dynamic blanking frequencies shift to the left, that is, the frequency difference ω_{sob} - ω_{dyn} decreases. At the same time, the frequency of natural oscillations decreases, but much slower. At high frequencies, the transmission coefficient of the vibration amplitude after the dynamic blanking mode tends to a limit value $|W(p)|_{p\to\infty} = \dfrac{m_2 i + m_2 i^2}{m_1 + m + m_2 i} < 1$ (the larger i, the larger the value$|W(p)|_{p\to\infty}$).

If $i \to \infty$ we get that$|W(p)|_{p\to 0} = \dfrac{m_2 i + m_2 i^2}{m_1 + m + m_2 i} = 1$.

The mathematical model of the system (86) can also be obtained by simplifying a more complex model (Figure 19), described not by two (y, y_2), but by three coordinates: $y_0, y, \varphi; y, y_1, y_0; y_1, y_2, y$.

Building a Complete Mathematical Model

Let's return to the calculation diagram shown in Figure 21. Then the expressions (81) and (82) for kinetic and potential energies can be converted to the form

$$T = \tfrac{1}{2}m_0\dot{y}_0^2 + \tfrac{1}{2}J_0\dot{\phi} + \tfrac{1}{2}m\dot{y}^2, \tag{88}$$

$$\begin{aligned} \Pi = {} & \tfrac{1}{2}k(y - z)^2 + \tfrac{1}{2}k_1(y_1 - z_1)^2 + \tfrac{1}{2}k_0(y_0 - z_2)^2 + \\ & + \tfrac{1}{2}k_2(y_2 - z_3)^2 + \tfrac{1}{2}k_{01}(y - y_1)^2, \end{aligned} \tag{89}$$

wherein $y_0 = ay_1 + by_2$, $\varphi = c(y_2 - y_1)$.

In the diagram (Figure 18), the following designations are adopted: m_0- the mass of the intermediate body (can turn into m_1 and m_2 connected by a lever); m - weight of the object; k, k_1, k_0, k_2 - elastic elements of the intermediate body resting on the base; z - z_3 - kinematic disturbances. The coordinates of vols. A_1 and A_2 (Figure 19) are determined as follows: $y_{A1} = y_1$ and $y_{A2} = y$. To obtain joints between the elements, the following conditions must be met

$$y_1 - y = 0, \; y_{10} = 0, \tag{90}$$

Let's take the coordinate values:

$$y_{10} = y_0 - z_2. \tag{91}$$

1. consider the movement of the system in the coordinates y_0, φ. Let's accept the designations

$$a = \frac{l_2}{l_1 + l_2};\ b = \frac{l_1}{l_1 + l_2};\ c = \frac{1}{l_1 + l_2}. \tag{92}$$

When the protection object moves in the coordinate system y_0, y_2 and y_1, the expression for kinetic energy can be reduced to the form:

$$T = \frac{1}{2}m_0(a\dot{y}_1 + b\dot{y}_2)^2 + \frac{1}{2}J_0c^2(\dot{y}_2 - \dot{y}_1)^2 + \frac{1}{2}m\dot{y}^2. \tag{93}$$

The expression for the potential energy in this case takes the form

$$\begin{aligned} \Pi = \frac{1}{2}k(y - z)^2 + \frac{1}{2}k_1(y_1 - z_1)^2 + \frac{1}{2}k_0(ay_1 + by_2 - z_2)^2 + \\ + \frac{1}{2}k_2(y_2 - z_3)^2 + \frac{1}{2}k_{01}(y - y_1)^2. \end{aligned} \tag{94}$$

The system of differential equations of motion of the system can be represented in the form:

$$\begin{aligned} \ddot{y}_1(m_0a^2 + J_0c^2) + y_1(k_1 + k_{01}k_0a^2) + \ddot{y}_2(m_0ab - J_0c^2) + \\ + y_2k_0ab = -k_{01}yk_1z_1 + k_0az_2, \end{aligned} \tag{95}$$

$$\begin{aligned} \ddot{y}_1(m_0ab - J_0c^2) + y_1(-k_0ab) + \ddot{y}_2(m_0b^2 + J_0c^2) + \\ + y_2(k_0b_2 + k_2) = k_0bz_2 + k_2z_3, \end{aligned} \tag{96}$$

$$y_1k_{01} + \ddot{y}m + y(k + k_{01}) = kz. \tag{97}$$

Table 6 shows the coefficients of the equations in the coordinates y_1, y_2 and y.

Table 6. Values of equation coefficients (95-97) in coordinates y_1, y_2 and y

a_{11}	a_{12}	a_{13}
$(m_0a^2 + J_0c^2)p^2 + k_1 + k_0a^2$	$(m_0ab - J_0c^2)p^2 + k_0ab$	$-k_{01}$
a_{21}	a_{22}	a_{23}
$(m_0ab - J_0c^2)p^2 + k_0ab$	$(m_0b^2 + J_0c^2)p^2 + k_0b^2 + k_2$	0
a_{31}	a_{32}	a_{33}
$-k_{01}$	0	$mp^2+ k+ k_{01}$

The generalized forces in y, y_2, and y_1 are:

$$Q_{y_1} = k_1z_1 + k_0az_2;\; Q_{y_2} = k_0bz_2 + k_2z_3;\; Q_y = kz. \tag{97'}$$

Equations (95 - 97) describe the movement in the coordinate system reflecting the vertical movements of the mass-inertial elements of the IBD. Such a mathematical model can be called basic. The peculiarity of the matrix is that $=a_{13}$, $-k_{01}= a_{23}$; $a_{32}=$ 0; it depends on the features of dynamic interactions determined by the structure of the IBD (Figure 18), and the choice of the system of generalized coordinates.

Let's move on to the coordinate system, y_0 and φ; we write down a number of relations $y= y_{11}+ y_1= y_{11}+ y_0-l_1\varphi$ and get an expression for the kinetic and potential energies:

$$T = \frac{1}{2}m_0(\dot{y}_0)^2+\frac{1}{2}J_0\dot{\phi}^2 + \frac{1}{2}m(\dot{y}_{11} + \dot{y}_0-l_1\dot{\phi})^2 \tag{98}$$

$$\begin{aligned}\Pi = {} & \frac{1}{2}k(y_{11} + y_0 - l_1\phi - z)^2 + \frac{1}{2}k_1(y_0 - l_1\phi - z_1)^2 \\ & +\frac{1}{2}k_0(y_0 - z_2)^2 + \frac{1}{2}k_2(y_0 + l_2\phi - z_3)^2 + \frac{1}{2}k_{01}(y_{11})^2.\end{aligned} \tag{99}$$

We write the equations of motion of the system in coordinates φ, y_0, y_{11}:

$$\begin{aligned}&\ddot{\phi}(J_0 + ml_1^2)+\ddot{\phi}(kl_1^2 + k_1l_1^2 + k_2l_2^2)+\ddot{y}_0(-ml_1)+y_0(-kl_1 + k_1l_1 + k_2l_2)+ \\ &+\ddot{y}_{11}(-ml_1) + y_{11}(-kl_1) = -kl_1z + k_1l_1z_1 + k_2l_2z_3,\end{aligned} \tag{100}$$

$$\begin{aligned}&\ddot{\phi}(-ml_1)+\phi(-kl_1 + k_1l_1 + k_2l_2)+\ddot{y}_0(m + m_0)+y_0(k + k_1 + k_0 + k_2)+ \\ &+\ddot{y}_{11}(m) + y_{11}(k) = kz - k_1z_1 + k_0z_2 - k_2z_3,\end{aligned} \tag{101}$$

$$\ddot{\phi}(-ml_1)+\phi(-k_1l_1)+\ddot{y}_0(m)+y_0(k) + \ddot{y}_{11}(m) + y_{11}(k + k_{01}) = kz. \tag{102}$$

The values of coefficients of equations (100) – (102) in coordinates φ, y_0 and y_{11} respectively are given in Table 7.

Table 7. Coefficients of equations (100) – (102) in coordinates φ, y_0 and x

a_{11}	a_{12}	a_{13}
$(ml_1^2 + J_0)p^2 + kl_1^2 + k_1l_1^2 + k_2l_2^2$	$(-ml_1)p^2 - kl_1 + k_1l_1 + k_2l_2$	$(-ml_1)p^2 - kl_1$
a_{21}	a_{22}	a_{23}
$(-ml_1)p^2 - kl_1 + k_1l_1 + k_2l_2$	$(m_0 + m)p^2 + k + k_1 + k_0 + k_2$	$(m)p^2 + k$
a_{31}	a_{32}	a_{33}
$(-ml_1)p^2 - k_1l_1$	$(m)p^2 + k$	$(m)p^2 + k + k_{01}$

Generalized forces by coordinates φ, y_0 and $\ddot{y}_{11}$ have the form:

$$Q_\varphi = -kl_1z + k_1l_1z_1 + k_2l_2z_3;\ Q_{y_0} = kz + k_1z_1 + k_0z_2 - k_2z_3;\ Q_{y11} = kz. \quad (103)$$

The mathematical model in the coordinate system φ, y*0* and y*1*$_1$ differs from the previous model (coordinates y_1, y_2 and *y)* in that there are no zero cells (Table 7), that is, there are connections between all partial systems. As for the coordinate, $y_{11} = y - y_1$ it can be "zeroed" by an assumption with the formation $k_{01} \to \infty$ of an articulation or kinematic pair. Under these conditions, it is possible to join the solids m and m_1. Physically, this means that the mass m is attached to the IBD element in volume A with mass-inertial parameters m0, J_0 and changes them (both the total mass and the moment of inertia). The movement of the system will be described in this case by the coordinates y_0 and $_\varphi$. The necessary data to obtain the transfer functions can be obtained by excluding the column and row containing y11 *(*in fact, the variable is eliminated, and the order of the matrix is reduced by one).

For further calculations, enter the coordinate system y_0, y_{11} and y_1. Then the expression for kinetic energy in the coordinates y_0, y_{11} and y_1 is converted to the form:

$$T = \tfrac{1}{2}m_0(\dot{y}_0)^2 + \tfrac{1}{2}J_0\dot{\phi}^2 + \tfrac{1}{2}m(\dot{y}_1 + \dot{y})^2. \quad (104)$$

We replace the coordinate system with respect to the expression for kinetic energy, assuming that $y = y_1 + y_{11}$; $y_{11} = y - y_1$, $\varphi = c(y_2 - y_1)$; $y_0 = ay_1 + by_2$; $y_2 = \dfrac{y_0 - ay_1}{b}$.
We will write

$$\phi = cy_2 - cy_1 = \frac{c \cdot (y_0 - ay_1)}{b} - cy_1 = \frac{cy_0 - cay_1 - bcy_1}{b} = \tfrac{c}{b}(y_0 - y_1),$$

down and find $\varphi = a_0(y_0 - y_1)$ where $a_0 = \tfrac{c}{b}$, $y_0 = ay_1 + by_2$;

$$y_2 = \frac{y_0 - ay_1}{b} = \tfrac{1}{b}y_0 - \tfrac{a}{b}y_1 = a_{10}y_0 - a_{01}y, a_{10} = \tfrac{1}{b}, a_{01} = \tfrac{a}{b}$$

We write the expression for the potential energy in expanded form in coordinates y_0, y_1 and y_{11}:

$$\Pi = \frac{1}{2}k(y_1 + y_{11} - z)^2 + \frac{1}{2}k_1(y_1 - z_1)^2 + \frac{1}{2}k_0(y_0 - z_2)^2 + \\ + \frac{1}{2}k_2(a_{10}y_0 - a_{01}y_1 - z_3)^2 + \frac{1}{2}k_{01}(y_{11})^2, \tag{105}$$

The equations of motion of the system in the coordinates y_0, y_{11} and y_1 are as follows:

$$\ddot{y}_1(Ja_0^2 + m) + y_1(k + k_1 + k_2a_{01}^2) + \ddot{y}_0(-Ja_0^2) + y_0(-k_2a_{01}a_{10}) + \\ + \ddot{y}_{11}(-m) + y_{11}(k) = kz - k_2a_{01}z_3 + k_1z_1\ ; \tag{106}$$

$$\ddot{y}_1(-Ja_0^2) + y_1(-k_2a_{01}a_{10}) + \ddot{y}_0(m_0 + J_0a_0^2) + \\ + y_0(k + k_2a_{10}^2) = k_0z_2 + k_2a_{01}z_3\ ; \tag{107}$$

$$\ddot{y}_1(-m) + y_1(k) + \ddot{y}_{11}(m) + y_{11}(k + k_{01}) = -kz. \tag{108}$$

Table 8 shows the values of coefficients of equations (107) - (108) in coordinates y_0, y_{11} and y_1.

Table 8. Values of coefficients of equations (107) - (108) in coordinates y_1, y_0 and y_{11}

a_{11}	a_{12}	a_{13}
$(m + J_0a_0^2)p^2 + k + k_1 + k_2la_{01}^2$	$(-J_0a_0^2)p^2 - k_2a_{01}a_{10}$	$(-m)p^2+k$
a_{21}	a_{22}	a_{23}
$(-J_0a_0^2)p^2 - k_2a_{01}a_{10}$	$(m + J_0a_0^2)p^2 + k + k_2a_{10}^2$	0
a_{31}	a_{32}	a_{33}
$(-m)p^2+k$	0	$(m)p^2+k+k_{01}$

Generalized forces by coordinates y_0, y_{11} and y_1 have the form:

$$Q_{y11} = -kz;\ Q_{y_0} = k_0z_2 + k_2a_{01}z_3;\ Q_{y_1} = kz - k_2a_{01}z_3 + k_1z_1. \tag{109}$$

In this coordinate system y_0, y_{11} and y_1, it is also possible to introduce a joint along the coordinate y_{11}. To obtain a transfer function of a system that has two degrees of freedom y_0 and y_1, it is necessary to exclude the corresponding columns and row. Note that if $y_{11} = 0$, then $y_1 = y$, the movement of the system with one hinge at point *A* will be described by the coordinates y_0 and y_1. In this case, the lever has an elastic support for its center of rotation. The case with two joints is of the greatest interest.

Consider the coordinate system y_0, y_{11}, and y_1. Introducing a number of relations ($y_{10}= y–z_2$), we write down the expressions for the kinetic energy of the system:

$$T = \frac{1}{2}m_0(\dot{y}_{10} + \dot{z}_2)^2 + \frac{1}{2}J_0 a_0{}^2(\dot{y}_{10} + \dot{z}_2 - \dot{y} + y_{11})^2 + \frac{1}{2}m(\dot{y})^2, \tag{110}$$

where $a_0 = \frac{c}{b}$.

The expression for the potential energy of the system in the coordinates y_0, y_{11} and y_1 has the form:

$$П = П_1 + П_2 + П_3 + П_4 + П_5, \tag{111}$$

where

$$\Pi_1 = \frac{1}{2}k(y - z)^2 = \frac{1}{2}k(y^2 - 2yz + z^2) \ ;$$
$$\Pi_2 = \frac{1}{2}k_1(y^2 - 2y_{11} + y_{11}{}^2 - 2z_1 y + 2z_1 y_{11} + z_1^2) \ ;$$
$$\Pi_3 = \frac{1}{2}k_0(y_{10})^2 \ ;$$
$$\Pi_4 = \frac{1}{2}k_2(a_{10}y_{10} - a_{01}y + a_{01}y_{11} + z_0)^2,$$

where $z_0 = a_{10}z_2 - z_3$. Let's make a number of auxiliary calculations and write down the equations of motion of the system in the coordinates y_0, y_{11} and y_1:

$$\begin{aligned} &\ddot{y}_{10}(m_0 + J_0 a^2) + y_{10}(k_0 + k_2 a_{10}^2) + \ddot{y}(-J a_0^2) + y(-k_2 a_{10} a_{01}) + \\ &+\ddot{x}(J_0 a_0^2) + x(k_2 a_{10} a_{01}) = -k_2 a_{10} z_0 - m_0 \ddot{z}_2 - J_0 a^2 \ddot{z}_2 \ ; \end{aligned} \tag{112}$$

$$\begin{aligned} &\ddot{y}_{10}(-J_0 a_0^2) + y_{10}(-k_2 a_{10} a_{01}) + \ddot{y}(m + J a_0^2) + y(k + k_1 + k_2 a_{01}^2) + \ddot{x}(-J_0 a_0^2) + \\ &+x(-k_1 - k_2 a^2{}_{01}) = J_0 a_0{}^2 \ddot{z}_2 + kz + k_1 z_1 + k_2 a_{01} z_0 \ ; \end{aligned} \tag{113}$$

$$\begin{aligned} &\ddot{y}_{10}(J_0 a_0{}^2) + y_{10}(k_2 a_{10} a_{01}) + \ddot{y}(-J a_0^2) + y(-k_1 - k_2 a^2{}_{01}) + \ddot{x}(-J_0 a_0^2) + \\ &+x(k_1 + k_2 a^2{}_{01} + k_{01}) = -J_0 a_0^2 \ddot{z}_2 - k_1 z_1 - k_2 a_{01} z_0. \end{aligned} \tag{114}$$

Table 9 shows the coefficients of equations (112) - (114).

Table 9. Values of coefficients of equations (112) - (114) in coordinates y_1, y_0 and y_{11}

a_{11}	a_{12}	a_{13}
$(m_0 + J_0 a_0{}^2)p^2 + k_0 + k_2 a_{10}^2$	$(-J_0 a_0{}^2)p^2 - k_2 a_{10} a_{01}$	$(J_0 a^2)p^2 + k_2 a_{10} a_{01}$
a_{21}	a_{22}	a_{23}
$(-J_0 a_0{}^2)p^2 - k_2 a_{10} a_{01}$	$(m + J_0 a_0{}^2)p^2 + k + k_2 a^2{}_{01} + k_1$	$(-J_0 a_0{}^2)p^2 - k_1 - k_2 a^2{}_{01}$
a_{31}	a_{32}	a_{33}
$(J_0 a_0{}^2)p^2 + k_2 a_{10} a_{01}$	$(-J_0 a_0{}^2)p^2 - k_1 - k_2 a^2{}_{01}$	$(J_0 a_0{}^2)p^2 + k_1 + k_2 a^2{}_{01} + k_{01}$

Generalized forces by coordinates y_0, y_{11} and y_1 have the form

$$Q_{y_{10}} = -(m_0 + J_0 a_0{}^2)\ddot{z}_2 - k_2 a_{10} z_0 \ ;$$
$$Q_y = (J_0 a_0{}^2)\ddot{z}_2 + kz + k_1 z_1 + k_2 a_{01} z_0; \ Qy_{11} = -(J_0 a_0{}^2)\ddot{z}_2 - k_1 z_1 - k_2 a_{01} z_0. \quad (115)$$

When $z_0 = a_{10} z_2 - z_3$ we get the following expressions for generalized forces:

$$Q_{y_{10}} = -(m_0 + J_0 a_0{}^2)\ddot{z}_2 - k_2 a_{10} z_0 = -(m_0 + J_0 a_0^2)p^2 - k_2 a_{10}^2 z_2 - k_2 a_{10} z_3 \ ; \quad (116)$$

$$\begin{aligned} Q_y &= -(J_0 a_0{}^2)\ddot{z}_2 + kz + k_1 z_1 + k_2 a_{01} z_0 \\ &= J_0 a_0^2 p^2 + kz + k_1 z_1 + k_2 a_{10} a_{01} z_2 - k_2 a_{01} z_3, \end{aligned} \quad (117)$$

$$Q_{y11} = -(J_0 a_0{}^2)\ddot{z}_2 - k_1 z_1 - k_2 a_{10} z_0$$

$$= -J_0 a_0^2 p^2 - k_1 z_1 - k_2 a_{10} a_{01} z_2 + k_2 a_{10} z_3. \quad (118)$$

In this coordinate system, it is possible to enter two joints: by coordinate y_{10} and by coordinate y_{11}. Using the matrix (Table 9) and excluding the corresponding rows and columns, we obtain the equation of motion for a system with two joints.

$$(m_0 + J_0 a_0^2)p^2 + k + k_2 a_{01}^2 + k_1 = J_0 a^2{}_0 \ddot{z}_2 + kz + k_1 z_1 + k_2 a_{01} z_0. \quad (119)$$

For simplification, we accept $k_2=0$; $k_1=0$, $z_1=0$, $z_3=0$ with this $z=z_2$, then equation (119) is converted to the form

$$\bar{y}(m_0 + J_0 a_0^2)p^2 + k = (J_0 a^2{}_0 p^2 + k)\bar{z}, \quad (120)$$

Where the transfer function of the system can be found

$$W(p) = \frac{y}{z_2} = \frac{J_0 a_0^2 p^2 + k}{m + J_0 a_0^2 + k}. \quad (121)$$

Comparison (121) and (87) shows that the structures of the transfer functions coincide, however, the moment of inertia of the solid $J_0 a^2_0$ does not yet reflect the features of the design scheme if it consists of two material points m_1 and m_2 connected by a rod.

To obtain a complete coincidence of the results, it is necessary to provide a more detailed calculation scheme. For output (87), the diagram shown in (Figure 19) was used. The peculiarity of this design scheme is to take into account the mass inertia properties of the linkage. The design diagram (Figure 21) indicates the masses m_1 and m_2; taking into account the features of their movement is an essential factor for the coincidence of expressions (121) and (87).

Figure 21. Design scheme of ATS in Figure 19, but with spaced masses m_1 and m_2

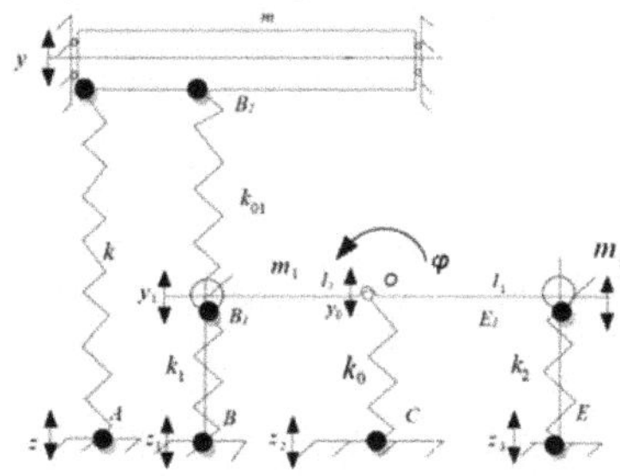

We will select the coordinate system y_0, y_{11} and y_1 for further calculations, assuming that $y_{10} = y_0 + z$, $y_{11} = y - y_1$, where $y_0 = ay_1 + by_2$.

Let's write down the expressions for the kinetic and potential energies of the system, taking into account the rotation of the B_1E_1 lever relative to the point O:

$$T = \frac{1}{2}m\dot{y}^2 + \frac{1}{2}m_1\dot{y}_1^2 + \frac{1}{2}m_2\dot{y}_2^2, \quad (122)$$

$$\Pi = \frac{1}{2}k(y - z)^2 + \frac{1}{2}k_1(y_1 - z_1)^2 + \frac{1}{2}k_2(y_2 - z_3)^2 + \\ + \frac{1}{2}k_{01}(y - y_1)^2 + \frac{1}{2}k_0(y_0 - z_2)^2. \quad (123)$$

Let's make a number of transformations:

$$y_2 = \frac{y_0 - ay_1}{b} = \frac{a_0 y_0 - iy_1}{b},\ i = \frac{a}{b} = \frac{l_2}{l_1};\ a_0 = \frac{1}{b};$$

$$y_2 = a_0 y_{10} + a_0 z_2 - iy + i y_{11} . y_0 = y_{10} + y_{11},$$

and we get (122), (123) in the form:

$$T = \frac{1}{2} m\dot{y}^2 + \frac{1}{2} m_1 (\dot{y} - \dot{x})^2 + \frac{1}{2} m_2 (a_0 \dot{y}_{10} + a_0 \dot{z}_2 - i\dot{y} + i\dot{y}_{11})^2, \quad (124)$$

$$\Pi = \frac{1}{2} k(y - z)^2 + \frac{1}{2} k_1 (y - y_{11} - z_1)^2 + \frac{1}{2} k_2 (a_0 y_{10} - iy + i y_{11} + z_0)^2 + \\ + \frac{1}{2} k_{01} (y_{11})^2 + \frac{1}{2} k_0 (y_{10})^2, \quad (125)$$

where $z_0 = a_0 z_2 - z_3$. Having made a number of auxiliary calculations similar to those given above in the derivation of Lagrange equations of the II kind, we obtain the equations of motion

$$\ddot{y}(m_1 + m_2 i^2 + m) + y(k + k_1 + k_2 i^2) + \ddot{y}_{11}(-m_1 - m_2 i^2) + y_{11}(-k_1 - k_2 i^2) + \\ + \ddot{y}_{10}(-m_2 i a_0) + y_{10}(-k_2 a_0 i) = m_2 a_0 i z_2 + kz + k_1 z_1 + k_2 z_0 i \ ; \quad (126)$$

$$\ddot{y}(-m_1 - m_2 i^2) + y(-k_1 - k_2 i^2) + \ddot{y}_{11}(m_1 + m_2 i^2) + y_{11}(k_1 + k_{01} + k_2 i^2) + \\ + \ddot{y}_{10}(m_2 i a_0) + y_{10}(k_2 a_0 i) = -m_2 a_0 i \ddot{z}_2 - k_1 z_1 - k_2 z_0 i \ ; \quad (127)$$

$$\ddot{y}(-m_2 i a_0) + y(-k_2 i^2 a_0) + \ddot{y}_{11}(m_2 i a_0) + y_{11}(k_2 i a_0) + \\ + \ddot{y}_{10}(m_2 a_0^2) + y_{10}(k_2 a_0^2 + k_0) = -m_2 a_0^2 \ddot{z}_2 - k_2 a_0 z_0. \quad (128)$$

Table 10 shows the coefficients of equations (126) - (128).

Table 10. Coefficients of equations (126) ÷ (128) in y, y_{11} and y_{10} coordinates

a_{11}	a_{12}	a_{13}
$(m_1 + m + m_2 i^2)p^2 + k + k_1 + k_2 i^2$	$(-m_1 - m_2 i^2)p^2 - k_1 - k_2 i^2$	$-m_2 i a_0 p^2 - k_2 a_9 i$
a_{21}	a_{22}	a_{23}
$(-m_1 - m_2 i^2)p^2 - k_1 - k_2 i^2$	$(m_1 + m_2 i^2)p^2 + k_1 + k_{01} + k_2 i^2$	$m_2 i a_0 p^2 + k_2 a_0 i$
a_{31}	a_{32}	a_{33}
$-m_2 i a_0 p^2 - k_2 a_0 i$	$m_2 i a_0 p^2 + k_2 a_0 i$	$m_2 a_0^2 p^2 + k_2 a_0^2 + k_0$

Generalized forces in this case have the form:

$$Q_y = m_2 a_0 i z_2 + kz + k_1 z_1 + k_2 z_0 i, \\ Q_x = -a_0 i m_2 \ddot{z}_2 - k_1 z_1 - k_2 i z_0, \\ Q_{y_{10}} = -m_2 a_0^2 \ddot{z}_2 - k_2 a_0 z_0. \quad (129)$$

Excluding columns and rows from the matrix by coordinates y_{11} and y_{10}, we find the equation of motion for the system with the coordinate y

$$\ddot{y}\left(m_1 + m + m_2 i^2\right) + y\left(k + k_1 + k_2 i^2\right) = m_2 a_0 i \ddot{z}_2 + kz + k_1 z_1 + k_2 z_0 i. \tag{130}$$

To obtain the transfer function "input offset," $y\ z_2$ we assume that $z=z_2$, $z_1=0$, $z_3=z_2$, $k_1=0$, $k_2=0$.

$$W(p) = \frac{\bar{y}}{\bar{z}_2} = \frac{m_2(i+1)a_0 i p^2 + k}{(m + m_1 + m_2 i^2)p^2 + k}. \tag{131}$$

Note that $\frac{1}{b} = \frac{l_1 + l_2}{l_1} = 1 + i$, then

$$W(p) = \frac{\bar{y}}{\bar{z}_2} = \frac{m_2 i(i+1)p^2 + k}{\left(m + m_1 + m_2 i^2\right)p^2 + k}. \tag{132}$$

Expression (87) and (131) match. A similar result in determining the transfer function (131) can be obtained at $k_{01}\ \infty$ and $k_0\ \infty$.

Thus, by selecting the system of generalized coordinates accordingly, it is possible to construct a mathematical model of a lever-linked mechanical system, simplifying the original high-order ones. In this case, the system receives fewer degrees of freedom than in the original situation. Articulation is possible between two bodies when connecting two bodies into a kinematic pair of a rotational type (V class). However, it is also possible to connect a solid body with another body with the loss of the possibility of relative movement. Mathematical model of system with joints is obtained on the basis of use of matrices of coefficients of equations of motion in coordinates, selection of which corresponds to selected type of joints. The formation of the coordinate system presupposes the introduction of coordinates of relative motion characterizing a possible joint, followed by "zeroing" of the generalized coordinate.

A similar result, as could be seen, can be obtained not only by excluding (or "zeroing") the coordinates of relative motion, but also by increasing the stiffness of the springs in relative relationships (in our case, $k_{01}\ \infty$, $k_0\ \infty$). Attention to the details of the above transformations is due to the fact that the separation of the joint or kinematic pair makes it possible to connect the two mating points with an elastic element.

The studies carried out are devoted to the search and development of methods and means for assessing and the possibilities of changing the dynamic state of mechanical oscillatory systems. In solving the problems of machine dynamics, in which the design schemes are often precisely mechanical oscillatory systems with one, two (or more) degrees of freedom, over the past years, there have been trends in the development of more general approaches based on the use of structural interpretations of mathematical models of systems described by ordinary linear and differential equations with constant coefficients. Structural analogues of differential equations obtained by known methods have the potential to search for newdesign solutions related to the expansion of the set of typical elementary links, the possibility of constructing more complex structures that interact with each other as quasi-elements according to the rules of structural transformations of circuit theory and automatic control theory. The basis for evaluating dynamic properties in cases of periodic perturbations close to harmonic are the transfer functions of the system. The subject of research with this approach is the peculiarities of establishing dependencies on input influences (vibrations of various types) and coordinates that determine the dynamic state of the protected objects. In the theory and practice of vibration protection systems, more attention is paid to the coordinates and speeds of points of protection objects in various versions of tasks for assessing and monitoring dynamic states. To a lesser extent, attention is paid to the force reactions between the elements of the system that occur in mechanical oscillatory systems, as well as in contacts with support surfaces.

With all the development of approaches related to the kinetostatics of machines and mechanisms, as well as issues of assessing the dynamic state of machines, their components and parts, mechanical oscillatory systems have their own specifics. The reason is that oscillatory systems differ from generally accepted ideas about mechanisms as mechanical chains consisting of solids connected by kinematic pairs. In mechanical oscillatory systems, traditional representations are associated with the perception of a traditional set of typical elements of mechanical oscillatory systems in the form of elastic elements (springs), dissipative devices (damper) and mass inertia elements in the form of material points or solids.

A comparative analysis of the technological foundations of building mathematical models, calculation schemes, the peculiarities of using various elements in the composition of oscillatory systems gives reason to believe that it is possible to develop some general approaches that reflect systemic ideas about the properties of oscillatory systems in general and its individual elements, in particular.

The possibilities of comparing a mathematical model of a mechanical oscillatory system, written in the form of a corresponding system of linear differential equations, in the form of a dynamically equivalent structural scheme of an automatic control system, make it possible to propose new methods for describing the dynamic properties of mechanical oscillatory systems and methods for determining the features of states associated with determining the conditions for the occurrence of certain modes, including the assessment of dynamic and static reactions.

In most machines, weight forces, as well as special constant forces applied to certain points or units of machines, form static reactions, which can be considered unchanged under the action of periodic forces (with a zero average component). The movements of machine elements lead to the appearance of dynamic reactions, which are summed up with static ones. The sum of static and dynamic reactions can be considered as a complete reaction. In calculations related to determining the possibilities of reliable operation, not only the value of the complete reaction is important, but also its sign. Changing the sign of the complete reaction can lead to contact opening, collisions and other undesirable effects.

The presented work proposes a method for determining static and dynamic reactions, which is based on the transformations of the initial structural model of the mechanical oscillatory system. The idea of the method is based on the idea that an elastic element in an oscillatory system is a feedback chain for an object making periodic movements of the relative position of static equilibrium. In this approach, the elastic element or spring is identified with a typical element of structural diagrams of automatic control systems. Dampers and other devices can be considered the same typical elements. Each of these typical elements has elementary transfer functions, respectively, amplifying, differentiating link, differentiating link of the second kind, etc.

Such typical links are similar to others, since the input to such links is displacement (some coordinate), and the output is force (elastic force, resistance force, inertial force, etc.).

More complex structures can be formed from such typical elements, which have the property of quasi-elements. The construction of complex structures follows the rules of parallel and sequential connections of springs.

Note that the advantage of evaluating the dynamic properties of systems based on the use of structural diagrams is that basic information can be obtained using only elastic elements (meaning the transfer function). With regard to other properties of the system, the dampers and motion conversion devices are simply connected in parallel to the respective contact points of the resilient members. Which, in particular, is observed when the springs and the damper are connected in parallel.

Returning to the essence of the proposed method for determining force reactions in an oscillatory mechanical system, it consists in the fact that the object of protection on the basis of the obtained system of differential equations of motion is distinguished and presented in the form of a second-order integrating link, and the released part of the mechanical system is organized into feedback. In this case, the coupling, in the physical sense, is some generalized spring (quasi-spring) having a well-defined dynamic stiffness. This makes it easy to determine dynamic coupling responses. Static reactions can also be determined on the basis of this approach if we assume that the mass of the object of protection is zero. In practical calculations, it should be taken into account that all inertial forces become zero.

The second section describes the methodological foundations of the method for determining static and dynamic reactions using numerous examples. The evidentiary basis of the method relies on the use of the Dalembert principle.

The methodological base for determining the reaction is based on algorithms for structural transformations of structural schemes, which serves not only as a means of obtaining analytical expressions for various reactions, but also allows us to disclose some poorly understood dynamic properties of systems that contain solids and devices for transforming motion.

As an example, theoretical justifications for building mathematical models of lever dynamic vibration dampers are given. Important, in our opinion, for subsequent applications is the idea of building compacts, which can include typical links, leverage mechanisms and other mechanisms, ultimately forming a quasi-element or a system of quasi-elements interacting according to the rules of structural transformations.

Structural formations of several typical elements, built according to certain rules, have quasi-spring properties. The latter, in their more complex interactions, form structures according to the same rules.

In systems containing solids, transformations are possible either with the introduction of coordinates of relative motion and their subsequent "zeroing" or it is possible to use the effect of a limit transition to an elastic element with infinitely high rigidity. In both cases, a joint is introduced into the system: in this case, the number of degrees of freedom is reduced, and a lever mechanism appears in the system, which brings new dynamic properties to the system.

REFERENCES

Bakalov, A. N. (2006). *Theory of chains*. Nauka.

Bolshakov, R. S., & Eliseev, S. V. (2013). Methodological foundations of problems of determining bond reactions in mechanical oscillatory systems. IrGUPS. (Deposited manuscript in VINITI 29.11.2013 No. 366)

Buryan, Y. A., Sorokin, B. N., & Galuza, Y. F. (2011). Active hydromechanical system for damping car vibrations. *Omsk Scientific Bulletin*, *103*, 122–126.

Eliseev, A. V., Selvinsky, V. V., & Eliseev, S. V. (2015). *Dynamics of vibration interactions of elements of technological systems, taking into account non-holding ties*. Nauka.

Eliseev, S. V. (2018). Applied system analysis and structural mathematical modeling (dynamics of transport and technological machines: connectivity of movements, vibration interactions, leverage) (A. I. Artyunin, Ed.). IrGUPS.

Eliseev, S. V., & Artyunin, A. I. (2016). *Applied theory of oscillations in problems of dynamics of linear mechanical systems*. Nauka.

Gardner, F., & Bairns, J. L. (1949). *Transients in linear systems with concentrated constants*. GITTL.

Khomenko, A. P., Eliseev, S. V., & Bolshakov, R. S. (2014). Features of the interaction of partial systems in a vibration protection circuit with two degrees of freedom: Leverage in dynamic damping of vibrations. Modern Technologies. Systems Analysis. *Modeling*, *3*(43), 8–19.

Kiryukhin, A. V., Tikhonov, V. A., Chistyakov, A. G., & Yablonsky, V. V. (2012). Active vibration protection - purpose, principles, condition. Active vibration protection and noise insulation of pipelines and experimental research. *Problems of Mechanical Engineering and Automation*, (4), 102–110.

Veitz, V. L., Kachura, A. E., & Martynenko, A. M. (1971). Dynamic calculations of machine drives. *Mechanical Engineering (New York, N.Y.)*.

ADDITIONAL READING

Artyunin, A. I., Eliseev, S. V., Kaimov, E. V., & Eliseev, A. V. (2014). Sensor for determining boundary parameters of interaction of bodies in vibration systems. Utility patent RU 148250 U1.

Eliseev, A. V. (2013). Peculiarities of interaction of a material particle with a vibrating surface depending on an additional force with a non-holding bond. *International Journal of Applied and Basic Research*, (3), 9–15.

Eliseev, A. V. (2013). Features of the implementation of the multiple toss mode in a model problem with non-holding coupling in the presence of viscous friction. *Science in Central Russia*, (2S), 42–47.

Eliseev, A.V. (2014). Technology for assessing the properties of dynamic interaction in contacts of composite solids. Scientific problems of transport in Siberia and the Far East, (1-2), 179-183.

Eliseev, A. V. (2015). Dynamics of vibration interactions of process system elements taking into account non-retaining connections. In V. S. Balbarov (Ed.), Problems of mechanics of modern machines: Materials of the VI International Conference (p. 93-98). Moscow: Publishing house.

Eliseev, A.V., Kuznetsov, N.K., & Moskovskikh, A.O. (2019). Machine dynamics. System representations, structure diagrams, and element relationships. Moscow: Publishing house.

Eliseev, A. V., Mamaev, L. A., & Sitov, I. S. (2015). Some approaches to justification of inertial excitation scheme in process vibration machines. Systems. Methods. *Technologies*, (4), 15–24.

Eliseev, A.V., Selvinsky, V.V., & Eliseev, S.V. (2015). Dynamics of vibration interactions of process system elements taking into account non-retaining connections. Novosibirsk: Publishing house.

Eliseev, A. V., & Vyong, K. Ch. (2016). Some control capabilities of the one-dimensional vibration field of the process machine. Modern technology. Systems analysis. *Modeling*, (1), 33–41.

Eliseev, S. V., & Eliseev, A. V. (2012). Modes of tossing a material particle on a vibrating surface in a model problem with not-holding ties. Modern technology. Systems analysis. *Simulation*, (3), 86–96.

Eliseev, S. V., & Eliseev, A. V. (2013). Determination of the coefficient of viscous friction for the mode of multiple tossing of a material particle in a model problem with not-holding ties. Systems. Methods. *Technologies*, (1), 22–27.

Eliseev, S. V., & Eliseev, A. V. (2013). Study of the interaction of a material particle with a vibrating surface in the presence of a viscous friction force in a model problem with not-holding ties. Modern technology. Systems analysis. *Modeling*, (1), 69–77.

Eliseev, S. V., & Eliseev, A. V. (2013). Peculiarities of gap occurrence in mechanical system with non-retaining connection under pulse action. Modern technology. Systems analysis. *Simulation*, (2), 36–45.

Eliseev, S. V., & Eliseev, A. V. (2013). Generalized approaches to the problems of determining contact reactions in solids under static loads, taking into account non-holding bonds. Modern technology. Systems analysis. *Simulation*, (4), 51–60.

Eliseev, S. V., & Eliseev, A. V. (2014). Defines the contact reactions in composite solids under dynamic loads with respect to non-holding bonds. Modern technology. Systems analysis. *Simulation*, (1), 45–54.

Eliseev, S.V., Eliseev, A.V., Bolshakov, R.S., & Khomenko, A.P. (2021). Methodology of system analysis in tasks of assessment, formation and control of dynamic state of technological and transport machines. Novosibirsk: Publishing house.

Eliseev, S. V., Kashuba, V. B., Pnev, A. G., Eliseev, A. V., & Sitov, I. S. (2014). Non-retaining bonds in dynamic interactions of loose medium and vibrating surface: Scientific and methodological substantiation of vibration hardening process technology. Systems. Methods. *Technologies*, (3), 17–31.

Khomenko, A.P., Eliseev, S.V., Artyunin, A.I., Eliseev, A.V., Bolshakov, R.S., & Kaimov, E.V. (2013). Elastic elements in mechanical systems. Structural interpretations. Deposited manuscript No. 230-V 2013.

Kletschkowski, T. (2023). Theoretical and non-dimensional investigations into vibration control using viscoelastic and endochronic elements. *Vibration*, *6*(4), 1030–1047. DOI: 10.3390/vibration6040060

Lewis, R. C., & Wrisley, D. L. (1950). A system for the excitation of pure natural modes of complex structure. *Journal of the Aeronautical Sciences*, *17*(11), 705–723. DOI: 10.2514/8.1780

Moheimani, S. O. R., Halim, D., & Fleming, A. J. (2003). *Spatial control of vibration: Theory and experiments*. World Scientific Publishing Co Pte. Ltd. DOI: 10.1142/5246

Peng, L., Jiang, H., Chen, X., Liu, D., Feng, H., Zhang, L., Zhao, Y., & Liu, C. (2019). A review on the advanced design techniques and methods of vibrating screen for coal preparation. *Powder Technology*, *347*, 136–147. DOI: 10.1016/j.powtec.2019.02.047

Sitov, I. S., & Eliseev, A. V. (2012). Theoretical foundations of the processes of interaction of a material particle with a vibrating surface with non-retaining bonds. Systems. Methods. *Technology*, (4), 19–29.

Chapter 3
Some Applications of Structural Mathematical Modeling Methods

ABSTRACT

The application of structural mathematical modeling methods in various fields is considered, including the problems of vibration isolation and protection of technical objects. Attention is paid to the concept of a lever and its manifestation in mechanical oscillatory systems. Dynamic interactions in mechanical systems are described, where partial systems reflect the properties of solids making angular vibrations. In such cases, the lever is realized in the form of levers of the first and second kind, their various forms of simple and complex connections. Examples of the use of two-wire groups in problems of dynamics of flat mechanisms are given. Peculiarities of lever development in mechanical chains and mechanisms are discussed, which requires development of specific methods for construction of mathematical models of technical objects and analysis of their dynamic properties.

INTRODUCTION

As was shown in the previous chapters, leverage and mechanisms are widely represented in various structural and technical forms characteristic of solving many problems of machine dynamics, including in the tasks of vibration isolation and vibration protection of technical objects. In conventional mechanical oscillatory systems, leverage manifests itself in specific forms, which is quite simply disclosed

DOI: 10.4018/979-8-3373-0447-2.ch003

using structural mathematical models of the system and the corresponding apparatus for frequency analysis of dynamic systems.

In a more visual form, leverage manifests itself when considering dynamic interactions in mechanical oscillatory systems, where partial systems reflect the properties of solids making angular vibrations. In such cases, leverage is realized in the form of levers of the first and second kinds, their various forms of simple and complex connections. Mechanical oscillating systems as design diagrams of technical systems and vibration protection systems, in particular, can include various structural formations from standard elementary links, including mechanisms consisting of solid bodies connected by certain kinematic pairs. Most often, such formations are considered in the form of separate blocks having zero degrees of freedom or the so-called Assur groups.

Two-lead groups are widely used in problems of dynamics of flat mechanisms. (Dimentberg & Kolesnikov, 1980) reflected the issues of evaluation of dynamic properties of vibration protection systems, in which linkage is implemented by articulated linkage mechanisms. As for the display of dynamic properties of this kind, it is necessary to take into account the peculiarities of the movement of such structural formations, since the vibration protection system is considered, as a rule, in states of small oscillations relative to the position of static equilibrium. Similar cases are typical for the dynamics of vehicles and vibration-type technological machines.

Although, as shown in the previous chapters, linkage is also characteristic of conventional mechanical oscillating systems, a wide variety of linkage is distinguished by mechanical systems that include devices for transforming motion, as well as mechanical chains and mechanisms of various nature. Such approaches are often implemented in the designs of suspensions of transport machines, in the development of vibration technological machines, during the creation of which the issues of creating certain spatial structures of the vibration field are solved.

Taking into account the peculiarities of leverage introduced by mechanical chains and mechanisms, it is necessary to develop quite specific techniques for building mathematical models of technical objects and analyzing their dynamic properties. Further materials use some structural and technical solutions protected by patents for utility models.

ADDITIONAL MASSES IN THE LINKAGE STRUCTURE

Mechanisms in the structures of vibration protection systems are most often considered in modes of small vibrations relative to the position of static equilibrium or steady motion. In such cases, the mechanisms implement the functions of superimposing additional connections that create certain conditions in solving specific problems of the dynamics of machines and equipment. In the future, the methodological foundations of approaches to the construction of mathematical models of mechanical vibration systems, in particular, vibration protection systems, in the structure of which there are mechanisms as additional elements of the system, are developing.

Additional Masses Location Options

A generalized approach is being developed related to structural interpretations of mechanisms as part of mechanical oscillatory systems using concepts about devices with motion transformation. As such, such a determination can be attributed to any mechanism, since the transmission of motion in a mechanical circuit is somehow related to the transformation of motion. A motion conversion device (MCD) can be called a non-self-braking screw mechanism with the functions of a typical second-order differentiating link. The screw mechanism can, in a certain formulation of tasks, be interpreted as a kind of lever in which a constant ratio connects the parameters of the rotational and translational motion of the elements of the kinematic pair. Figure 1*a* - *e* show options for the arrangement of mechanisms in the basic design scheme (Figure 1*a*), where the protection object with mass M is supported by an elastic element with rigidity k and a generalized mechanism designated MCD. In turn, Figure 1*b* shows one of the interpretations of the MCD in the form of a screw mechanism. In subsequent studies, hinge-lever mechanisms are also used (Lurie, 1968).

1. In this case, harmonic disturbances are considered: power Q (t) and kinematic z (t); uses the coordinate system associated with the fixed basis. Small movements of the system relative to the position of static equilibrium are assumed, there are no resistance forces, and the system has linear properties.

Figure 1. Vibration protection systems with devices for motion conversion: a - basic calculation scheme with a supporting elastic element and a device for motion conversion (MCD); b - MCD in the form of a non-self-braking mechanism; c - MCD in the form of a hinge-lever mechanism with additional weight and extension of the lower lever; d - MCD with bent lower lever; e - MCD - with additional weight on extended upper lever; f - variable speed drive with curved upper lever

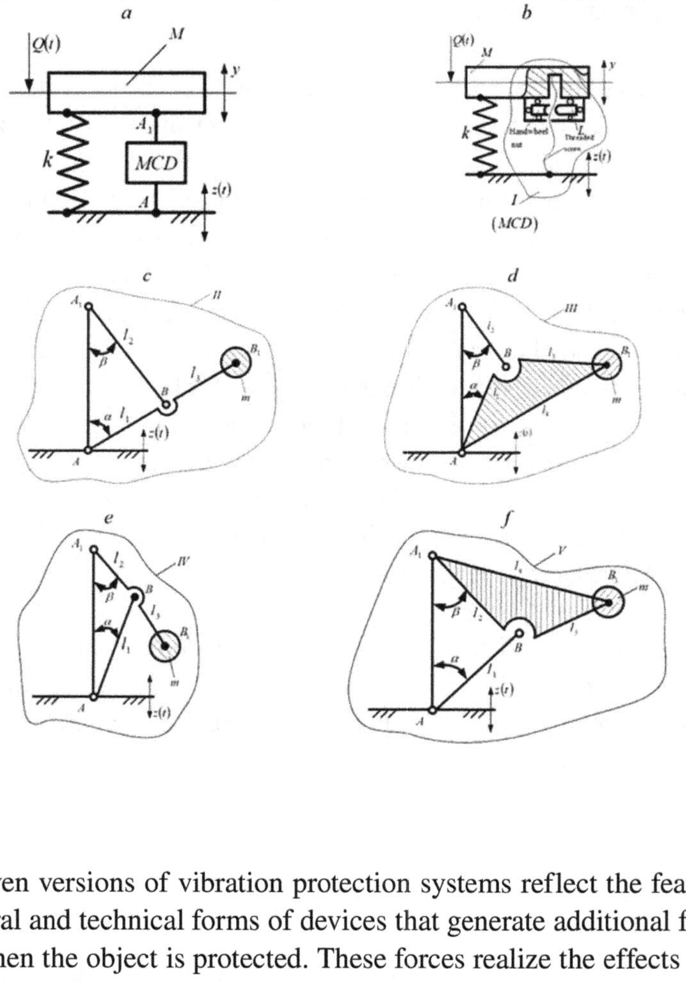

2. The given versions of vibration protection systems reflect the features of the structural and technical forms of devices that generate additional forces of inertia when the object is protected. These forces realize the effects of dynamic interactions of elements, which is manifested through changes in the frequency characteristics of systems and the creation of certain dynamic modes, in particular, dynamic damping modes.

The system (Figure 1*a*, *b*) uses a supporting elastic element with stiffness k; object of protection with mass M, and mass inertia parameters (L and m) introduced by motion conversion devices are also taken into account.

Figure 1*b* shows a vibration protection system with a device in the form of a non-self-braking screw mechanism (circuit *I*). At relative motion of protection object *M*, elastic and additional inertial forces are generated as a result of relative motions of elements. In this case, *L* is a parameter characterizing the reduced mass of the MCD, depending on the moment of inertia of the flywheel nut, the radius of the screw and the angle of inclination of the helical line. A feature of the diagram in Figure 1*b* is the fact that the axis of rotation of the flywheel nut is in the plane of movement *y* of the protected object. Resistance forces in the screw mechanism (as well as for subsequent cases in Figure 1*c* - *e*) are not taken into account. In (Eliseev, Lukyanov, Reznik, & Khomenko, 2006) the dynamics of mechanical oscillatory systems with MCD according to the diagram in Figure 1*b* is presented in sufficient detail, which, to a certain extent, initiated attention to taking into account the dynamic features created by other mechanisms.

Flat hinge-lever mechanisms (Figure 1*c* - *e*) differ in that the axes of the rotary hinges are located perpendicular to the axis of the plane of motion *y* of the protected object. Loops II ÷ V in Figure 1, *c* - *e* show options for the location of additional loads with a mass of *m* to ensure the tuning of vibration protection systems in the simplest ways, which is achieved, for example, by simply changing the position of the additional mass *m* (l_3 changes).

In addition, the diagram of Figure 1*d* and Figure 1*e* show the use of curved arms. The latter is connected with possibility of additional masses transfer to service area of adjustment mechanisms or movement. Upper and lower levers in Figure 1c - *e* are considered weightless rigid rods. It is assumed that external disturbances can be power (they are applied directly to the object of protection *M*) or kinematic, which is associated with base vibrations. Harmonic forms of vibrations and small movements of all elements of vibration protection systems relative to static equilibrium position are considered. Given in Figure 1, *the* parameters of vibration protection systems: the angles of inclination of the levers α and β, the lengths of the links l_1 ÷ l_4, the additional mass *m* determine the configuration of the vibration protection system and, therefore, its dynamic properties. When tuning the vibration protection system, these parameters can change in one way or another, which, in particular, refers to the study of the effects arising from various types of external disturbance.

Mathematical Models of the System. Forced Oscillations

When analyzing the systems in Figure 1, it is necessary to take into account the peculiarities of the action of power and kinematic external disturbances.

Force disturbance ($Q \neq 0$, $z = 0$)**.** The diagram in Figure 1c. Point O_1 in Figure 2 is the instantaneous velocity center. The velocity of point *B* can be found from the relations:

$$\dot{y}_B = \dot{y} \cdot a \tag{1}$$

$$a = \frac{i \cdot \cos\beta}{\sin\alpha \cdot (\cos\alpha + i \cdot \cos\beta)} \tag{2}$$

In this case, the gear ratio $i = \frac{l_2}{l_1}$ of the lengths of the upper and lower rods.

Figure 2. Kinematic diagram of the system shown in Figure 1c

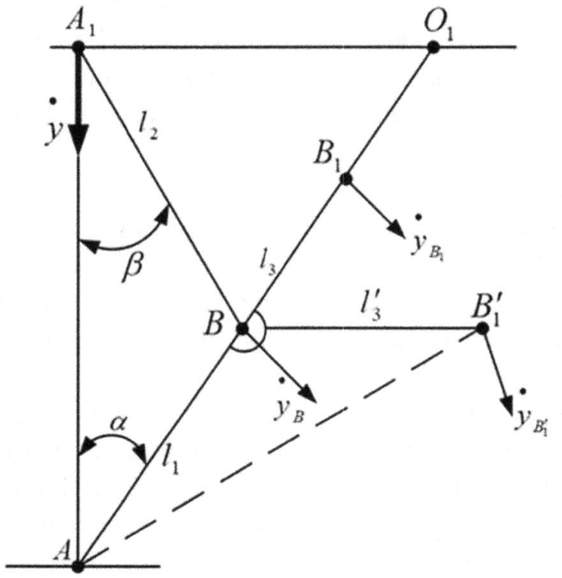

1. Determination of kinematic parameters. The angular velocity of the lower arm relative to point A can be found

$$\omega_1 = \frac{\dot{y}_B}{AB} = \frac{\dot{y}_B}{l_1} = \frac{y \cdot a}{l_1} = \dot{y} \cdot \frac{i \cdot \cos\beta}{l_1 \cdot \sin\alpha \cdot (\cos\alpha + i \cdot \cos\beta)} \tag{3}$$

The speed of point B_1 in this case is determined by the expression:

$$\dot{y}_{B_1} = \omega_1 \cdot (l_1 + l_3) = \dot{y} \cdot \frac{i \cdot (l_1 + l_3) \cdot \cos\beta}{l_1 \cdot \sin\alpha \cdot (\cos\alpha + i \cdot \cos\beta)} = \dot{y} \cdot b \tag{4}$$

$$b = \frac{i \cdot (l_1 + l_3) \cdot \cos\beta}{l_1 \cdot \sin\alpha \cdot (\cos\alpha + i \cdot \cos\beta)} \tag{5}$$

If the scheme uses a curved lever in connection with the position of the p.B'_1, then the velocity vector p.B'_1will be directed perpendicular to the AB_1. As for the length$A\,B'_1$, it can be defined as the length l_4 of the geometric relations

$$\left|\dot{y}_{B'_1}\right| = \dot{y} \cdot \frac{i \cdot l_4 \cdot \cos\beta}{l_1 \cdot \sin\alpha \cdot (\cos\alpha + i \cdot \cos\beta)} \tag{6}$$

If there is a curved lever, you can enter the parameter

$$b_1 = \frac{i \cdot l_4 \cdot \cos\beta}{l_1 \cdot \sin\alpha \cdot (\cos\alpha + i \cdot \cos\beta)} \tag{7}$$

Increasing the distance l_4 (or l_3) leads to an increase in the parameter b. To compile the equation of motion of the system, the diagram of which is shown *in* Figure 1, then we compose expressions for the kinetic and potential energies

$$\mathrm{T} = \frac{1}{2}M(\dot{y})^2 + \frac{1}{2}m\left(\dot{y}_{B'_1}\right)^2 \tag{8}$$

$$\Pi = \frac{1}{2} \cdot k(y - z)^2 \ (8')$$

The equation of motion of the system will take the form

$$\ddot{y} \cdot \left(M + m b_1^2\right) + ky = Q \tag{9}$$

From (9), you can go to the structural diagram of the system (Figure 3*a* - d) using the transformations of the methodological methods described in the previous sections.

Figure 3. Structural diagrams of vibration protection systems corresponding to equation (8): a - the system has reduced weight; $(M + mb_1^2)$ *b - MCD is an additional feedback on acceleration as per relation to the protected object; c - MCD as a parallel link; d - elastic element and MCD can be connected according to the rules of parallel connection of springs*

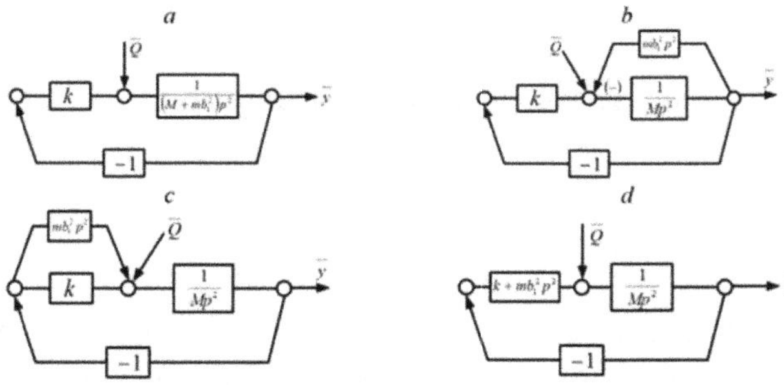

Thus, in case of power disturbance, the MCD is transformed into a link with the transfer function of a typical second-order differentiating element. The methods of introducing such an additional bond correspond to the rules of transformation in the structural theory of vibration protection systems. The transfer function of the system can be determined from equation (9) or directly from the structural diagrams in Figure 3*a* - *d*; in this case

$$W(p) = \frac{\bar{y}}{\bar{Q}} = \frac{1}{(M + mb_1^2) \cdot p^2 + k} \tag{10}$$

From (10) you can find the natural frequency of the system:

$$\omega_{\text{sob}}^2 = \frac{k}{M + mb_1^2} \tag{11}$$

The introduction of the TPC may be considered as a method of controlling the natural frequency. For these purposes, the parameters of the levers, the additional mass *m* and the angles of installation of the links can vary. If the α setting angle is small, then the reduced mass:

$$M_{\text{pr}} = M + mb_1^2 \tag{12}$$

can take larger values by increasing the parameter b_1.

Perturbation caused by the motion of the datum surface. In case of kinematic disturbance, the initial system (Figure 1*c*) will behave differently, since the movement of the base ($z \neq 0$, $Q = 0$) forms additional forces of inertia. These forces can be thought of as the inertia forces of portable motion. In particular, the speed of point B will be the sum of the speeds of motion created by the oscillations of the base, $\left(\dot{y}_{B_{prt}}\right)$as well as the movement of the object of protection$\left(\dot{y}_{B_{rlv}}\right)$. Similarly, the speeds of pp. B_1 andB'_1. For kinematic calculations, let's build a schematic diagram, as shown in Figure 4.

Figure 4. Kinematic diagram for calculation of parameter ratios at kinematic disturbance

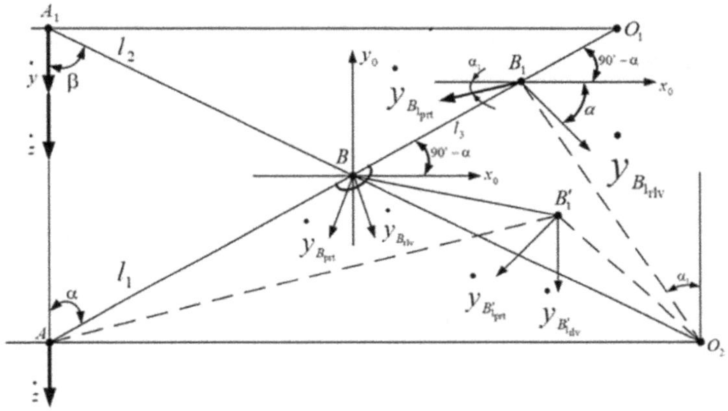

2. Determination of point speeds. Find the speed of point B, while:

$$\bar{\dot{y}}_B = \bar{\dot{y}}_{B_{rlv}} + \bar{\dot{y}}_{B_{prt}} \tag{13}$$

$\dot{y}_{B_{rlv}}$ can be defined from expression (1)

$$\left|\dot{y}_{B_{rlv}}\right| = y \cdot a$$

wherein a is used in accordance with Expression (2).

The velocity vector $\dot{y}_{B_{rlv}}$ is perpendicular to the AO_1 (O_1 is the instantaneous center of velocities at,)$\dot{z} = 0$. $\dot{y} \neq 0$In turn, the velocity vector is perpendicular to the $\dot{y}_{B_{prt}}$ A_1O_2 (O_2 is the instantaneous center of velocities at $\dot{z} \neq 0$, $\dot{y} = 0$). The magnitude $\dot{y}_{B_{rlv}}$ is given by

$$\dot{y}_{B_{rlv}} = \frac{\dot{z} \cdot BO_2}{AO_2} \tag{14}$$

We find that $AO_2 = (l_1 \cdot \cos\alpha + l_2 \cdot \cos\beta)$, $BO_2 = A_1O_2 - l_2$,$BO_2 = l_1\frac{\cos\alpha}{\cos\beta}$,

$$A_1O_2 = \frac{l_1\cos\alpha + l_2\cos\beta}{\cos\beta}$$

and get

$$\dot{y}_{B_{пер}} = \frac{\dot{z} \cdot l_1 \cdot \cos\alpha}{\sin\beta \cdot (l_1\cos\alpha + l_2\cos\beta)} = \frac{\dot{z} \cdot \cos\alpha}{\sin\beta \cdot (\cos\alpha + i\cos\beta)} \tag{15}$$

In this case,

$$\dot{y}_{B_{пер}} = \dot{z} \cdot a_1 \tag{16}$$

where

$$a_1 = \frac{\cos\alpha}{\sin\beta \cdot (\cos\alpha + i\cos\beta)} \tag{17}$$

Since, by entering $a = \frac{i \cdot \cos\beta}{\sin\alpha \cdot (\cos\alpha + i\cos\beta)}$ the relations $(\cos\alpha + i\cos\beta) = \frac{i \cdot \cos\beta}{\sin\alpha \cdot a}$, $(\cos\alpha + i\cos\beta) = \frac{\cos\alpha}{\sin\beta \cdot a_1}$, we find

$$\frac{i \cdot \cos\beta}{\sin\alpha \cdot a} = \frac{\cos\alpha}{\sin\beta \cdot a_1}$$

$$\frac{a_1 \cdot i \cdot \cos\beta}{\sin\alpha} = \frac{\cos\alpha \cdot a}{\sin\beta}$$

$$a_1 = a\frac{\cos\alpha \cdot \sin\alpha}{\sin\beta \cdot i \cdot \cos\beta}.$$

The velocity vector of the point B, is projected on the coordinate axis y_0 and x_0 in accordance with Figure 4 (in projections on the axis y_0 and x_0), then

$$\dot{y}_{B_{x_0}} = \dot{y}_{B_{rlv_{x0}}} - \dot{y}_{B_{prt_{x0}}} = \dot{y}_{B_{rlv}} \cdot \cos\alpha - \dot{y}_{B_{prt}} \cdot \cos\beta \tag{18}$$

$$\dot{y}_{B_{y_0}} = -\dot{y}_{B_{rlv}} \cdot \sin\alpha - \dot{y}_{B_{prt}} \cdot \sin\beta \tag{19}$$

while

$$\dot{y}_{B_{prt}} = a_1 \cdot \dot{z} \quad (20)$$

$$\dot{y}_{B_{rlv}} = a \cdot \dot{y} \quad (21)$$

where $a = \frac{i \cdot \cos\beta}{\sin\alpha \cdot (\cos\alpha + i\cos\beta)}$; $a_1 = \frac{\cos\alpha}{\sin\beta \cdot (\cos\alpha + i\cos\beta)}$.
It follows from (18) that

$$\dot{y}_{B_{x_0}} = a \cdot \dot{y} \cdot \cos\alpha - a_1 \cdot \dot{z} \cdot \cos\beta \quad (22)$$

$$\dot{y}_{B_{y_0}} = -a \cdot \dot{y} \cdot \sin\alpha - a_1 \cdot \dot{z} \cdot \sin\beta \quad (23)$$

then one can obtain

$$\dot{y}_B = \sqrt{\begin{array}{c} a^2 \cdot (\dot{y})^2 \cos^2\alpha - 2aa_1\dot{y}\dot{z} \cdot \cos\alpha \cdot \cos\beta + a_1^2 \cdot (\dot{z})^2 \cos^2\beta + a^2 \cdot (\dot{y})^2 \sin^2\alpha + \\ +2aa_1\dot{y}\dot{z} \cdot \sin\alpha \cdot \sin\beta + a_1^2 \cdot (\dot{z})^2 \sin^2\beta \end{array}}$$

$$V_B^2 = a^2 \cdot (\dot{y})^2 + a_1^2 \cdot (\dot{z})^2 + 2aa_1\dot{y}\dot{z} \cdot (\sin\alpha \cdot \sin\beta - \cos\alpha \cdot \cos\beta) \quad (24)$$

If we accept that $b_2 = \sin\alpha \bullet \sin\beta - \cos\alpha \bullet \cos\beta$ then in this case

$$(\dot{y}_B)^2 = a^2 \cdot (\dot{y})^2 + a_1^2 \cdot (\dot{z})^2 + 2aa_1\dot{y}\dot{z} \cdot b_2 \quad (25)$$

Peculiarities of Dynamic Properties of the System Under Kinematic Disturbances

Let's build a mathematical model of the system under kinematic perturbation. Find expressions for the kinetic and potential energies of the system

$$\begin{aligned} \mathrm{T} &= \frac{1}{2}M(\dot{y} - \dot{z})^2 + \frac{1}{2}m \cdot (\dot{y}_B)^2 \\ &= \frac{1}{2}M(\dot{y} - \dot{z})^2 + \frac{1}{2}m \cdot \left[a \cdot (\dot{y})^2 + a_1 \cdot (\dot{z})^2 + 2aa_1b_2\dot{y}\dot{z}\right] \end{aligned} \quad (26)$$

$$\Pi = \frac{1}{2}k \cdot (y - z)^2 \quad (27)$$

Using the Lagrange equation of the second kind and the Laplace transform, we obtain the expressions for the transfer function under input action $\bar{z}$ and output -$\bar{y}$

$$W(p) = \frac{\bar{y}}{\bar{z}} = \frac{aa_1b_2mp^2 + k}{(M + a^2m)p^2 + k} \quad (28)$$

where b_2= cosα•cosβ – sinα•sinβ.

The natural oscillation frequency of the system does not change compared to the case of power disturbance; however, two dynamic effects can be detected in the system. The first is the dynamic damping mode at frequency

$$\omega^2_{\text{dyn}} = \frac{k}{a a_1 b_2 m} \tag{29}$$

Secondly, with $p \to \infty$ ($p = j\,\omega$ -complex variable$j = \sqrt{-1}$), the transfer function (28) transforms to the form:

$$\left| W(p) \right|_{p \to \infty} = \frac{\overline{y}}{\overline{z}} = \frac{a a_1 b_2 m}{M + a^2 m} \tag{30}$$

If α = β, then: $b_2 = \cos^2\alpha - \sin^2\alpha$ At $\alpha = 45^o$ $b_2 = 0$. If α <45°, then: b2> 0. At α> 45° b_2 <0.

Thus, a mechanism with a mass at point B can be interpreted as an additional connection created by a second-order differentiating link. Link capabilities are determined by the parameters m, a, a_1, b_2. If the attachment point of the additional mass is at point B_1, then the velocity of point B_1 is determined by analogy with (4)

$$\overline{\dot{y}}_{B_1} = \overline{\dot{y}}_{B_{1_{\text{rlv}}}} + \overline{\dot{y}}_{B_{1_{\text{prt}}}} \tag{31}$$

If you design (31) on the y_0, x_0 axis, you get the following relations:

$$\left| \overline{\dot{y}}_{B_{1_{\text{rlv}}}} \right| = \omega_2 \cdot A B_1$$

$$\omega_2 = \frac{\dot{y}_{B_{1_{\text{rlv}}}}}{l_1} \tag{32}$$

$$\left| \overline{\dot{y}}_{B_{1_{\text{rlv}}}} \right| = \frac{\dot{y}_{B_{\text{rlv}}}}{l_1} \cdot (l_1 + l_3)$$

$$\left| \overline{\dot{y}}_{B_{1_{\text{rlv}x_y}}} \right| = \frac{\dot{y}_{B_{1_{\text{rlv}}}}}{l_1} \cdot (l_1 + l_3) \cdot \cos\alpha \tag{33}$$

Here $\overline{\dot{y}}_{B_{1_{\text{rlv}}}}$ is the velocity vector $\perp$ O_2B_1.

$$\dot{y}_{B_{1_{per.s.}}} = \dot{y}_{B_{1_{per}}} \cdot \cos\alpha_1,$$

$$\left|\dot{y}_{B_{1_{per}}}\right| = \frac{\dot{z} \cdot O_2B_1}{AO_2},$$

$$AO_2 = \left(\frac{l_1\cos\alpha + l_2\cos\beta}{tg\beta}\right).$$

In this case, the value O_2B_1 is determined according to the scheme in Figure 4. From (28) it follows that the sign b_2 determines the type of amplitude-frequency characteristic of the system. If $b_2 > 0$, then the numerator (28) will be positive for any p-value. In the event that $p \to \infty$, "locking" is possible, while

$$|W(p)|_{p\to\infty} = \frac{maa_1b_2p^2 + k}{(M + ma^2)p^2 + k} \to \frac{maa_1}{M + ma^2} \tag{34}$$

The situation is similar at $b_2 < 0$ with $p \to \infty$. However, in this case, a dynamic damping mode is possible in the system at the frequency defined by the expression (29). Figure 5 shows the relationship between the α and β angles at $b_2 = 0$.

Figure 5. Plot of α and β at $b_2 = 0$

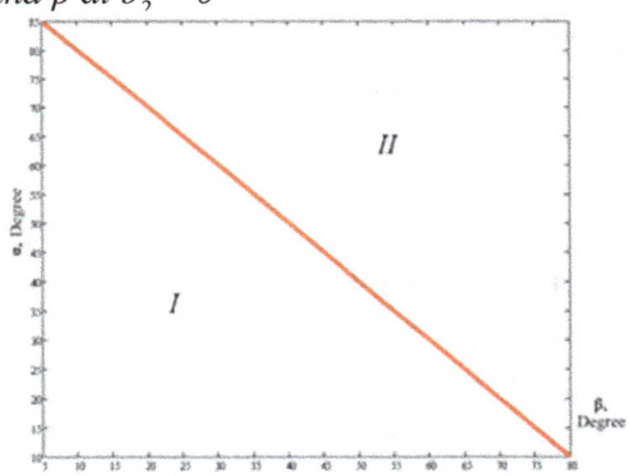

The peculiarity of the relationship between the parameters α and β in Figure 5 is such that the graph divides the space of parameters α and β into two regions indicated by symbols *I* and *II*. In this case, pairs of values from region *I* correspond to $b_2 > 0$, and pairs of values α and β from region *II* correspond to $b_2 < 0$. In the graph itself, for which condition $b_2 = 0$ is met, pairs of α and β values determine the configuration of the system links such that the amplitude-frequency characteristic of the system will have the same appearance as under external force.

Thus:

$$\dot{y}_{B_{1_{x_0}}} = \dot{y}_{B_{1_{rlv}}} \cdot \cos\alpha - \dot{y}_{B_{1_{prt}}} \cdot \cos\alpha_1$$

$$\dot{y}_{B_{1_{y_0}}} = - \dot{y}_{B_{1_{rlv}}} \cdot \sin\alpha - \dot{y}_{B_{1_{prt}}} \cdot \sin\alpha_1 \tag{35}$$

$$\left(\dot{y}_{B_1}\right)^2 = \left[\dot{y}_{B_{1_{rlv}}} \cdot \cos\alpha - \dot{y}_{B_{1_{prt}}} \cdot \cos\alpha_1\right]^2 + \left[\dot{y}_{B_{1_{rlv}}} \cdot \sin\alpha + \dot{y}_{B_{1_{prt}}} \cdot \sin\alpha_1\right]^2 \tag{36}$$

Let's enter a series of relations

$$\dot{y}_{B_{1_{rlv}}} = \frac{\dot{y}_{B_{rlv}}}{l_1} \cdot (l_1 + l_3); \dot{y}_{B_{rlv}} = a \cdot \dot{y}; \dot{y}_{B_{prt}} = a_1 \cdot \dot{z}; \dot{y}_{B_{1prt}} = \frac{\dot{y}_{B_{prt}} \cdot B_1O_2}{BO_2}$$
$$= \frac{\dot{z} a_1 \cdot B_1O_2}{BO_2}; b_3 = \frac{B_1O_2}{BO_2};$$

$$\frac{\dot{z}}{AO_2} = \frac{\dot{y}_{B_{rlv}}}{BO_2}; \dot{y}_{B_{prt}} = \dot{z} \cdot \frac{BO_2}{AO_2}; \dot{y}_{B_{1prt}} = \dot{z} \cdot b_3 \cdot a_1; BO_2 = A_1O_2 - l_2;$$
$$BO_2 = l_1 \cdot \frac{\cos\alpha}{\cos\beta}. \tag{37}$$

We get that

$$\dot{y}_{B_{prt}} = z \cdot \frac{l_1 \cdot \cos\alpha}{\cos\beta \cdot (l_1 \cdot \cos\alpha + l_2 \cdot \cos\beta) \cdot tg\beta}$$
$$= z \cdot \frac{l_1 \cdot \cos\alpha}{\sin\beta \cdot (l_1 \cdot \cos\alpha + l_2 \cdot \cos\beta)} \tag{38}$$

Further construction of the equation of motion and the transfer function of the system, taking into account the peculiarities of the location of additional masses, is carried out in the same way as for the location of the mass at point *B*. It is possible to consider the joint action of additional masses, which is carried out according to the above method. The introduction of additional links, which are mechanical chains in the configuration of a mechanism with a zero degree of freedom or a double-lead Assur group, expands the capabilities of vibration protection systems. The latter is achieved by choosing the configuration of the mechanical system, depending on the angles of installation of the levers, as well as their lengths. The choice of parameters can significantly affect the properties of the system through changes in amplitude-frequency characteristics. Thus, the type of external influence is essential for determining the capabilities of the system. Kinematic action differs in that it generates additional forces of interaction determined by portable movement of elements created by vibrations of support surface. Under certain conditions, the influence of such effects can be compensated by the configuration and features of dynamic interactions between system elements. At force action in the system there are also additional connections by absolute acceleration. The developed approach and the proposed method of building mathematical models make it possible to assess the possibilities of finding and developing new structural and technical solutions related to the removal of additional masses to various zones of the MCD, which can be used in the tasks of rational design of vibration protection systems.

MOTION CONVERSION DEVICES IN LEVER STRUCTURES

Similar in dynamic content tasks in which methodological positions are developed in the search and development of methods and means for changing the state of oscillatory systems are tasks for the analysis and dynamic synthesis of vibration protection systems for technical objects. The proximity of dynamics problems determines, to a significant extent, the feasibility of developing generalized approaches related, among other things, to attempts to introduce new typical elements and new types of connections into consideration.

Features of Building Mathematical Models

A vibration protection system (according to the patent (Khomenko, Eliseev, Artyunin, Eliseev, & Kaimov, 2014)) is considered, which consists of a main unit consisting of a protection object *M* with a spring k_0 and a special mechanical oscillating circuit formed by articulated linkage mechanisms. Mass inertia elements *m* are fixed in hinges *A* and *B*. In addition, the points are interconnected by an elastic element with stiffness *k* (points *A* and *B*) (Figure 6).

Figure 6. Design diagram of vibrating machine system with adjustable elastic mass parameters

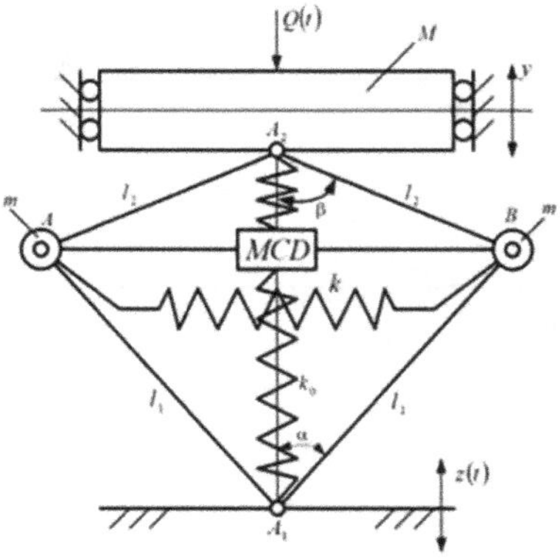

Hinge-lever mechanisms are symmetrical and have correspondingly links in the form of weightless rigid rods l_1 and l_2. The α and β angles are determined by the configuration of the system with respect to the static equilibrium position. All motions of system elements are considered as small, resistance forces in relative motions of elements are considered small. In addition to the elastic element with rigidity *k*, points *A* and *B* are connected to each other by a motion conversion device (MCD); *L* is the reduced mass of the motion conversion device, which forms inertial interactions between vols. *A* and *B*. As such a device, for example, the above-mentioned non-self-braking screw mechanism can be used. The resulting inertial forces are determined by the parameters of the relative motion of vols. *A* and *B*. It is assumed that the mass of the device for transforming motion (*L*) is small and it is not taken into account when determining the kinetic energy of the system. As external disturbances, the force harmonic effect *Q is* considered, which is applied to the object of protection (*M*), as well as kinematic - in the form of a known law of harmonic motion of the base or support surface *z(*t). When considering the peculiarities of the movement of the system elements (Figure 6), it is taken into account that one of the external influences is simultaneously implemented in it: power or kinematic. As for the joint action of two perturbations, the problem can be solved in a linear formulation based on the superposition method. However, in this case, it is necessary to establish certain relationships that determine the conditions for the interaction of two factors.

This section illustrates the capabilities of the developed method of building mathematical models for vibration protection systems, which include mechanisms that create additional contours for the formation of dynamic interactions, new physical effects, and dynamic state modes.

Force disturbance of the system *($Q \neq 0$, $z(t) = 0$)*. Schematic diagram of kinematic relations in relative motions of system elements is given in Figure 7.

Figure 7. Schematic diagram of kinematic relations at force disturbance

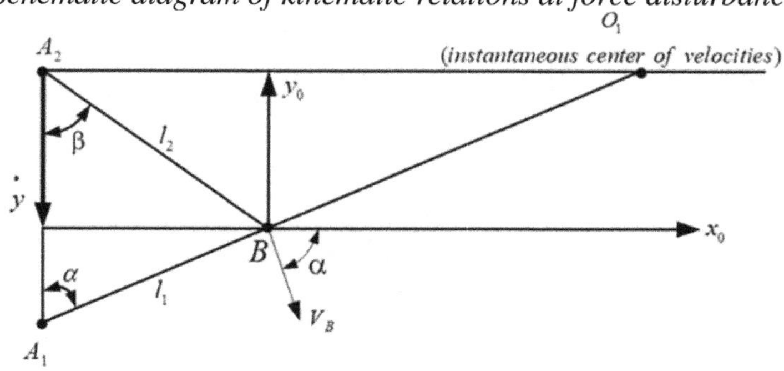

In this case, the velocity vector $\dot{y}_B$will be $\perp O_1B$. Let's enter the auxiliary coordinate system x_0By_0 (Figure 7), which has parallel axes with the coordinate system of the fixed basis. From the kinematic relations in Figure 7 it follows that

$$\overline{V}_B = \dot{y} \cdot a \tag{39}$$

$$a = \frac{i \cdot \cos\beta}{\sin\alpha \cdot (\cos\alpha + i \cdot \cos\beta)} \tag{40}$$

Here, $i = l_2/l_1$ is the linkage ratio.

To determine the horizontal deviations of vols. A and B in relative offsets in the direction of the x_0 axis, you can enter the following relations:

$$x_{0B}= y{\bullet}a{\bullet}\cos\alpha \tag{41}$$

The total deformation of the springs k in this case will be determined

$$x_{rlv\,AB} = y \cdot 2a \cdot \cos\alpha \tag{42}$$

The speeds of relative movements of vols. A and B are related by the relations:

$$\dot{x}_{\mathrm{rlv}AB} = \dot{y} \cdot 2a \cdot \cos\alpha \tag{43}$$

Kinetic and potential energies at force disturbance can be represented by

$$\mathrm{T} = \frac{1}{2}M \cdot (\dot{y})^2 + 2 \cdot \frac{1}{2}m \cdot V_B^2 + \frac{1}{2}L \cdot \left(\dot{x}_{\mathrm{rlv}_{AB}}\right)^2 \tag{44}$$

$$\Pi = \frac{1}{2}k_0 y^2 + \frac{1}{2}k x_{\mathrm{rlv}_{AB}}^2 \tag{45}$$

Given (42), (43), the resulting expressions (44) and (45) can be converted to the form

$$\mathrm{T} = \frac{1}{2}M \cdot (\dot{y})^2 + 2 \cdot \frac{1}{2}m \cdot (\dot{y} \cdot a)^2 + \frac{1}{2}L \cdot (2\dot{y}a \cdot \cos\alpha)^2 \tag{46}$$

$$\Pi = \frac{1}{2}k_0 y^2 + \frac{1}{2}k(2\dot{y}a \cdot \cos\alpha)^2 \tag{47}$$

and get the equation of motion of the system in the form

$$\ddot{y} \cdot (M + 2ma^2 + 2La^2\cos^2\alpha) + y \cdot \left(k_0 + 2ka^2 \cdot \cos^2\alpha\right) = Q \tag{48}$$

After Laplace transformations, the transfer function of the original system will be determined

$$W(p) = \frac{\bar{y}}{\bar{Q}} = \frac{1}{[M + 2a^2 \cdot (m + L \cdot \cos^2\alpha)] + k_0 + 2ka^2\cos^2\alpha} \tag{49}$$

To assess the possibilities of changing the dynamic properties of the system, Figure 8*a* - *e* show options for constructing structural diagrams.

The element with the transfer function $2ma^2p^2$ reflects the mass inertia properties of the attached masses in rotational hinges *A* and *B*. Elastic and dissipative elements, as well as mass inertia elements that are not objects of vibration protection, are considered as elements of a generalized spring. Such elements in structural interpretations are inverse negative associations with respect to the object of protection. Variants of such bonds, which are shown in Figure 8, could be expanded, which ultimately creates certain opportunities in the tasks of dynamic synthesis of vibration protection systems.

The transfer function of the system under force disturbance, represented by the expression (39), makes it possible to determine the frequency of natural oscillations:

$$\omega^2_{\text{соб}} = \frac{k_0 + 2ka^2\cos^2\alpha}{M + 2a^2(m + L\cos^2\alpha)} \tag{50}$$

From (50) it follows that the natural frequency of the system depends significantly on the configuration of the mechanisms *(*angles α and β*),* the ratio of the lengths of the links l_1 and l_2, as well as the mass inertia properties of the additional masses and the device for converting motion.

Figure 8. Variants of the mutual arrangement of elements in the structural diagram of the mechanical oscillatory system in Figure 6a - a device for converting motion as a negative feedback on the absolute deviation of the object ($M + ma^2$); b - device for conversion as parallel link k_0 in the basic circuit; c - device for converting motion and spring $2ka^2 \cos^2 \alpha$ as a parallel chain for k_0 in the basic circuit; d - system elements in parallel negative feedback with respect to the object of protection ($M + 2ma^2$); e - system elements in parallel spring; f - generalized structural diagram

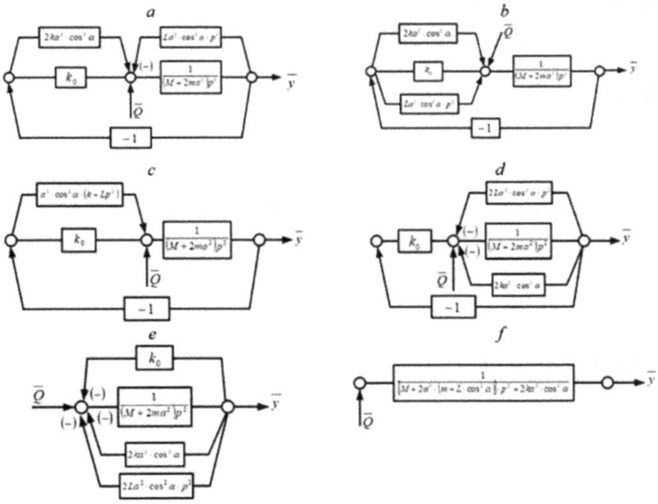

In case of power disturbance, the initial design scheme is essentially reduced to a conventional mechanical oscillatory system. The parameters of such a system have the ability to adjust by changing the shape of the mechanisms and the conditions of dynamic interaction between their elements.

Kinematic External Disturbance of the System

In this case, the schematic diagram of the kinematic relations (Figure 9) will differ from the diagram in Figure 7 by using another instantaneous center of velocities O_2 and the idea that points *A* and *B* will participate in two types of movement: relative, which is associated with the movement of the object of protection *M*, and portable - created by the movement of the base (z (t) $\neq$ 0).

Figure 9. Schematic diagram of kinematic relations at kinematic disturbance z (t)

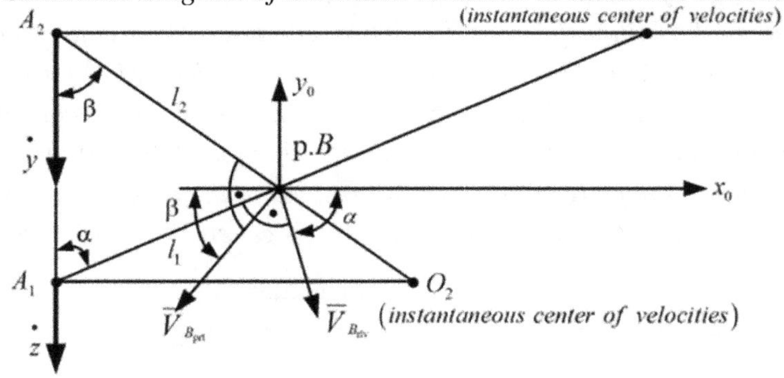

Using the diagram in Figure 9, we find that

$$V_{B_{prt}} = \dot{z} \cdot a_1 \tag{51}$$

where coefficient a_1 is given by:

$$a_1 = \frac{\cos\alpha}{\sin\beta \cdot (\cos\alpha + i \cdot \cos\beta)} \tag{52}$$

Full speed of a point *of B* at $\overline{V}_{B_{prt}} \perp A_2O_2$, $\overline{V}_{B_{rlv}} \perp A_1O_1$, can be found through projections to axes of coordinates $_{\text{of}}\, y_0 \mathrm{B} x_0$

$$V_{B_{x_0}} = V_{B_{rlv_{x_0}}} + V_{B_{prt_{x_0}}} = \dot{y} \cdot a \cdot \cos\alpha - \dot{z} \cdot a_1 \cdot \cos\beta \tag{53}$$

$$V_{B_{y_0}} = V_{B_{rlv_{y_0}}} + V_{B_{prt_{y_0}}} = -\dot{y} \cdot a \cdot \sin\alpha - \dot{z} \cdot a_1 \cdot \sin\beta. \tag{54}$$

Converting (53) and (54) to

$$V_B^2 = \left(\dot{y}a\cos\alpha - \dot{z} \cdot a_1 \cdot \cos\beta\right)^2 + \left(\dot{y} \cdot a \cdot \sin\alpha + \dot{z} \cdot a_1 \cdot \sin\beta\right)^2 \tag{55}$$

after a series of calculations we get

$$V_B^2 = (\dot{y}a)^2 + \left(\dot{z}a_1\right)^2 - 2\dot{y}\dot{z}a\,a_1 \cdot (\sin\alpha \cdot \sin\beta - \cos\alpha \cdot \cos\beta) \tag{56}$$

Let us assume that

$b = \sin\alpha \bullet \sin\beta - \cos\alpha \bullet \cos\beta$ (56′)

then:

$$V_B^2 = (\dot{y}a)^2 + (\dot{z}a_1)^2 - 2\dot{y}\dot{z}aa_1 \cdot b$$

(56″)

To determine the speeds of relative movement of the TPD, we use the horizontal components of the speeds of movement of vols. *A* and *B*. As for the horizontal displacements of pp. *A* and *B*, which form the deformation of the elastic element *k*, then this displacement will be

$$x_{sum} = 2\cdot(ya\cdot\cos\alpha - za_1\cdot\cos\beta) \qquad (57)$$

Let's write down the expressions for the kinetic and potential energies of the system according to Figure 6 at kinematic disturbance

$$\begin{aligned} T &= \tfrac{1}{2}M\cdot(\dot{y})^2 + 2\cdot\tfrac{1}{2}\cdot m\cdot\left[(\dot{y}a)^2 + (\dot{z}a_1)^2 - 2\dot{y}\dot{z}aa_1\cdot b\right] + \\ &\quad + 2\cdot\tfrac{1}{2}\cdot L\cdot(\dot{y}\cdot a\cdot\cos\alpha - \dot{z}\cdot a_1\cdot\cos\beta)^2. \end{aligned} \qquad (58)$$

$$\Pi = \tfrac{1}{2}k_0\cdot(y-z)^2 + 2\cdot\tfrac{1}{2}k\cdot(ya\cdot\cos\alpha - za_1\cos\beta)^2 \qquad (59)$$

we obtain the equation of motion of the system under kinematic disturbance:

$$\begin{aligned} &\ddot{y}\cdot(M + 2ma^2 + 2L\cdot a^2\cdot\cos^2\alpha) + y\cdot(k_0 + ka^2\cdot\cos\alpha) = \\ &= \ddot{z}\cdot(2L\cdot aa_1\cdot\cos\alpha\cdot\cos\beta - 2maa_1b) + z\cdot(k_0 + 2kaa_1\cdot\cos\alpha\cdot\cos\beta). \end{aligned} \qquad (60)$$

After the Laplace transforms (60) will take the form

$$\bar{y}\cdot\begin{bmatrix}(M + 2ma^2 + 2L\cdot a^2\cdot\cos^2\alpha)\cdot p^2 \\ +k_0 + ka^2\cdot\cos\alpha\end{bmatrix} = \bar{z}\cdot\begin{bmatrix}(2L\cdot aa_1\cdot\cos\alpha\cdot\cos\beta - 2maa_1b)\cdot p^2 + \\ +k_0 + 2kaa_1\cdot\cos\alpha\cdot\cos\beta\end{bmatrix}$$

(61)

The transfer function of the system at kinematic disturbance from the bearing surface (Figure 6) will be determined

$$W(p) = \frac{\bar{y}}{\bar{z}} = \frac{2maa_1\cdot\left(\frac{L}{m}\cos\alpha\cdot\cos\beta - b\right)\cdot p^2 + k_0 + 2kaa_1\cdot\cos\alpha\cdot\cos\beta}{\left[(M + 2ma^2 + 2L\cdot a^2\cdot\cos^2\alpha)\cdot p^2 + k_0 + ka^2\cdot\cos\alpha\right]} \qquad (62)$$

From (62) we can draw a conclusion about the dynamic properties of the system. In particular, a dynamic damping mode is possible in the system, as well as "locking" at high frequencies, at certain parameter ratios.

Evaluation of the Dynamic Properties of the System at Kinematic Disturbance

From (62) it follows that the system has a natural frequency:

$$\omega^2_{\text{sob}} = \frac{k_0 + 2ka^2 \cdot \cos^2\alpha}{M + 2a^2 \cdot (m + L \cdot \cos^2\alpha)} \tag{63}$$

which matches the expression (50). In contrast to the force disturbance in this case, the system can determine the frequency of dynamic damping of vibrations

$$\omega^2_{\text{dyn}} = \frac{k_0 + 2kaa_1 \cdot \cos\alpha \cdot \cos\beta}{2maa_1 \cdot \left(\frac{L}{m} \cdot \cos\alpha \cdot \cos\beta - b\right)} \tag{64}$$

At $p \to \infty$, the system, having kinematic disturbance, is "locked" and the ratio of vibration amplitudes at the input and output of the system takes the value:

$$\underset{p\to\infty}{W(p)} = \frac{y}{z} = \frac{2maa_1 \cdot \left(\frac{L}{m}\cos\alpha \cdot \cos\beta - b\right)}{M + 2ma^2 + 2L \cdot a^2 \cdot \cos^2\alpha}. \tag{65}$$

Consider the relation $\frac{L}{m}$ cosα•cosβ–*b* that, taking into account (56 ′), can be denoted as

$$R = \frac{L}{m}\cos\alpha \cdot \cos\beta + \cos\alpha \cdot \cos\beta - \sin\alpha \cdot \sin\beta \tag{66}$$

or:

$$R = \left(\frac{L}{m} + 1\right)\cos\alpha \cdot \cos\beta - \sin\alpha \cdot \sin\beta \; (66')$$

Figure 10 shows the relationship between the *R* function, the α and β angles, and *the L/m* parameter values.

Figure 10. Dependencies between the function R, the angles of α and β, as well as the L/m parameter: a - a family of curves R with fixed values of the angle α and L/m; curve 1 - α = 15º, L/m = 0.1; curve 2 - α = 30º, L/m = 0.5; curve 3 - α = 45º, L/m = 1; curve 4 - α = 60º, L/m = 5; curve 5 - α = 75º, L/m = 0; b - dependences between angles α and β in case of R = 0 at fixed values of L/m: curve 1 - L/m = 0.1; curve 2 - L/m = 0.5; curve 3 - L/m = 1; curve 4 - L/m = 5; curve 5 - L/m = 0; c - dependence between α angle and L/m value at fixed value of β angle: curve 1 - β = 15º; curve 2 - β = 30º; curve 3 - β = 45º; curve 4 - β = 60º

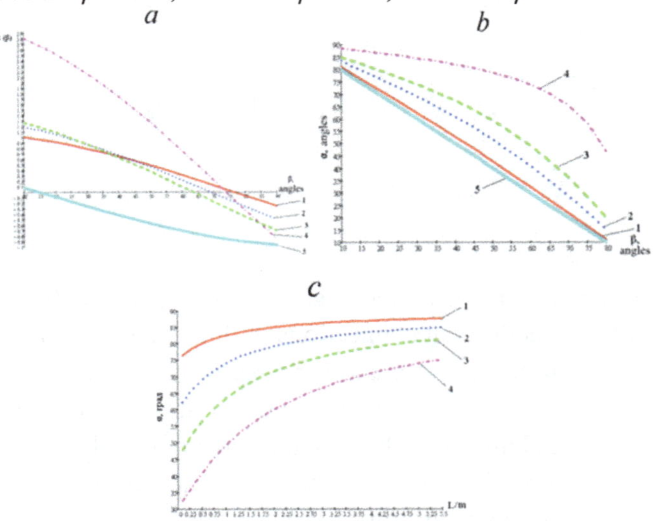

The function R can take a critical value of zero. If $R = 0$, then there is a condition

$$\left(\frac{L}{m} + 1\right) = \frac{\sin\alpha \cdot \sin\beta}{\cos\alpha \cdot \cos\beta} = tg\alpha \cdot tg\beta \tag{67}$$

If the condition is met, it determines the type of transfer function (62) at which the dynamic damping mode is excluded. Since the system in question is related to the tasks of finding and developing methods and means for describing vibration disturbances, it is of interest to compare the amplitude-frequency characteristics at different values of the R parameter, which is shown in Figure 11.

Figure 11. Comparison of types of amplitude-frequency characteristics at different values of parameter R: curves 1,2 - corresponds to R> 0; curve 3 - R = 0; curve 4 - R <0

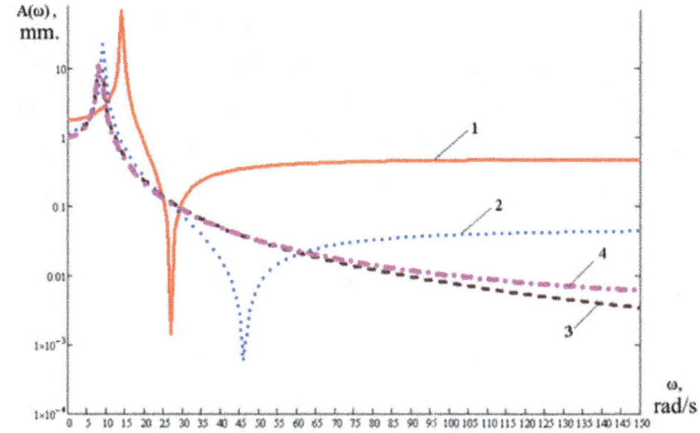

Amplitude-frequency characteristics significantly depend on the ratio of parameters of the introduced additional connections in that dynamic modes of vibration protection systems can be selected, and their parameters can be adjusted by appropriate selection of values that characterize the configuration of the system and the presence of certain elements in the structure. In a physical sense, this situation reflects the manifestations of dynamic interactions created by portable inertia forces that create conditions for compensating for external influences.

The dynamic properties introduced by additional links depend on the type of external perturbations. At kinematic disturbance it is possible to create modes of partial compensation of external disturbances due to portable inertia forces. Motion conversion devices, as well as additional masses, can act as tuning elements.

Some Structural and Technical Forms Using Linkage

The introduction of lever mechanisms into the structures of oscillatory systems is associated with the need to take into account a number of features that arise when choosing the location of the points of support of the levers, their configuration and forms of use of the connecting elements. A system is considered (according to the patent (Khomenko et al., 2015)), which has an object of protection with a mass of m_0, which makes (Figure 12) vertical movements (y_0) relative to the position of static equilibrium.

Figure 12. Design diagram of the robotic device suspension unit

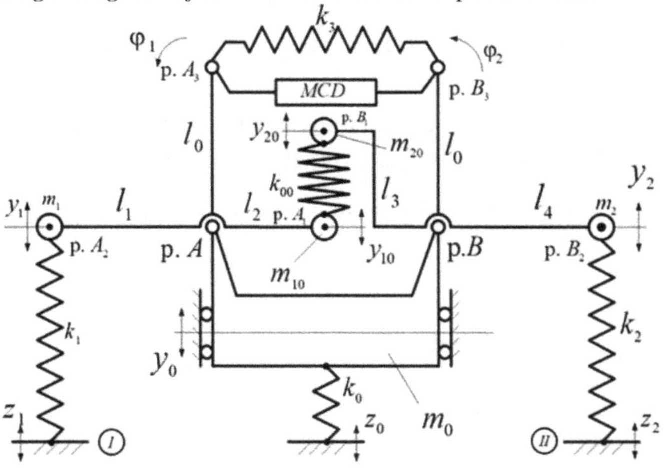

In pp. A and B, two lever mechanisms of reverse T-shape are attached to the object of protection. Each lever mechanism has mass inertia elements fixed respectively in pp. A_2, A_1 (m_1, m_{10}) and pp. B_2, B_1 (m_2, m_{20}). In pp. A_3, B_3 the system elements are fixed: a spring with rigidity k_3 and a motion conversion device (MCD) in the form of a screw non-self-braking mechanism. In addition, the points A_1 and B_1 are connected by an elastic member of rigidity k_{00}. In turn, the object is also supported by a spring with rigidity k_0 .

Mass inertia elements m_1 and m_2 rest on support surfaces *I* and *II,* the law of motion of which is known (z_1, z_2). In addition, the element k_0 rests on the support surface with the law of motion $z_0(t)$. It is assumed that the system makes small vibrations without taking into account resistance forces. Coordinate systems y_0, φ_1, φ_2 and y_0, y_1, y_2 are used to describe the motion. All necessary geometric parameters are shown in Figure 12.

Kinematic perturbations at the stage of preliminary study are considered harmonic. The relationship between the motion parameters z_0, z_1, z_2 is discussed separately.

Among the tasks to be solved, it is supposed to develop a mathematical model of the system and assess the possibilities of changing the dynamic state by structural transformations and choosing the tuning parameters of the original design.

Building a Mathematical Model of the System

We write down the expressions for kinetic and potential energies under the assumption that the system makes small vibrations relative to the position of static equilibrium

$$T = \frac{1}{2} \cdot (m_0 + m_1 + m_2 + m_{10} + m_{20}) \cdot (\dot{y})^2 + \frac{1}{2} \cdot L \cdot (\dot{\varphi}_2 - \dot{\varphi}_1)^2 \cdot l^2 + \\ + \frac{1}{2} \cdot (\dot{\varphi}_1)^2 \cdot (m_1 l_1^2 + m_{10} l_2^2) + \frac{1}{2} \cdot (\dot{\varphi}_2) \cdot (m_2 l_4^2 + m_{20} l_3^2), \tag{68}$$

where φ 1 and φ 2 are angles of lever devices rotation relative to *the object* m0; L - reduced mass-inertial characteristic of MCD. In turn

$$\Pi = \frac{1}{2} k_0 \cdot (y_0 - z_0)^2 + \frac{1}{2} k_1 \cdot (y_1 - z_1)^2 + \frac{1}{2} k_2 \cdot (y_2 - z_2)^2 + \frac{1}{2} k_3 \cdot (\phi_2 - \phi_1)^2 l^2 + \frac{1}{2} \\ k_{00} \cdot (\phi_2 l_3 - \phi_1 l_2)^2. \tag{69}$$

Let's enter a series of relations

$$\left. \begin{aligned} y_{10} &= \phi_1 \cdot l_2, y_{20} = \phi_2 \cdot l_3 \\ y_1 &= y_0 + \phi_1 \cdot l_1, y_2 = y_0 + \phi_2 \cdot l_4 \end{aligned} \right\} \tag{70}$$

Taking into account (70) potential energy will be recorded

$$\Pi = \frac{1}{2} k_0 \cdot (y_0 - z_0)^2 + \frac{1}{2} k_1 \cdot (y_1 - \phi_1 l_2 - z_1)^2 + \frac{1}{2} k_2 \cdot (y_2 - \phi_2 l_4 - z_2)^2 + \\ + \frac{1}{2} k_3 \cdot (\phi_2 - \phi_1)^2 l^2 + \frac{1}{2} k_{00} \cdot (\phi_2 l_3 - \phi_1 l_2)^2. \tag{71}$$

Find the equations of motion in the coordinates y_0, φ $_1$ and φ $_2$

$$\ddot{y}_0 \cdot (m_0 + m_1 + m_2 + m_{10} + m_{20}) + y_0 \cdot (k_0 + k_1 + k_2) - \\ - \varphi_1 k_1 l_2 - \varphi_2 k_2 l_4 = k_0 z_0 + k_1 z_1 + k_2 z_2; \tag{72}$$

$$\ddot{\varphi}_1 \cdot (m_1 l_1^2 + m_{10} l_2^2 + L \cdot l^2) + \ddot{\varphi}_1 \cdot (k_1 l_2^2 + k_2 l^2 + k_{00} l_2^2) + \\ + \ddot{\varphi}_2(-L l^2) + \ddot{\varphi}_2(-k_3 l_2 - k_{00} l_2 l_3) - k_1 l_2^2 y_0 = -k_1 z_1 l_2; \tag{73}$$

$$\ddot{\varphi}_2 \cdot (m_2 l_4^2 + m_{20} l_3^2 + L \cdot l^2) + \varphi_2 \cdot (k_2 l_4^2 + k_3 l^2 + k_{00} l_3^2) + \ddot{\varphi}_1(-L l^2) + \\ + \varphi_1(-k_3 l^2 - k_{00} l_2 l_3) - k_2 l_4 y_0 = -k_2 z_2 l_4. \tag{74}$$

Figure 13 shows the structural diagram of the source system (Figure 12) in coordinates y_0, φ $_1$, φ $_2$.

The peculiarity of the system lies in the fact that between partial systems (φ_1) and (φ_2) the connection has an inertial-elastic type. At the frequency of external exposure

$$\omega^2_{\text{prt}} = \frac{k_{00} l_2 l_3 + k_3 l_2^2}{L l^2 p^2} \tag{75}$$

zeroing of inter-partial links φ_1 and φ_2 (as well as between y_1 and y_2) is possible.

Figure 13. Block diagram of the system with T-shaped levers (Figure 12) in coordinates y_0, φ_1, φ_2

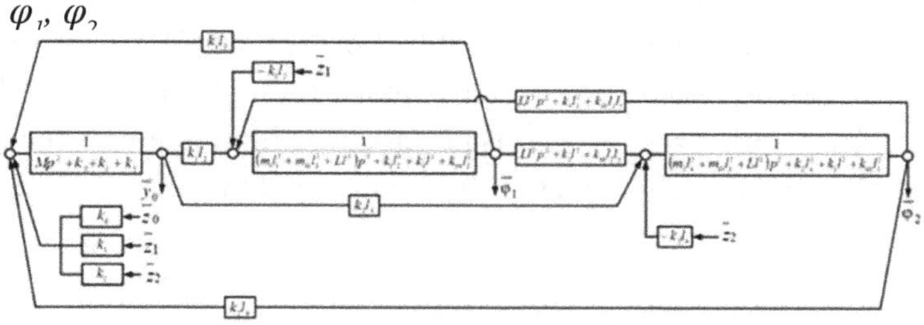

In other cases, that is, when considering y_0 and φ_1, as well as φ_2 and y_0, the links are elastic (no zeroing).

Evaluation of the Dynamic Properties of the System

Cramer's formulas (Druzhinsky, 1977) can be used to evaluate the dynamic properties of the system:

$$\bar{y}_0 = \frac{1}{A_0}[\bar{Q}_1 \cdot (a_{22}a_{33} - a_{23}a_{32}) + \bar{Q}_2 \cdot (a_{13}a_{32} - a_{12}a_{33}) + \bar{Q}_3 \cdot (a_{12}a_{23} - a_{13}a_{22})] \quad (76)$$

$$\bar{\varphi}_1 = \frac{1}{A_0}[\bar{Q}_1 \cdot (a_{23}a_{31} - a_{21}a_{33}) + \bar{Q}_2 \cdot (a_{11}a_{33} - a_{13}a_{31}) + \bar{Q}_3 \cdot (a_{13}a_{21} - a_{11}a_{23})] \quad (77)$$

$$\bar{\varphi}_2 = \frac{1}{A_0}[\bar{Q}_1 \cdot (a_{21}a_{32} - a_{22}a_{31}) + \bar{Q}_2 \cdot (a_{12}a_{31} - a_{11}a_{32}) + \bar{Q}_3 \cdot (a_{11}a_{22} - a_{12}a_{21})] \quad (78)$$

where A_0 is the characteristic equation of the system shown in Figure 13.

$$A_0 = a_{11}a_{22}a_{23} - a_{11}a_{23}^2 - a_{22}a_{13}^2 - a_{33}a_{12}^2 + 2a_{12}a_{23}a_{31} \quad . \quad (79)$$

Using formulas (67) - (69), one can find the transfer functions that determine the interaction between the input external perturbations from the set $\bar{z}_0$, $\bar{z}_1$, $\bar{z}_2$the output signals. If there are several signals at the $\bar{y}_0$,$\bar{\varphi}_1$,$\bar{\varphi}_2$input to the partial system, then the superposition principle can be used. It is of particular interest to consider a situation in which some relationships are established between external signals $z_{0,}$ z_1, z_2. For example, we can assume that the system rests simultaneously on one surface, then we can assume that with the corresponding $z=z_0=z_1=z_2$ features of determining transfer functions based on expressions (67) - (69). If $k_0 = 0$, then the object m0 is not directly connected to the support surface, contact is made through the elastic elements k_1 and k_2 (Figure 12). With $z = z_1 = z_2$, it is possible to obtain the transfer function at the input signal (or generalized external influence) and the output $\bar{z}$signal$\bar{y}_0$:

$$W_1(p) = \frac{\bar{y}_0}{\bar{z}} = \frac{(a_{13}a_{32} - a_{12}a_{33}) + (a_{13}a_{21} - a_{11}a_{23})}{A_0} \tag{80}$$

Since a_{23} in accordance with the block diagram in Figure 13 looks like:

$$a_{23} = Ll^2p^2 + k_3 l_2^2 + k_{00} l_2 l_3 \tag{81}$$

then the numerator (80) at

$$a_{11} = (m_0 + m_1 + m_2 + m_{10} + m_{20}) \cdot p^2 + k_0 + k_1 + k_2 \tag{82}$$

is a biquad equation relative to the square of the frequency of external exposure. If we exclude the contact $_{\text{of m0}}$ with the support surface, assuming $k_0 = 0$, then with respect to the object of protection, we can expect the appearance of two modes of dynamic damping of vibrations under the action of vibrations from the base ($z = z_1 = z_2$). The situation, in principle, will not change if the condition $z = z_0 = z_1 = z_2$ ($k_0 \neq 0$) is met. However, in this case, the parameters (that is, the dynamic blanking frequencies) will change. The amplitude-frequency characteristics of the system by variable $\bar{y}_0$will be of a normal nature. With the characteristic equation A*0*, we can expect the presence of three resonant frequencies with a corresponding increase in vibration amplitudes along the coordinate *y0* In this case, the two dynamic blanking frequencies will be located between the three resonant frequencies. The suspension circuit (Figure 12) has the ability to change the dynamic state of the protection object m_0 by appropriate selection of adjustment parameters. Such parameters can be the lengths of the links of the levers, the rigidity of the elastic elements k_3 and $k_{00,}$ as well as the mass inertial parameters of the MCD (meaning L - the reduced mass of the MCD and l - the length of the lever arm). It was noted above that if the ratio (67) is observed, partial bonds can be decoupled.

In suspensions with lever links, in the presence of several input signals, local dependencies between external influences are possible. For example, the relations $z = z_0 = z_1 = z_2$ reflect the possibility of vertical oscillations of all support surfaces simultaneously. However, it is quite possible that movements on individual surfaces (z_0, z_1, z_2) are considered as fragments of some general movement. Such a movement can be a swing of a common surface relative to a certain center, etc. In the spectral theory of suspension of transport systems, more complex functional relationships between the actions of several input signals are considered. Simplifications can be obtained by considering the possibilities of linear relations in the parameters of simultaneously acting disturbances. Note also that the emerging problem of assessing the dynamic state of the system under the action of several force factors can be considered quite "natural" under the action of an external force shifted relative to the center of gravity, with vibrations, for example, of a solid body on elastic supports, making flat vibrations. The use of tuning parameters k_3 and k_{00} in this scheme (Figure 12) has quite definite disadvantages. In particular, at $k_{00} \to \infty$, communication between elements m_{10} and m_{20} is blocked, which simultaneously leads to blocking of the relative movement of the levers (in relation to the protected object m_0). In this regard, suspensions with the possibilities of gear engagement of lever devices have great capabilities.

Features of the dynamic properties of the system in coordinates y_0, y_1, y_2. To build a mathematical model of the system, a number of relations are introduced

$$\phi_1 = \frac{y - y_0}{l_1} = a \cdot (y - y_0), \phi_2 = b \cdot (y - y_0), a = \frac{1}{l_1}, b = \frac{1}{l_4}. \tag{83}$$

Let's write down the expressions for kinetic and potential energies:

$$\begin{gathered} \mathrm{T} = \frac{1}{2}M \cdot (\dot{y}_0)^2 + \frac{1}{2}Ll^2 \cdot (b\dot{y}_2 - b\dot{y}_0 - a\dot{y}_1 + a\dot{y}_0)^2 + \\ +\frac{1}{2} \cdot (m_1 l_1^2 + m_{10} l_2^2) \cdot a^2 \cdot (\dot{y}_1 - \dot{y}_0)^2 + \frac{1}{2} \cdot (m_2 l_4 + m_{20} l_3^2) \cdot b^2 \cdot (\dot{y}_2 - \dot{y}_0)^2. \end{gathered} \tag{84}$$

$$\begin{gathered} \Pi = \frac{1}{2}k_0 \cdot (y_0 - z_0)^2 + \frac{1}{2}k_1 \cdot (y_1 - z_1)^2 + \frac{1}{2}k_2 \cdot (y_2 - z_2)^2 + \\ +\frac{1}{2}k_3 l^2 \cdot [by_2 - ay_1 + y_0 (a - b)^2] + \frac{1}{2}k_{00} \cdot [y_2 \cdot bl_3 - y_1 \cdot al_2 + y_0 \cdot (al_2 - bl_3)]^2. \end{gathered} \tag{85}$$

Let's enter a number of symbols

$$a - b = a_0, bl_3 = a_1, al_2 = a_2 \tag{86}$$

then (85) is converted to the form

$$\begin{gathered} \Pi = \frac{1}{2}k_0 \cdot (y_0 - z_0)^2 + \frac{1}{2}k_1 \cdot (y_1 - z_1)^2 + \frac{1}{2}k_2 \cdot (y_2 - z_2)^2 + \\ +\frac{1}{2}k_3 l^2 \cdot [by_2 - ay_1 + y_0 a_0^2] + \frac{1}{2}k_{00} \cdot [y_2 \cdot a_1 - y_1 \cdot a_2 + y_0 \cdot a_3]^2. \end{gathered} \tag{87}$$

Let's build a system of equations of motion in coordinates $_{y0}$, $_{y1}$, $_{p2}$, taking

$$M = m_0 + m_1 + m_2 + m_{10} + m_{20} \tag{88}$$

which will amount to

$$\ddot{y}_0 \cdot \left[M + Ll^2 a_0^2 + \left(m_1 l_1^2 + m_{10} l_2^2\right) \cdot a^2 + \left(m_2 l_4^2 + m_{20} l_3^2\right) \cdot b^2\right]$$
$$+y_0 \cdot \left[k_0 + k_3 l^2 a_0^2 + k_{00} a_3^2\right] +$$

$$+\ddot{y}_1 \cdot \left[-a_0 a L l^2 - \left(m_1 l_1^2 + m_{10} l_2^2\right) a^2\right] + y_1 \cdot \left[-k_3 l^2 a - k_{00} a_3 a_2\right] +$$

$$+\ddot{y}_2 \cdot \left[L l^2 a_0 b - \left(m_2 l_4^2 + m_{20} l_3^2\right) b^2\right] + y_2 \cdot \left[k_3 l^2 b + k_{00} a_3 a\right] = k_0 z_0. \tag{89}$$

$$\ddot{y}_0 \cdot \left[-L l^2 a a_0 - \left(m_1 l_1^2 + m_{10} l_2^2\right) \cdot a^2\right] + y_0 \cdot \left[-k_3 l^2 a a_0 - k_{00} a_3 a_2\right]$$
$$+\ddot{y}_1 \cdot \left[L l_1 a_1^2 - \left(m_1 l_1^2 + m_{10} l_2^2\right) a^2\right] +$$

$$+y_1 \cdot \left[k_1 + k_3 l^2 a^2 + k_{00} a_2^2\right] + \ddot{y}_2 \cdot \left[-L l^2 a b\right] + y_2 \cdot \left[-k_3 l^2 a b - k_{00} a_1 a_2\right]$$
$$= k_1 z_1. \tag{90}$$

$$\ddot{y}_0 \cdot \left[L l^2 b a_0 - \left(m_2 l_4^2 + m_{20} l_3^2\right) \cdot b^2\right] + y_0 \cdot \left[k_3 l^2 b a_0 + k_{00} a_3 a_1\right] +$$
$$+\ddot{y}_1 \cdot \left[-L l^2 b a\right] + y_1 \cdot \left[-k_3 l^2 a b - k_{00} a_1 a_2\right] + \tag{91}$$
$$+\ddot{y}_2 \cdot \left[L l^2 b^2 + \left(m_2 l_4^2 + m_{20} l_3^2\right) \cdot b^2\right] + y_2 \cdot \left[k_2 + k_3 l^2 b^2 + k_{00} a_1^2\right] = k_2 z_2.$$

Analysis of the system properties using (89) - (91) can be built on the transfer function of the form

$$\underset{(z_0=z_1=z_2=z)}{W_2(p)} = \frac{\bar{y}_0}{\bar{z}} = \frac{k_0 \cdot \left(a_{22} a_{33} - a_{23}^2\right) + k_1 \left(a_{13} a_{32} - a_{12} a_{23}\right) + k_2 \left(a_{12} a_{23} - a_{13} a_{33}\right)}{A_0} \tag{92}$$

From (92), conditions for the occurrence of dynamic damping modes on the coordinate $\bar{y}_0$at different parameter ratios can be obtained. Since the observed coordinate y_0 in the coordinate system y_0, φ_1, φ_2 and$\bar{y}_0$,$\bar{y}_1$, $\bar{y}_2$is the same, the dynamic blanking frequencies will be the same.

However, in the analysis (92), the values of the coefficients a_{ij} should be taken from the system of equations (89) - (91). As for the characteristic equation, its general construction scheme remains unchanged, however, when moving from the coordinate system y_0, φ_1, φ_2 to the system y_0, y_1, constant factors will form between the characteristic equations that determine the coordination of the dimensions of the state variables (linear and angular).

Taking into Account the Features of the Gear Connection of Lever Sectors

A system with two degrees of freedom is considered, as shown in Figure 14.

Figure 14. Design diagram of a gear system in contact with two lever sectors

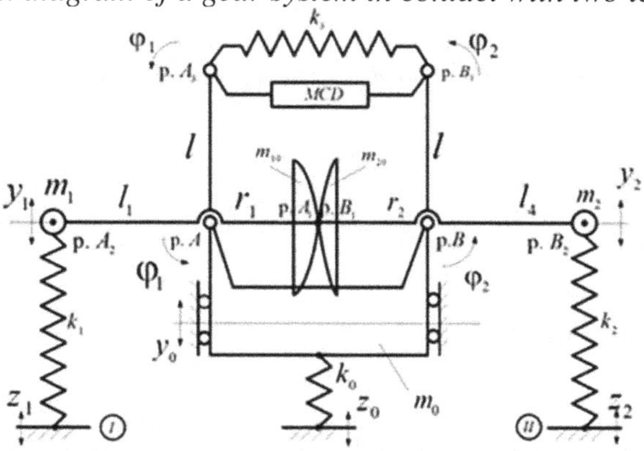

In the circuit (Figure 14), gear engagement is implemented, which ensures the connection of angular speeds of rotation of φ_1 and φ_2, $\phi_1 r_1 = \phi_2 r_2$; $\phi_2 = \phi_1 \frac{r_1}{r_2}$ wherein the rotations occur in different directions. The expression for the potential energy is

$$\Pi = \frac{1}{2}k_0 \cdot (y_0 - z_0)^2 + \frac{1}{2}k_1 \cdot (y_1 - z_1)^2 + \frac{1}{2}k_2 \cdot (y_2 - z_2)^2 + \frac{1}{2}k_3 \cdot (\phi_2 + \phi_1)^2 l^2 \quad (93)$$

The last term of the expression (93) can be transformed:

$$\phi_1 + \phi_2 = \phi_1 + \phi_1 \frac{r_1}{r_2} = \phi_1\left(\frac{r_1 + r_2}{r_2}\right) = \phi_1(1 + i) \quad (93')$$

Then (93) can be written in the form

$$\Pi = \frac{1}{2}k_0 \cdot (y_0 - z_0)^2 + \frac{1}{2}k_1 \cdot (y_1 - z_1)^2 + \frac{1}{2}k_2 \cdot (y_2 - z_2)^2 + \frac{1}{2}k_3 \cdot \phi_1^2 \cdot (1 + i)^2 l^2 \quad (94)$$

Lever sectors have mass inertia properties

$$J_1 = m_1 l_1^2 + m_{10} r_1^2 \quad (95)$$

$$J_2 = m_2 l_4^2 + m_{20} r_2^2 \quad (96)$$

The expression for the kinetic energy of the system is

$$\mathrm{T} = \frac{1}{2}M \cdot (y_0)^2 + \frac{1}{2}(J_1 + J_2 i^2) \cdot (\dot{\varphi}_1)^2 + \frac{1}{2}L l^2 (1 + i)^2 \cdot (\dot{\varphi}_1)^2 \quad (97)$$

Since $y_1 = y_0 - \phi_1 l_1; y_2 = y_0 - \phi_2 l_4$ (97) can be written

$$\mathrm{T} = \frac{1}{2}(m_1 + m_2 + m_{10} + m_{20} + m_0) \cdot (\dot{y}_0)^2 + \frac{1}{2}J_1 (\dot{\varphi}_1)^2 + \frac{1}{2}J_2 (\dot{\varphi}_1)^2 i^2 + \frac{1}{2}L l^2 \dot{\varphi}_1 (1 + i)^2 \quad (98)$$

Let's enter a series of relations

$$y_1 = y_0 + \phi_1 l_1; y_2 = y_0 + \phi_1 l_4 i; y_1 - y_2 = \phi_1 (l_1 - l_4 i); \phi_1 = \frac{y_1 - y_2}{(l_1 - l_4 i)} \quad (99)$$

which allows (94) to lead to the form

$$\Pi = \frac{1}{2}k_0 \cdot (y_0 - z_0)^2 + \frac{1}{2}k_1 \cdot (y_0 + \phi_1 l_1 - z_1)^2 + \frac{1}{2}k_2 \cdot (y_0 + \phi_1 l_4 - z_2)^2 + \frac{1}{2}k_3 \cdot \phi_1^2 \cdot (1 + i)^2 l^2 \quad (100)$$

Let's write the equations of motion of the system in coordinates y_0 and φ_1:

$$\ddot{y}_0 M + y_0 [k_0 + k_1 + k_2] + \varphi_1 [k_1 l_1 + k_2 l_4 i] = k_0 z_0 + k_1 z_1 + k_2 z_2 \quad (101)$$

$$\begin{aligned} &\dot{\varphi}_1 \cdot [J_1 + J_2 i^2 + L l^2 (1 + i)^2] + \varphi_1 \cdot [k_1 l_1^2 + k_4 l_4^2 i^2 + k_3 l^2 (1 + i)^2] + \\ &+ y_0 \cdot [k_1 l_1^2 + k_2 l_4 i] = k_1 l_1 z_1 + k_2 l_4 i z_2. \end{aligned} \quad (102)$$

Coefficients of equations (101), (102) are given in Table 1.

Table 1. Coefficients of the system of equations in coordinates y_0, φ_1

a_{11}	a_{12}
$Mp^2 + k_1 + k_2 k_0$	$k_1 l_{1'} + k_2 l_4 i$
a_{21}	a_{22}
$k_1 l_{1'} + k_2 l_4 i$	$J_1 + J_2 i^2 + Ll^2(1+i)^2 p^2 + k_1 l_1^2 + k_2 l_4^2 i^2 + k_3 l^2 (1+i)^2$
Q''_1	Q''_2
$k_0 z_0 + k_1 z_1 + k_2 z_2$	$k_1 l_1 z_{1'} + k_2 l_4 i z_2$

Note:, Q''_1 - Q''_2 generalized forces corresponding to coordinates y_0 and φ

Figure 15 shows the structural diagram of the system in coordinates l_0 and φ. In this system, inter-partial connections are elastic.

Figure 15. Block diagram of the system (Figure 12) with toothed sectors

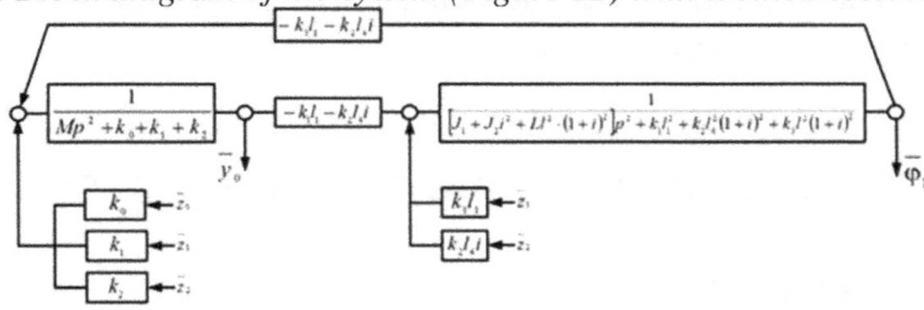

Let $z_1 = 0$, $z_2 = 0$. Then the transfer function of the system (Figure 15) will take the form:

$$W_1(p) = \frac{\bar{y}_0}{\bar{z}_0}$$

$$= \frac{k_0 \cdot \{[J_1 + J_2 i^2 + Ll^2 \cdot (1+i)^2]p^2 + k_1 l_1^2 + k_2 l_4^2 (1+i)^2 + k_3 l^2 (1+i)^2\}}{[a_{11}] \cdot [a_{22}] - (k_1 l_1 + k_2 l_4 i)^2} \tag{103}$$

From (103), the dynamic blanking frequency ($z_0 = z$; $z_1 = 0$; $z_2 = 0$):

$$\omega^2 = \frac{k_1 l_1^2 + k_2 l_4^2 (1+i)^2 + k_3 l^2 (1+i)^2}{J_1 + J_2 i^2 + Ll^2 \cdot (1+i)^2} \tag{104}$$

In a similar manner, other dynamic quenching modes can be found arising from various combinations of external influences. System tuning capabilities are determined by system parameters: stiffness k_3, lever arm lengths (l), gear ratio i and mass inertia properties of toothed sectors. The system has two resonant frequencies (or natural frequencies), which can be found from the characteristic equation (103). Thus, the mechanical linkage system considered in the dynamic interactions with the vibratingbase has an extended set of adaptation properties. When providing elastic connection k_{00} between adjustment masses (m_1, m_2, m_{10}, m_{20}), the system can have two modes of dynamic suppression. The spring constant k_3 and the inertial properties of the MCD (via Ll^2) can act as adjustment parameters.

With the introduction of gear transmission between lever sectors, the system loses one degree of freedom, however, adaptation capabilities in suppressing external kinematic influences can be supported by the introduction of gear ratios between lever sectors and the choice of other adjustment parameters. The method developed by the author for constructing and analyzing the dynamic properties of the original vibration protection system is based on taking into account the leverage implemented by the leverage sectors. Lever sectors can be considered as a new element of mechanical oscillatory systems, having expanded capabilities of dynamic actions, in particular, due to the ease of application and use of active means by applying controlled moments of forces to change the dynamic state.

TRANSPORT SUSPENSIONS. MATHEMATICAL MODELS

Select Coordinate System

Passive suspensions with an expanded set of links are widespread in the dynamics of transport systems. However, leverage and its capabilities have not yet received proper coverage, although developments in this direction are being carried out quite intensively. The relaxation suspensions used in practice have a number of characteristic features - a series connection "damper - elastic element." Details such as the location of the element in a series connection circuit are also taken into account. Figure 16 shows the design diagram of the suspension k_0 - the main spring rate; M - weight of the protected object, bp - damping element (b - coefficient of viscous friction); k - spring stiffness factor in relaxation chain, $p = j\,\omega$ - complex variable.

Figure 16. Design diagram of suspension with relaxation chain

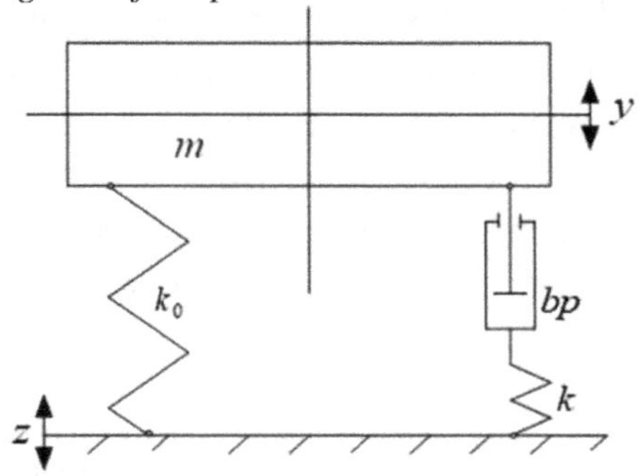

Motion of protection object m at kinematic disturbance is determined by coordinate in y conditionally fixed coordinate system. As shown in (Eliseev, Moskovskikh, Bolshakov, & Savchenko, 2012), the equivalent (or reduced) stiffness of two consecutive typical elements (*bp* is the first-order differentiating link, and k is the amplifying link) is determined by the formula

$$k_{eqv} = \frac{bp \cdot k}{bp + k} \tag{105}$$

The sequence of the elements does not matter. The reduced spring rate can be found for the diagram in Figure 16 as follows:

$$k_{pr} = k_0 + k_{eqv} = k_0 + \frac{pbk}{pb + k} = \frac{kk_0 + bp(k + k_0)}{bp + k}. \tag{106}$$

In this case, instead of a purely elastic support, the object rests on a structure that is some combination of three elementary links (two springs k_0 and k, as well as a differentiating link - bp). At zero frequency ($p = 0$), the object rests on an elastic element with rigidity k_0; at $p \to \infty$, the object "works" with an elastic element of rigidity $k_0 + k$. Thus, the relaxation circuit allows in the system in Figure 16 to turn the usual spring k_0 into a passive link with the transfer function of a general aperiodic link (106). The properties of such a link are well studied (Kiryukhin, Tikhonov, Chistyakov, & Yablonsky, 2012; Ryabov, Chernyshov, Eliseev, & Upyr, 2010). Relaxation suspensions of this type make it possible to ensure smooth travel, for which it is necessary to minimize the amplitude of vibrations. One way to improve the dynamic properties of relaxation suspensions is to control damping, which involves turning off the damper at times when the damping force prevents the object from moving to the equilibrium position. Such approaches allow you to control the processes of development of oscillations and can level resonant phenomena. At the same time, the physics of the process is more complex, since the manifestations of movement are of an integration nature, which, in turn, is determined by taking into account other factors related to the specifics of the design of the suspension. Examples of design solutions for relaxation suspensions are given in Figure 17. In further calculations, it is assumed that the center of mass of the levers, interacting links coincides with the attachment point to the vibration source, and the "masses" of the damper *bp* and spring k_2 are quite small.

Figure 17. Examples of design: a - series connection with spring k_2; b - accounting of damper elasticity; c - elastically damping leaf spring

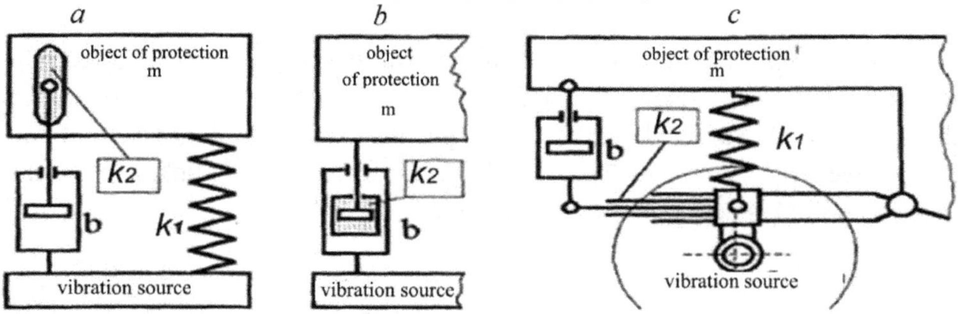

Figure 18 shows diagrams of stores, which include linkage.

Figure 18. Diagrams of lever relaxation suspensions: a - mechanical; b - hydro-mechanical

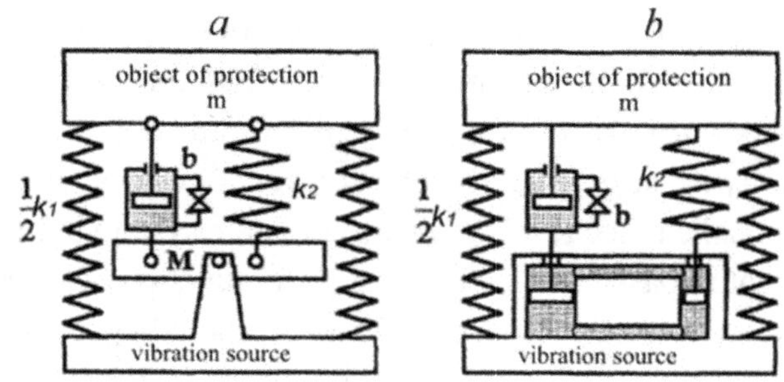

The presence of such connections implemented by lever mechanisms allows you to change the dynamic properties of suspensions, create the effects of dynamic damping and locking at high frequencies.

Task Statement. Building a Mathematical Model

The design diagram of the transport suspension is shown in Figure 19 and is a system with three degrees of freedom and a linkage.

Figure 19. Design diagram of suspension with linkage

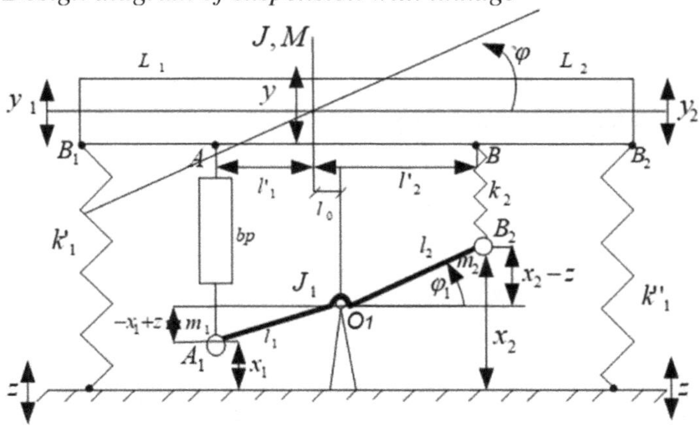

It is assumed that the movement of the lever does not lead to noticeable deviations of the lines of action of the forces of the elastic and damping elements from the vertical. In Figure 19, a number of designations are adopted: O - center of gravity of the object; O_1 - lever attachment point; M, J - mass inertia parameters of the protection object; m_1, m_2, J_1 - mass inertial parameters of lever suspension; $L_1 = OB_1$, $L_1 = OB_1$, $l_0 = OO_1$, $l_1 = O_1A_1$, $l_2 = O_1A_2$ $l'_1 = OA$, $l'_2 = OB$, z - kinematic disturbance; k'_1, k''_1- elastic elements of the object resting on the base; bp - damper; k_2 - stiffness of elastic suspension element; y, φ - generalized coordinates of the motion of the center of mass and rotation around the center of mass; $y_{1,}$ y_2 - coordinates of protection object movement relative to fixed reference frame; φ_1 - angle of lever rotation in motion relative to the base; x_1, x_2 are the coordinates of the relative motion (to the base). We write a number of auxiliary relations: $y_1 = y - L_1$, $y_2 = y + L_2\varphi$, $\frac{l_2}{l_1} = i$. The diagram (Figure 17) uses a lever of the second kind, so the relative speeds of points A_1 and A_2 will be directed in different directions. If we take the direction of rotation of the lever φ_1 counterclockwise as positive, then the relative position of points A_1 and A_2 is determined respectively as $z - x_1$ and $x_2 - z$ (Figure 19). For preliminary calculations, it can be assumed that $l_1 \approx l'_1$and. We also $l_2 \approx l''_2$denote the distances to the attachment points A and B at the object of protection.

Consider the suspension motion in the coordinate system y, φ and φ_1 and write down the expressions for the kinetic and potential energies:

$$T = \frac{1}{2}M\dot{y}^2 + \frac{1}{2}J\dot{\varphi}^2 + \frac{1}{2}J_1\dot{\varphi}_1{}^2 \tag{107}$$

$$\Pi = \frac{1}{2}k'_1(y_1 - z)^2 + \frac{1}{2}k''_1(y_2 - z)^2 + \frac{1}{2}k_2(y_B - x'_2)^2\ ; \tag{108}$$

The energy dissipation function (Kopylov, 2011) can be taken as:

$$\Phi = \frac{1}{2}b(\dot{y}_A - \dot{x}'_1)^2 \tag{109}$$

We introduce a number of auxiliary relations:

$$\varphi = \frac{1}{2}\frac{y_2 - y_1}{L_1 + L_2} = c(y_2 - y_1), c = \frac{1}{L_1 + L_2}, \varphi = \frac{z - x_1}{l_1}, \varphi_1 = \frac{x_2 - z}{l_2}, x_2 = \varphi_1 + l_2 = i(z - x_1); \tag{110}$$

we will assume that

$$x_2 = \varphi_1 l_2, x'_1 = y_1 - z - x_1 = y_1 - z - l_1\varphi, x'_1 = y_1 - z + \varphi_1 l_1, x'_2$$

$$= y_2 - z + l_2\varphi_1 \tag{111}$$

Then (108) can be converted to the form:

$$\Pi = \frac{1}{2}k'_1(y_1 - z)^2 + \frac{1}{2}k''_1(y_2 - z)^2 + \frac{1}{2}k_2(y + l_2\varphi - y_2 + z - l_2\varphi_1)^2. \tag{112}$$

Since $y_2 = y_0 + l_2\,\varphi$, then (112) is written as

$$\Pi = \frac{1}{2}k'_1(y_1 - z)^2 + \frac{1}{2}k''_1(y_2 - z)^2 + \frac{1}{2}k_2[\varphi(l_1 + l_2) + (z - l_1\varphi_1)]^2. \tag{113}$$

We introduce into the expression for the potential energy of the system the coordinates of the relative motion of the elastic elements:

$$\Pi = \frac{1}{2}k'_1(y_1 - z)^2 + \frac{1}{2}k''_1(y_2 - z)^2 + \frac{1}{2}k_2(y_B - x'_2)^2 \tag{114}$$

where $y_B = y_0 + l_2\,\varphi, x'_2 = y_2 - z + l_2\varphi_1$.
In turn, the scattering function will take the form

$$\Phi = \frac{1}{2}b(\dot{y}_A - \dot{x}'_1)^2$$

where

$$\dot{y}_A = \dot{y}_0 - l_1\dot{\varphi};x'_1 = \dot{y}_1 - \dot{z} - l_1\dot{\varphi}_1 \qquad (115)$$

Having made a number of intermediate calculations, we write down the equations of motion of the system:

$$\ddot{y}(M)+y(k'_1 + k''_1)+\varphi(-k'_1L_1 + k''_1L_2)= k'_1z + k''_2z \; ; \qquad (116)$$

$$\varphi(-k_2l_2(l_2 - L_2))+\ddot{\varphi}_1(J_1)+\varphi_1(k_2l_2{}^2)= k_2l_2z.$$

Values of coefficients of equations (116) in *coordinates* y, φ_1 are given in Table 2.

Note that in the y, φ, φ_1 coordinates, the system is under the action of kinematic disturbance and does not contain direct information about the action of portable inertia forces. Of greater interest is a coordinate system in which the motion of an object is considered relative to the base. In this case, you can enter the coordinate $y_0 = y_{01} + z$, then $y_{01} - z = y$, and then enter y in all other relations.

Table 2. Values of coefficients of equations (116) in coordinates y, φ, φ_1

a_{11}	a_{12}	a_{13}
$Mp^2 + (k''_1 + k'_1)$	$-k'_1L_1 + k''_1L_2$	0
a_{21}	a_{22}	a_{23}
$(-k'_1L_1 + k''_1L_2)$	$Jp^2 + b(L_1 - l_1)^2p + k''_1L_1{}^2 + k''_1L_2^2 + k_2(l_2 - L_2)^2$	$bl_1(L_1 - l_1)p - k_2(l_2 - L_2)$
a_{31}	a_{32}	a_{33}
0	$pbl_1(L_1 - l_1) - k_2l_2(l_2 - L_2)$	$J_1p^2 + k_2l_2{}^2 + bl_1{}^2p$
Q_1	Q_2	Q_3
$(k'_1 + k''_1)z$	$-bp(L_1 - l_1) - k'_1L_1z + k''_1L_2z - k_2(l_2 - L_2)z$	$k_2l_2z - bl_1pz$

Note: Q_1, Q_2, Q_3 - generalized forces corresponding to y, φ, φ_1 coordinates are determined from (116)

Accounting for Portable Forces of Inertia

Let's present the design diagram of the trailer suspension (Figure 19) in the coordinate system, which allows us to proceed to a more convenient mathematical model in the future. For these purposes, an additional degree of freedom y_{01} is introduced. Figure 20 shows a calculation scheme that uses the coordinate system y_1, y_2, $y'_1y'_2$as well as y_0, φ, y_{01}, φ_1. Such a system has four degrees of freedom. Base oscillations are considered as external impact. Other symbols are shown in Figure 20. match Figure 19 (but enter k_{01} and k_1 instead of *bp*).

Figure 20. Design diagram of trailer suspension

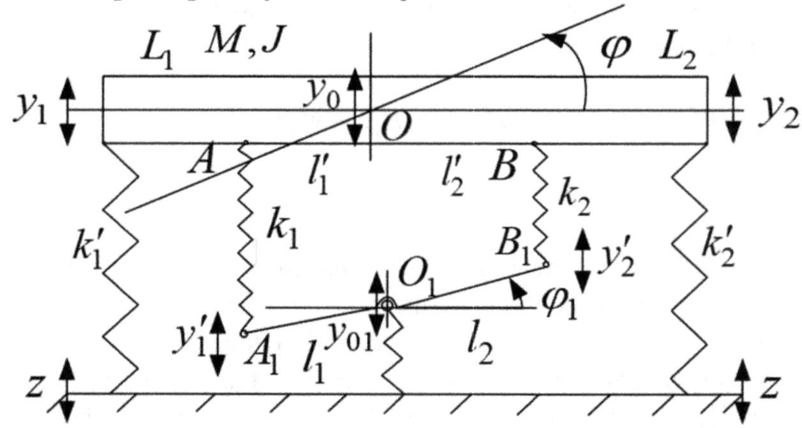

We write expressions for kinetic and potential energies

$$T = \tfrac{1}{2}M\dot{y}_0{}^2 + \tfrac{1}{2}J\dot{\varphi}^2 + \tfrac{1}{2}m_1(\dot{y}'_1)^2 + \tfrac{1}{2}m_2(\dot{y}'_2)^2 + \tfrac{1}{2}J_1\dot{\varphi}_1{}^2 \tag{117}$$

$$\begin{aligned}\Pi &= \tfrac{1}{2}k'_1\left(y_1 - z\right)^2 + \tfrac{1}{2}k'_2\left(y_2 - z\right)^2 + \tfrac{1}{2}k_1\left(y_A - y'_1\right)^2 \\ &+ \tfrac{1}{2}k_2\left(y_B - y'_2\right)^2 + \tfrac{1}{2}k_{01}\left(y_{01} - z\right)^2.\end{aligned} \tag{118}$$

Let $y_{01} = (y'_{01})+z; y_0 = ay_1 + by_2 \varphi = c(y_2 - y_1); y_{01} = a_1 y'_1 + b_1 y'_2;$

$$\varphi_1 = c_1(y'_2 - y'_1);\ y_A = y_0 - l'_1\varphi; y_B = y_0 + l'_2\varphi; a = \frac{L_2}{L_1 + L_2}$$

$$b = \frac{L_1}{L_1 + L_2}; c = \frac{1}{L_1 + L_2}; a_1 = \frac{l_2}{l_1 + l_2}; b_1 = \frac{l_1}{l_1 + l_2}; c_1 = \frac{1}{l_1 + l_2} \tag{119}$$

We will enter $y'_1 = \frac{y_{01} - b_1 y'_2}{a_1} y'_2 = \frac{y_{01} - a_1 y'_1}{b_1}$, as well as

$$y_{01} = a_1 y'_1 + b_1 y'_2,\ y'_1 = \frac{y_{01} - b_1 y'_2}{a_1} = a_0 y_{01} - i y'_2,$$

where$i = \frac{l_2}{l_1}, \frac{1}{b_1} = a_0$. Express through the variable

$$y'_0\, \varphi_1 = c_1(y'_2 - \frac{y_{01}}{a_1} + \frac{b_1 y'_2}{a_1}),$$

where

$$y_1 = y_0 - l_1\varphi;\ y_2 = y_0 + l_2\varphi \tag{120}$$

Define $y'_1 = a_0 y_{01} - iy'_2$ then:

$$\varphi = c_1(y'_2 - a_0 y_{01} + iy'_2) = c_1[(i+1)y'_2 - a_0 y_{01}] \tag{121}$$

and find the kinetic energy

$$T = \frac{1}{2}M\dot{y}_0{}^2 + \frac{1}{2}Jc^2(\dot{y}_2 - \dot{y}_1)^2 + \frac{1}{2}m_1\left[a_0\dot{y}_{01} - i(\dot{y}'_2)^2\right]^2 + \frac{1}{2}m_2(\dot{y}'_2)^2 + \frac{1}{2}J_1 c_1{}^2[\dot{y}'_2 - a_0\dot{y}_{01}], \tag{122}$$

where$i = \frac{l_2}{l_1}, a_0 = \frac{1}{b_1}$.
Let's enter a series of relations

$$a_2 = a + l'_1 c, b_2 = b - l'_1 c, a_3 = a - l'_2 c, b_3 = b + l'_2 c\ ; \tag{123}$$

in its final form, the equations of motion of the system will take the form:

$$\ddot{y}_1(Ma^2 + Jc^2) + y_1(k'_1 + k_1 a_2{}^2 + k_2 a_3{}^2) + \ddot{y}_2(Mab - Jc^2) + y_2(k_1 a_2 b_2 + k_1 a_2 i + k_2 a_3 b_3) + y(k_1 a_2 a_0) + y'_2(-k_2 a_3) = k'_1 z\ ; \tag{124}$$

$$\ddot{y}_1(Mab - Jc^2) + y_1(k_1 a_2 b_2 + k_2 a_3 b_3) + \ddot{y}_2(Mb^2 + Jc^2) + y_2(k'_2 + k_2 b_2{}^2 + k_2 b_3{}^2) + y_{01}(-k_2 a_0 b_2) + y'_2(k_1 b_2 i - k_2 b_3) = k'_2 z\ ; \tag{125}$$

$$\ddot{y}_1(ma_0{}^2 + Jc^2 a_0{}^2) + y_1(-k_1 a_2 a_0) + y_2(-k_1 a_0 b_2) + \ddot{y}_{01}(m_1 a_0{}^2 + Jc_1{}^2 a_0) + y_{01}(k_1 a_0{}^2 + k_{01}) + \ddot{y}'_2[-m_1 a_0 i - Jc^2(1+i)] + y'_2(-k_1 a_0 i) = k_{01} z\ ; \tag{126}$$

$$y_1(k_1 a_2 i - k_2 a_3) + y_2(k_1 b_2 i - k_2 b_3) + \ddot{y}_{01}[-m_1 i - Jc_1 a_0(1+i)] - y_{01}(-k_1 a_0 i) + \ddot{y}'_2[m_1 i^2 + Jc_1{}^2(1+i)^2] + y'_2(k_1 i^2 + k_2) = 0. \tag{127}$$

Table 3 presents the coefficients of equations (124) - (127).

Table 3. Values of equation coefficients in coordinate system y_1, y_2, y_{01}, y'_2

a_{11}	a_{12}	a_{13}	a_{14}
$(Ma^2+Jc^2)p^2$ $+k'_1+k_1a_2^2+k_2a_3^2$	$(Mab-Jc^2)p^2$ $+k_1a_2b_2+k_1a_2i+k_2a_3b_3$	$-k_1a_0a_2$	$-k_2a_3$
a_{21}	a_{22}	a_{23}	a_{24}
$(Mab-Jc^2)p^2$ $+k_1a_2b_2+k_2a_3b_3$	$(Mb^2+Jc^2)p^2$ $+k_2'+k_2b_2^2+k_2b_3^2$	$-k_2a_0b_2$	$k_1b_2i-k_2b_3$
a_{31}	a_{32}	a_{33}	a_{34}
$-k_1a_2a_0$	$-k_1a_0b_2$	$(m_1a_0^2+Jc_1^2a_0)p^2$ $+k_1a_0^2+k_{01}$	$-[m_1a_0i+Jc^2(1+i)]p^2-$ $-k_1a_0i$
a_{41}	a_{42}	a_{43}	a_{44}
$-k_2a_3$	$+k_1b_2i-k_2b_3$	$-[m_1i+Jc_1a_0(1+i)]$ $-k_1a_0i$	$[m_1i^2+Jc_1^2(1+i)^2]p^2+$ $+k_1i^2+k_2$
Q_1	Q_2	Q_3	Q_4
$Q_1=k'_1z$	$Q_2=k'_2z$	$Q_3=k_{01}z$	$Q_4=0$

Note: Q_1-Q_4- generalized forces by coordinatesy_1, y_2, y_{01}, y'_2.

Using the coordinates from Table 3, one can construct a system of transfer functions with respect to any of the coordinates $y_1, y_2, y_{01}, y'_2$from the base disturbance; apply the Cramer rule for these purposes. In this case, the necessary expressions for constructing the amplitude-frequency characteristics can be obtained, however, in this case, some dynamic properties can be determined from the coefficient matrix. In particular, it can be concluded that the system has four natural frequencies, and therefore four possible resonance peaks on the amplitude-frequency characteristics. Between coordinates, intercoordinate y_1, y_2, y_{01}, y'_2connections at certain frequencies can be "canceled," which suggests the possibility of two dynamic blanking modes. Given the fact that generalized forces have an effect at three coordinates, a more complex picture of the interdependence of movements between partial systems can be y_1, y_2, y_{01} expected. However, these comments are only preliminary, since the subsequent transformation of generalized coordinates involves the introduction of joints that reduce the total number of degrees of freedom.

To take into account possible joints at points A and O_1, it is necessary to enter appropriate coordinates reflecting the relative movement, which in subsequent actions could be considered as disappearing small. Assuming that, $y''_2 = y_A - y'_2a$, we rewrite the $y_{01} = y'_{01} + z$expression for kinetic energy in the form

$$T = \frac{1}{2}M\left(a\dot{y}_1 + b\dot{y}_2\right)^2 + \frac{1}{2}Jc\left(\dot{y}_2 - \dot{y}_1\right)^2 + \frac{1}{2}m_1\left[a_0\dot{y}_{01} - i(\dot{y}'_2)\right]^2 + \\ + \frac{1}{2}m_2\left(\dot{y}'_2\right)^2 + \frac{1}{2}J_1c_1\left[\dot{y}'_2(1+i) - a_0\dot{y}_{01}\right]^2. \tag{128}$$

The potential energy of the system in this case will take the form

$$\Pi = \tfrac{1}{2}k'_1\left(y_1 - z\right)^2 + \tfrac{1}{2}k'_2\left(y_2 - z\right)^2 + \tfrac{1}{2}k_1\left(y_A - y'_1\right)^2 + \\ + \tfrac{1}{2}k_2\left(y_B - y'_2\right)^2 + \tfrac{1}{2}k_{01}\left(y_{01} - z\right)^2. \qquad (129)$$

Let's move on to a coordinate system $y_1, y''_1, y'_{01}, y''_2$in which the expressions for kinetic and potential energy can be written as:

$$T = \tfrac{1}{2}m\left(\dot{y}_1 a + \dot{y}_2 b\right)^2 + \tfrac{1}{2}Jc^2\left(\dot{y}_2 - \dot{y}_1\right)^2 + \\ + \tfrac{1}{2}m_1\left(\dot{y}'_1\right)^2 + \tfrac{1}{2}m_2\left(\dot{y}'_2\right)^2 - \tfrac{1}{2}J_1 c_1{}^2\left(\dot{y}'_2 - \dot{y}'_1\right)^2, \qquad (130)$$

$$\Pi = \tfrac{1}{2}k'_1\left(y_1 - z\right)^2 + \tfrac{1}{2}k'_2\left(y_2 - z\right)^2 + \\ + \tfrac{1}{2}k_1\left(y''_1\right)^2 + \tfrac{1}{2}k_2\left(y''_2\right)^2 + \tfrac{1}{2}k_{01}\left(y''_{01}\right)^2, \qquad (131)$$

where$y_{01} = y - z, y''_1 = y_A - y', y''_2 = y_B - y'_2, y_A = y_0 - l'_1\varphi,$

$y'_1 = y_{01} - l_1\varphi_1, y_A = a_2 y_1 + b_2 y'_2, y'_1 = y_{01} - l_1\varphi_1, y_B = y_0 + l'_2\varphi, y'_2$ $= y_{01} + l_2\varphi_1,$

$y_B = a'_2 y_1 + b'_2 y_2, y'_2 = y_{01} + l_2\varphi, y_{01} = y'_{01} + z, y_1 = y_0 - l_1\phi, y_2 = y_0 + l_2\varphi$

Enter the coordinatey''_2, (let$y_B - y'_2 = y''_2$), and y'_{01}denote by the relations$y_{01} = y'_{01} + z, y_B = a_3 y_1 + b_3 y_2$. We continue the further construction of the mathematical model of suspension in the coordinate systemy_1, y_2, y_0, y'_2. Such a coordinate system is of interest in that it allows you to build a model of a system that has an articulation at the point O, that is, at the center of gravity, which simplifies the further analysis of dynamic interactions. Let's write down the initial data for representing expressions for kinetic and potential energies. Let's move on to the coordinate system with the intention of entering. Then, doing, you can go to a simplified design scheme of suspension. Let's define the terms of the expression for kinetic energy:

$$T = T_1 + T_2 + T_3 + T_4 + T_5 \qquad (132)$$

where

$T_4 = \frac{1}{2}m_2\dot{y}_2 = \frac{1}{2}m_2(a_3\dot{y}_1 + b_3\dot{y}_2 - \dot{y}''_2); T_5 = \frac{1}{2}Jc_1{}^2[(i+1)^2a_3{}^2\dot{y}_1{}^2 + 2(1+i)^2a_3$
$b_3\dot{y}_1\dot{y}_2 +$

$+(1+i)^2b_3{}^2\dot{y}_2{}^2 + (i+1)^2(\dot{y}''_2)^2 + 2(i+1)a_0\dot{y}''_2\dot{y}_{01} + a_0{}^2\dot{y}_{01}{}^2 - 2(i+1)^2a_3\dot{y}_1\dot{y}''_2 -$

$-2(i+1)^2b_3\dot{y}''_2 - 2a_0a_3(1+i)\dot{y}_1\dot{y}_{01} - 2a_0(1+i)b_3\dot{y}_2\dot{y}_{01}]$.

Let's take $a_3(1+i) = a_6, b_3(1+i) = b_6$.Similarly, we detail the expressions for the potential energy:

$$\Pi = \Pi_1 + \Pi_2 + \Pi_3 + \Pi_4 + \Pi_5 \tag{133}$$

where

$\Pi_1 = \frac{1}{2}k'_1(y_1 - z)^2; \Pi_2 = \frac{1}{2}k'_2(y_2 - z)^2$;
$\Pi_3 = \frac{1}{2}k_1[y_1{}^2a_7{}^2 + 2a_7b_7y_1y_2 + y_2{}^2b_7{}^2 + y_{01}{}^2a_0{}^2 + 2a_0iy_{01}y''_2 +$
$+i^2(\dot{y}''_2)^2 - a_7y_1a_0y_{01} - y_2b_7a_0y_{01} - y_1a_7iy''_2 - y_2b_7i\dot{y}''_2]$;

$\Pi_4 = \frac{1}{2}k_2(\dot{y}''_2)^2; \Pi_5 = \frac{1}{2}k_{01}(y_{01} - z)^2$,

at the same time $a_7 = a_2 + ia_3$; $b_7 = b + ib_3$.
The equations of motion of the system in this case will take the form:

$$\ddot{y}_1(Ma^2 + Jc^2 + m_1i^2a_3{}^2 + m_2a_3{}^2 + J_1c_1{}^2a_6{}^2) + y_1(k'_1 + k_1a_7{}^2) + \ddot{y}_2(Mab - Jc^2 +$$
$$+m_1i^2a_3b_3 + m_2a_3b_3 - m_2a_3 + J_1c_1{}^2a_6b_6) + y_2(k_1a_7b_7) + \ddot{y}_{01}(-m_1a_0a_3 - J_1c_1{}^2a_0a_6) +$$
$$+y_{01}(-k_1a_7a_0) + \ddot{y}_2(-m_1i^2a_3 - m_2a_3 - J_1c_1{}^2a_6(1+i)) + \ddot{y}_2(-k_1a_7i) = k'_1z \; ; \tag{134}$$

$\ddot{y}_1(Mab - Jc^2 + m_1 i^2 a_3 b_3 + m_2 a_3 b_3 + J_1 c_1{}^2 a_6 b_6) + y_1(k_1 a_7 b_7) + \ddot{y}_2(Mb^2 + Jc^2 + m_1 i^2 b_3{}^2$
$+$

$+m_2 b_3{}^2 + J_1 c_1{}^2 b_6{}^2) + y_2(k'_2 + k_1 b_7{}^2) + \ddot{y}_{01}[-m_1 a_0 i b_3 - J_1 c_1{}^2 a_0 b_6) + y_{01}(-k_1 b_7 a_0) +$

$$+\ddot{y}''_2(-m_1 i^2 b_3 - J_1 c_1{}^2 b_6(1+i)] + y_2(-k_7 a_7 i) = k'_2 z \ ; \tag{135}$$

$\ddot{y}_1(-m_1 a_3 a_0 - J_1 c_1{}^2 a_0 a_6) + y_1(-k_1 a_7 a_0) + \ddot{y}_2(-m_1 a_0 b_3 i - J_1 c_1{}^2 a_0 b_6) +$

$+\ddot{y}_2(-k_1 b_7 a_0) + \ddot{y}_{01}(m_1 a_0{}^2) + y_{01}(k_1 a_0{}^2 + k_{01}) +$

$$+\ddot{y}_2(m_1 i a_0 + J_1 c_1{}^2(1+i) a_0) + \ddot{y}_2(k_1 a_0 i) = k_{01} z \ ; \tag{136}$$

$\ddot{y}_1[-m_1 i^2 a_3 - m_2 a_3 - J_1 c_1{}^2(1+i) a_6] + y_1(-k_1 a_7 i) + \ddot{y}_2[-m_1 i^2 b_3 - m_2 b_3 -$

$-J_1 c_1{}^2 b_6(1+i)] + y_2(-k_1 b_7 i) + \ddot{y}_{01}[m_1 i a_0 + J_1 c_1{}^2(1+i) a_0] + y_{01}(k_1 a_0 i) +$

$$+\ddot{y}''_2[m_1 i^2 + m_2 + J_1 c_1{}^2(1+i)] + y''_2(k_2 + k_1 i^2) = 0. \tag{137}$$

Table 4 shows the corresponding values of the system equation coefficients.

Table 4. Values of coefficients of equation system (134) - (137) coordinates y_1, y_2, y_0, y''_2

a_{11}	a_{12}	a_{13}	a_{14}
$(Ma^2 + Jc^2 + m_1 i^2 a_3{}^2$ $+m_2 a_3{}^2 + J_1 c_1{}^2 a_6{}^2) p^2$ $+k'_1 + k_1 a_7{}^2$	$(Mab - Jc^2 + m_1 i^2 a_3 b_3$ $+m_2 a_3 b_3 + J_1 c_1{}^2 a_6 b_6) p^2$ $+k_1 a_7 b_7$	$(-m_1 a_0 a_3$ $-J_1 c_1{}^2 a_0 a_6) p^2$ $-k_1 a_7 a_0$	$(-m_1 i^2 a_3$ $-m_2 a_3 - J_1 c_1{}^2 a_6 (a$ $+1)] p^2 k_1 a_7 i$
a_{21}	a_{22}	a_{23}	a_{24}
$(Mab - Jc^2 + m_1 i^2 a_3 b_3$ $+m_2 a_3 b_3 + J_1 c_1{}^2 a_6 b_6) p^2$ $+k_1 a_7 b_7$	$(Mb^2 + Jc^2$ $+m_1 i^2 b_3{}^2 + J_1 c_1{}^2 b_6{}^2) p^2$ $+k'_2 + k_1 b_7{}^2$	$(-m_1 a_0 i b_3$ $-J_1 c_1{}^2 a_0 b_6) p^2$ $-k_1 b_7 a_0$	$(-m_1 i^2 b_3 - m_2 b_3$ $-J_1 c_1{}^2 b_6 \times (1+i)) p^2$ $+(-k_7 a_7 i)$
a_{31}	a_{32}	a_{33}	a_{34}
$(-m_1 a_3 a_0$ $-J_1 c_1{}^2 a_0 a_6) p^2$ $-k_1 a_7 a_0$	$(-m_1 a_0 b_3 i$ $-J_1 c_1{}^2 a_0 b_6) p^2$ $-k_1 b_7 a_0$	$(m_1 a_0)^2 p^2$ $+k_1 a_0{}^2 + k_{01}$	$(m_1 i a_0 + J_1 c_1{}^2$ $\times (i+1) a_0) p^2$ $+k_1 a_0 i$
a_{41}	a_{42}	a_{43}	a_{44}
$[-m_1 i^2 a_3 - m_2 a_3$ $-J_1 c_1{}^2(1+i) a_6]$ $p^2 - k_1 a_7 i$	$[-m_1 i^2 b_3 - m_2 b_3$ $-J_1 c_1{}^2 b_6(1+i)]$ $p^2 - k_1 b_7 i$	$[m_1 i a_0$ $+J_1 c_1{}^2(1+i) a_0]$ $p^2 + k_1 a_0 i$	$[m_1 i^2 + m_2$ $+J_1 c_1{}^2 \times (1+i)]$ $p^2 + k_2 + k_1 i^2)$

Q_1	Q_2	Q_3	Q_4
$Q_{y_1} = k'_1 z$	$Q_{y_2} = k'_2 z$	$Q_{y_{01}} = k_{01} z$	$Q_{y''_2} = 0$

Note: $Q_1 - Q_4$- generalized forces.

Further research boils down to finding transfer functions

$$W_1(p) = \frac{\bar{y}_1(p)}{\bar{z}} \text{ and } W_2(p) = \frac{\bar{y}_2(p)}{\bar{z}}$$

what can be done using Kramer formulas (Druzhinsky, 1977). Changing the system of generalized coordinates leads to a change in cross-links between partial systems. In this case, it is possible to "neutral" the bonds between all partial systems, which can create conditions for reducing dynamic interactions in the suspension.

Since a system with four degrees of freedom is used and there are no joints with the base in the system, no peculiarities in frequency characteristics should be expected compared to traditional forms of systems. However, the mathematical model gives an idea of the possible spectrum of dynamic properties of the suspension in a wide frequency range (dynamic damping modes, natural vibration frequencies). Further investigation can be continued in the coordinate system, the choice of which implies the possibility of forming joints of solids.

Influence of Coordinate System Selection

Let's move on to the coordinate system, $(y_{01} = y'_{01} + z)y''_2(y_B - y'_2)$. In this case

$$T = T_1 + T_2 + T_3 + T_4 + T_5 \tag{138}$$

Where

$$T_1 = \frac{1}{2}M\dot{y}_0^{\,2} = \frac{1}{2}M(a\dot{y}_1 + b\dot{y}_2)^2; T_2 = \frac{1}{2}Jc^2(\dot{y}_2 - \dot{y}_1)^2 \;;$$

$$T_3 = \frac{1}{2}m_1[(a_0\dot{y}'_{01} - i\dot{y}_1 a_3 - i\dot{y}_2 b_3 + i\dot{y}''_2) + a_0\dot{z}]^2 \;;$$

$$T_4 = \frac{1}{2}m_2\dot{y}_2^{\,2} = \frac{1}{2}m_2(a_3\dot{y}_1 + b_3\dot{y}_2 - \dot{y}''_2)^2 \;;$$

$$T_5 = \frac{1}{2}J_1 c_1^{\,2}[(1+i)a_3\dot{y}_1 + (1+i)b_3\dot{y}_3 - (i+1)\dot{y}''_2 - a_0\dot{y}_{01}]^2 \;;$$

$$T_5 = \frac{1}{2}J_1 c_1(a_6\dot{y}_1 + b_6\dot{y}_2 - (1+i)\dot{y}''_2 - a_0\dot{y}'_{01} - a_0\dot{z})^2.$$

In turn, the potential energy of the system is

$$\Pi = \Pi_1 + \Pi_2 + \Pi_3 + \Pi_4 + \Pi_5 \tag{139}$$

where

$$\Pi_1 = \frac{1}{2}k'_1(y_1 - z)^2; \Pi_2 = \frac{1}{2}k'_2(y_2 - z)^2; \Pi_3 = \frac{1}{2}k_1[y_1 a_7 + y_2 b_7 - iy''_2 - a_0 y_{01}]^2,$$

$$\Pi_4 = \frac{1}{2}k_2(y''_2)^2; \Pi_5 = k_{01}(y'_{01})^2$$

while doing so$a_7 = a_2 + ia_3, b_7 = b_2 + ib_3$.

The given diagram of equations output gives an idea of the influence at the mathematical model level of the values of the parameters of the composite elements of the suspension. At the same time parts of joints formation are opened. Equations of motion, more precisely coefficients of equations of motion, and generalized forces, are given in Table 5.

It follows from Table 5 that the influence of portable inertia forces when choosing the appropriate coordinates is reflected through the structure of generalized forces. You can see that the generalized forces contain terms formed by the motion of the base, which means inertial forces. The system can be simplified by introducing coordinate jointsy'_{01}, y''_2. To do this, it is necessary to transform the initial system of equations (Table 5) accordingly

Since the kinetic energy does not change, we write down general expressions for the potential energy in the coordinate systemy_1, y_2, y'_{01}, y''_2. Let us assume that

$$\Pi = \Pi_1 + \Pi_2 + \Pi_3 + \Pi_4 + \Pi_5,$$

where

$$\Pi_1 = \frac{1}{2}k'_1(y_1 - z)^2, \Pi_2 = \frac{1}{2}k'_2(y_2 - z)^2,$$

$$\Pi_3 = \frac{1}{2}k_1[y_1 a_7 + y_2 b_7 - iy''_2 - a_0 y'_{01} - a_0 z]^2,$$

$$\Pi_3 = \frac{1}{2}k_1(y_1 a_7 + y_2 b_7 - i\dot{y}''_2 - a_0 y_{01} - a_0 z)^2,$$

$$\Pi_4 = k_2(y''_2)^2, \Pi_5 = k_{01}(y'_{01})^2.$$

Table 5. Values of equation coefficients in coordinate system y_1, y_2, y'_{01}, y''_2

a_{11}	a_{12}	a_{13}	a_{14}
$(Ma^2 + Jc^2 + m_1 i^2 a_3^{\ 2}$ $+m_2 a_3^{\ 2} + J_1 c_1^{\ 2} a_6^{\ 2})p^2$ $+k'_1 + k_{01} a_7^{\ 2}$	$(Mab - Jc^2 + m_2 a_3 b_3$ $+m_1 i^2 a_3 b_3 + J_1 c_1^{\ 2} a_6 b_6)p^2$ $+k_{01} a_7 b_7$	$(-m_1 a_0 a_3 i$ $-J_1 c_1^{\ 2} a_0 a_6)p^2$ $-k_{01} a_7 a_0$	$(-J_1 c_1^{\ 2} a_6(1+i)$ $-m_1 i^2 a_3 - m_2 a_3)$ $p^2 - k_1 a_7 i$
a_{21}	a_{22}	a_{23}	a_{24}
$Mab - Jc^2 + m_1 i^2 a_3 b_3$ $+m_2 a_3 b_3 + J_1 c_1^{\ 2} a_6 b_6$ $+k_{01} a_7 b_7$	$(Mb^2 + Jc^2 + m_1 i^2 b_3^{\ 2}$ $+m_2 b_3^{\ 2} + J_1 c_1^{\ 2} b_6^{\ 2})+k'_2$ $+k_1 b_7^{\ 2}$	$(J_1 c_1^{\ 2} b_6 a_0$ $-m_1 i a_0 b_3)p^2$ $-k_1 b_7 a_0$	$(-m_1 i^2 b_3 - m_2 b_3$ $-J_1 c_1^{\ 2}(1+i) b_6)$ $p^2 - k_1 b_7 i)$
a_{31}	a_{32}	a_{33}	a_{34}
$(-m_1 a_0 i a_3$ $-J_1 c_1^{\ 2} a_6 a_0)p^2$ $-k_1 a_7 a_0$	$(-m_1 i a_0 b_3$ $+J_1 c_1^{\ 2} a_0 b_6)p^2$ $-k_{01} b_7 a_0$	$(m_1 a_0^{\ 2}$ $+J_1 c_1^{\ 2} a_0^{\ 2})p^2$ $+k_1 a_0^{\ 2} + k_{01}$	$[J_1 c_1^{\ 2}(i+1) a_0$ $+m_1 i a_0]p^2$ $+k_1 a_0 i$
a_{41}	a_{42}	a_{43}	a_{44}
$[-m_1 i^2 a_3 - m_2 a_3$ $-J_1 c_1^{\ 2} a_6(1+i)]$ $p^2 - k_1 a_7 i$	$[-m_1 i^2 b_3 - J_1 c_1^{\ 2} b_6(1+i)$ $-m_2 b_3]p^2 - k_1 b_7 i$	$[J_1 c_1^{\ 2}(1+i) a_0$ $+k_1 i a_0 + m_1 a_0 i]$ $p^2 + k_{01} a_0 i$	$[m_1 i^2 + J_1 c_1^{\ 2}$ $\times (1+i)^2]p^2$ $+k_2$
Q_1	Q_2	Q_3	Q_4
$Q_{1,} = m_1 a_0 a_3 z$ $+J_1 c_1^{\ 2} a_0 a_6 z$ $+k'_1 z + k_1 a_0 a_7$	$Q_2 = m_1 a_0 i b_3 z$ $+J_1 c_1^{\ 2} a_0 b_6 z$ $+zk'_2 + k_1 a_0 b_7 z$	$Q_3 = m_1 a_0 z$ $+J_1 c_1^{\ 2} a_0^{\ 2} z$	$Q_4 = a_0 iz$ $+J_1 c_1^{\ 2} a_0(1+i)z$

Note: $Q_1 - Q_4$- generalized forces.

The equations we need to consider the dynamics of the lever suspension, that is, the MCD, in which the joints $y'_{01} y''_2$ are introduced through and, have the form:

$$\ddot{y}_1(Ma^2 + Jc^2 + m_1 i^2 a_3^{\ 2} + m_2 a_3^{\ 2} + J_1 c_1^{\ 2} a_6^{\ 2}) + y_1(k'_1 + k_1 a_7^{\ 2}) +$$

$$+\ddot{y}_2(Mab - Jc^2 + m_1 i^2 a_3 b_3 + m_2 a_3 b_3 + +J_1 c_1 a_6 b_6) + y_2(k_1 a_7 b_7) =$$

$$= \ddot{z}(m_1 a_0 a_3 + J_1 c_1^{\ 2} a_0 a_6) + (k'_1 + k_1 a_0 a_7); \tag{140}$$

$$\ddot{y}_1(Mab - Jc^2 + m_1 i^2 a_3 b_3 + m_2 a_3 b_3 + J_1 c_1^{\ 2} a_6 b_6) + y_1(k_1 a_7 b_7) + \ddot{y}_2(Mb^2 +$$

$$+Jc^2 + m_1 i^2 b_3^{\ 2} + m_2 b_3^{\ 2} + +J_1 c_1^{\ 2} b_6^{\ 2}) + y_2(k'_2 + k_1 b_7^{\ 2}) = \ddot{z}(m_1 a_0 i b_3 +$$

$$+J_1 c_1^{\ 2} a_0 b_6) + k_2 + k_1 a_0 b_7. \tag{141}$$

Design diagram of suspension with two joints is a system with two degrees of freedom; for her it can be written that

$$a_{11} = (Ma^2 + Jc^2 + m_1 i^2 a_3{}^2 + m_2 a_3{}^2 + J_1 c_1{}^2 a_6{}^2)p^2 + k'_1 + k_1 a_7{}^2,$$

$$a_{12} = a_{21} = (Mab - Jc^2 + m_1 i^2 a_3 b_3 + m_2 a_3 b_3 + J_1 c_1{}^2 a_6 b_6)p^2 + k'_1 + k_1 a_0 a_7,$$

$$a_{22} = (Mb^2 + Jc^2 + m_1 i^2 b_3{}^2 + m_2 b_3{}^2 + J_1 c_1{}^2 b_6{}^2)p^2 + k'_2 + k_1 b_7{}^2. \qquad (142)$$

The values of generalized forces are determined by the right parts (141) and (142). The corresponding transfer functions can now be found:

$$W_1(p) = \frac{\bar{y}_1(p)}{\bar{z}} = W_2(p) = \frac{\bar{y}_2(p)}{\bar{z}}.$$

In general, the relationship between coordinates and generalized forces is given by the formulas:

$$\bar{y}_1 = \frac{\bar{Q}_1 a_{22} - \bar{Q}_2 a_{12}}{a_{11} a_{22} - a_{12}{}^2} \qquad (143)$$

$$\bar{y}_2 = \frac{-\bar{Q}_1 a_{12} - \bar{Q}_2 a_{11}}{a_{11} a_{22} - a_{12}{}^2}. \qquad (144)$$

Thus, the mathematical model of the suspension can be obtained from the matrix, the elements of which are presented in Table 5, by excluding two columns a_{i3}, a_{i4} and two rows of the matrix $3i$, $4i$ ($i = \overline{i,3}$). The system of equations (141), (142) is defined by the remaining matrix 2×2.

Mathematical models of vehicle suspensions include a fairly large number of elements, including lever links that form joints. The proposed approach to building mathematical models is based on the selection of systems of generalized coordinates, within which the focus is on the coordinates of relative movements. With the appropriate choice of coordinates of relative motion, their "neutralization" is possible, which leads to a decrease in the number of degrees of freedom of movement of the system, but allows you to obtain a mathematical model of the system with joints. In this case, the corresponding transformation of generalized forces takes place, taking into account the equality of virtual works on possible movements. Subsequent actions to evaluate dynamic properties are related to the use of the transfer functions of the system in dynamic synthesis problems. It is rational to search for parametric spaces in which at certain frequencies it would become possible to simultaneously dynamically cancel over several or all coordinates of the system.

REFERENCES

Dimentberg, F. M., & Kolesnikov, K. S. (Eds.). (1980). *Vibrations in technology: A reference book in 6 volumes* (Vol. 3). Mechanical Engineering.

Druzhinsky, I. A. (1977). Mechanical chains. *Mechanical Engineering (New York, N.Y.).*

Eliseev, S. V., Lukyanov, A. V., Reznik, Yu. N., & Khomenko, A. P. (2006). *Dynamics of mechanical systems with additional ties*. ISU.

Eliseev, S. V., Moskovskikh, A. O., Bolshakov, R. S., & Savchenko, A. A. (2012). Possibilities of integration of methods of circuit theory and theory of automatic control in machine dynamics problems. Science and Education: Electronic Scientific and Technical Publication, 6. http://technomag.edu.ru/doc/378699.html

Khomenko, A. P., Eliseev, S. V., Artyunin, A. I., Kaimov, E. V., & Eliseev, A. V. (2015). Device for damping oscillations (Patent No. 150331 RUS, IPC F16F 15/04). Russian Federation. Application No. 2014138832/11, filed September 25, 2014, and published February 10, 2015. Bulletin No. 4.

Kiryukhin, A. V., Tikhonov, V. A., Chistyakov, A. G., & Yablonsky, V. V. (2012). Active vibration protection - purpose, principles, condition. Active vibration protection and noise insulation of pipelines and experimental research. *Problems of Mechanical Engineering and Automation*, (4), 102–110.

Kopylov, Y. R. (2011). Dynamics of vibration shock hardening processes [Monograph]. CPI "Scientific Book."

Lurie, A. I. (1968). *Analytical mechanics*. Nauka.

Ryabov, I. M., Chernyshov, K. V., Eliseev, S. V., & Upyr, R. Y. (2010). Experimental studies of inertia-friction dampers in vehicle suspension systems (part 1). *Bulletin of Irkutsk State Technical University*, *2*(42), 146–154.

ADDITIONAL READING

Eliseev, A. V. (2020). Frequency function and damping function in the evaluation of dynamic processes in symmetric mechanical oscillatory systems. *Advanced Engineering Research (Rostov-on-Don)*, *20*(4), 360–369.

Eliseev, A. V., Kuznetsov, N. K., & Eliseev, S. V. (2021). Frequency energy function in assessing the dynamic states of technical objects. MAI Works, (118).

Eliseev, A. V., Kuznetsov, N. K., & Eliseev, S. V. (2021). New approaches to evaluating the dynamic properties of oscillatory structures: Frequency functions and motion connectivity. MAI Works, (120).

Eliseev, A. V., Kuznetsov, N. K., & Nikolaev, A. V. (2022). The concept of dynamic invariants for assessing the structural properties of mechanical oscillatory systems. Transport. *Mining and Construction Engineering: Science and Production*, (15), 18–30.

Eliseev, A. V., & Mironov, A. S. (2023). Dynamic invariant maps for evaluating motion modes in mechanical oscillatory systems. MAI Works, (128).

Eliseev, S. V., Artyunin, A. I., Kaimov, E. V., & Eliseev, A. V. (2014). Device for regulating elastic-dissipative properties of vibration protection systems (RU 142137 U1). Patent for utility model. Application No. 2014100299/11, filed January 9, 2014.

Eliseev, S. V., Artyunin, A. I., Khomenko, A. P., Kaimov, E. V., & Eliseev, A. V. (2013). Device for damping vibrations (RU 136112 U1). Patent for utility model. Application No. 2013135078/11, filed July 25, 2013.

Eliseev, S. V., & Eliseev, A. V. (2019). *Theory of oscillations: Structural mathematical modeling in problems of dynamics of technical objects*. Springer.

Eliseev, S. V., & Eliseev, A. V. (2020). Some applications of the methods of structural mathematical modeling. In *Studies in Systems* (Vol. 252, pp. 433–521). Decision and Control., DOI: 10.1007/978-3-030-36104-3_15

Eliseev, S. V., Eliseev, A. V., & Kashuba, V. B. (2013). Specifics of gap assessment in a model problem of a material particle bouncing with non-holding constraints. *Mechanical Engineering and Safety of Life Activities*, (1[15]), 50–56.

Eliseev, S. V., Sitov, I. S., & Eliseev, A. V. (2012). Motion of a material particle with bouncing: A case study with non-holding constraints. *Mechanical Engineering and Safety of Life Activities*, (3[13]), 53–58.

Jiang, Y. Z., He, K. F., Dong, Y. L., Yang, D.-L., & Sun, W. (2019). Influence of load weight on dynamic response of vibrating screen. *Shock and Vibration, 4232730*. Advance online publication. DOI: 10.1155/2019/4232730

Korendiy, V., Kachur, O., Predko, R., Kotsiumbas, O., Stotsko, R., & Ostashuk, M. (2023). Generating rectilinear, elliptical, and circular oscillations of a single-mass vibratory system equipped with an enhanced twin crank-type exciter. *Vibroengineering Procedia*, *51*, 8–14. DOI: 10.21595/vp.2023.23657

Krot, P., Shiri, H., Dąbek, P., & Zimroz, R. (2023). Diagnostics of bolted joints in vibrating screens based on a multi-body dynamical model. *Materials (Basel)*, *16*(17), 5794. DOI: 10.3390/ma16175794 PMID: 37687487

Peng, L., Jiang, H., Chen, X., Liu, D., Feng, H., Zhang, L., Zhao, Y., & Liu, C. (2019). A review on the advanced design techniques and methods of vibrating screen for coal preparation. *Powder Technology*, *347*, 136–147. DOI: 10.1016/j.powtec.2019.02.047

Vellingiri, V., & Udhayakumar, S. (2023). Effect of vibrator parameters and physical characteristics of parts on conveying velocity. Strojniški vestnik -. *Jixie Gongcheng Xuebao*, *69*, 352–363. DOI: 10.5545/sv-jme.2022.510

Chapter 4
Some Areas of Development of Methods Structural Mathematical Modeling

ABSTRACT

Mathematical modeling of mechanical oscillatory systems using structural analysis methods is considered. The main focus is on the interpretation of the dynamic interactions of the system elements using structural diagrams, which makes it possible to obtain new results in the theory of vibration technologies, transport dynamics and vibration protection. The main approaches to the transformation of mathematical models into structural analogues are presented, including the use of dynamically equivalent automatic control systems. This allows you to apply frequency analysis and dynamic synthesis methods to solve vibration protection problems. Examples of single-degree-of-freedom mechanical systems include models with springs, dampers, and mass-inertial links. Frequency equations and concepts of dynamic stiffness of elements and the system as a whole are analyzed. Particular attention is paid to resonant vibrations and their effect on the dynamic properties of the system.

CHARACTERISTIC FREQUENCY EQUATION: STRUCTURE, DYNAMIC STIFFNESS, FEATURES OF INTERACTION OF ELEMENTS SYSTEMS

Mathematical modeling is widely used in solving various problems related to the assessment of the dynamic state of technical objects (Eliseev, 2018; Lontsikh & Eliseev, 2014). Structural mathematical modeling methods are a fairly developed

DOI: 10.4018/979-8-3373-0447-2.ch004

direction defined by interdisciplinary space defined by the methods of system theory, automatic control theory, circuit theory, graph theory in their various forms of applications (Eliseev, Volkov, & Kukharenko, 1990; Eliseev & Khomenko, 2014). The dynamics of mechanical oscillatory systems with concentrated parameters was reflected in many works on the theory of linear and nonlinear oscillations, which, in a certain sense, predetermined the interest in the formation and peculiarities of connections arising in dynamic interactions between the elements of the system under periodic influences. A number of rather developed applications are known related to solving the problems of the theory of vibration technology, vibration technologies, problems of transport dynamics, vibration protection and vibration isolation of technical objects (Belokobylsky, Eliseev, & Sitov, 2013).

The interpretation of the dynamic interactions of elements of mechanical systems, displayed by structural diagrams of dynamically equivalent automatic control systems, became the basis for the formation of unconventional approaches, which initiated the introduction of a number of new questions into the consideration related to ideas of expanding ideas about the set of typical elements, methods of their connection, the possibilities of structural transformations and the identification of dynamic features (Eliseev, Kuznetsov, Bolshakov, et al., 2015; Khomenko, Eliseev, Artyunin, Bolshakov, Kaimov, Kinash, & Nguyen, 2015).

Questions about the relationship between characteristic frequency equations of linear mechanical oscillatory systems with concentrated parameters and ideas about possible structural forms, the implementation of connections between individual elements of the system or their structural analogues are considered.

One of the directions of structural mathematical modeling is the use of dynamically equivalent automatic control systems as mathematical models of structural schemes. The approach is based on dynamic analogies arising from interactions of elements of mechanical oscillatory systems of various nature and corresponding interactions between elements of the automatic control system. The commonality of interactions in systems is determined by one-to-one relationships with the original mathematical model in the form of a system of ordinary differential equations with constant coefficients. The approach is generalized and can be extended to systems with distributed parameters (Eliseev, 1978; Halperin, 1968; Shatalov, 1962) .

Figures 1, *a-e* show different interpretations in representations of the original physical model as a mechanical oscillatory system with one degree of freedom. Such models are widely used in the tasks of machine dynamics and protection of equipment and equipment from vibrations, shocks, etc. The mechanical system in Figure 1, *a* consists of the so-called typical mechanical elements in the form of a spring k, a damper bp and a mass-inertial link m. The mass inertial element can be considered as a control object (in the tasks of vibration protection - an object of protection), which predetermines the possibility of using the analogy between automatic control systems (Eliseev & Khomenko, 2014).

The mathematical model of the system (Figure 1, *a*) under force disturbance in the form of harmonic force Q has the form

$$m\ddot{y} + b\dot{y} + ky = Q(t) \quad (1)$$

After the Laplace transform with respect to (1) under zero initial conditions, differential equation (1) is transformed to the form

$$mp^2\bar{y} + bp\bar{y} + k\bar{y} = \bar{Q} \quad (2)$$

where $p = j\,\omega$ is a complex variable; symbol <-> means Laplacian image.

1. By analogy with the basics of the theory of automatic control, the structure of the system in Figure 1, *a* is transformed into a structural diagram shown in Figure 1, b. Typical elements *m*, *b* and *k* in the structural diagram (Figure 1, *b*) correspond to typical links of the automatic control system (ACS), respectively, in the form of links with the transfer function of the integrating link of the second kind (object of protection) - mp^2, differentiating link of the first kind (damper) - *bp* and amplifying link (spring) - *k*. The block diagram in Figure 1, *b* shows not only the composition of the systems (the presence of standard links), but also the connections (forward, reverse, etc.). Typical links in the block diagram can be interconnected and form appropriate links, as shown in Figure 1, *c* and *d*.

The structural diagram of the initial mechanical oscillatory system in Figure 1, *a* can be presented in a generalized form, as shown in Figure 1, e.Note that the elements *bp* and *k* (Figure 1, *e*) can be translated into the "status" of negative feedback. In the physical sense, the negative feedback (Figure 1, *e*) is interpreted as the stiffness of the elastic element (*k*). Since a force is generated at the input to the adder (vol. *1* in Figure 1, *f*), $k\bar{y}$the negative feedback (-k) reflects the action of the elastic element to return the object (m) to its initial position in the presence of a deviation. The damper, as a link with a transfer function *bp,* in the physical sense displays the actions of a viscous resistance force. In this case, the negative feedback with the transfer function of the first-order differentiating link *bp*, as well as the spring k), forms a force effect.

Figure 1. Design and structural diagrams of a mechanical oscillatory system with one degree of freedom at power external disturbance

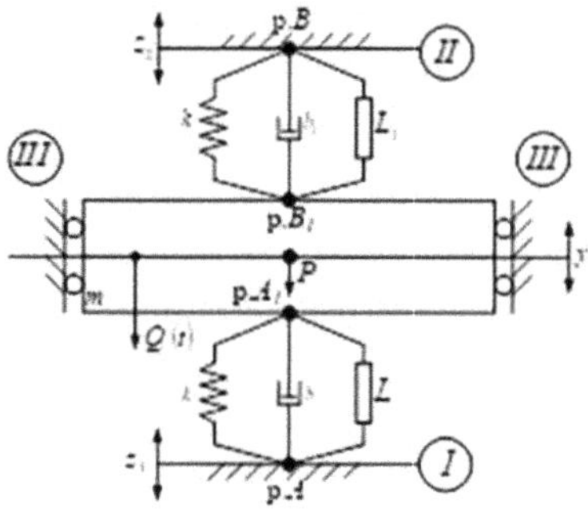

a - design diagram; *b* - block diagram showing the action of two elements; *c* - block diagram with a generalized element ($k + bp$); *d* - block diagram with elastic and damping elements as negative feedback; *e* - system block diagram in generalized form; *f* - diagram reflecting the features of transferring an element from direct to feedback

Thus, in p.(1) in Figure *1, e* one can compose the kinetostatic equation

$$\bar{y}mp^2 + \bar{y}bp + \bar{y}k = \bar{Q} \tag{3}$$

Expression (3) is a form of representation of the Dalembert principle (Lurie, 1959).

2. For further research, it is important that (3) implies the possibility of representing elements (*bp*) and (*k*) as links with an input signal for which the offset is, $\bar{y}$and the output is a certain force factor. The latter can be interpreted as the possibility of considering these elements in the "status" of the spring with the corresponding rigidity. If for spring k the stiffness of the element does not depend on the ω (that is, on the frequency of the disturbance), then for element *bp* - the "stiffness" of the element depends on the frequency of the ω. In this case, the stiffness can be called dynamic. In systems with several degrees of freedom in vibration protection tasks, one of the mass-inertial elements is selected as an object of protection, which is displayed by the type of transfer function of the second-order integrating link. As for the rest of the mass inertia elements (they can be called intermediate), they take the form of differentiating links of the second order with the properties of entering the connections between themselves and the support surfaces according to the same rules, then the springs and the damper.

The translation of the original mathematical model into a structural analogue and the corresponding transformations create the possibility of using advanced methods of frequency analysis and dynamic synthesis of systems for various applications, including vibration protection systems (Druzhinsky, 1977; Eliseev, Moskovskikh, Bolshakov, & Savchenko, 2012; Eliseev, 1978).

The objective of the study is to develop an approach to assessing the dynamic properties of mechanical oscillatory systems based on the use of a frequency characteristic equation and concepts about the dynamic rigidities of elements, their structural formations and the system as a whole.

DEVELOPMENT OF STRUCTURAL REPRESENTATIONS OF MECHANICAL SYSTEMS

The transition to block diagrams, as shown in Figure 1, predetermines the use of transfer functions (Figure 1, *e*). In its simplest form (Figure 1, *a*, *b*), the transfer function of the system under force disturbance $\overline{Q}$is

$$W(p) = \frac{\overline{y}}{\overline{Q}} = \frac{1}{mp^2 + bp + k} \tag{4}$$

The denominator of the transfer function (4) is used as a frequency characteristic equation

$$mp^2 + bp + k = 0 \tag{5}$$

When substituting $p = j\omega$ equation can be solved by determining the natural frequency. In the simplest case (b=0), the natural frequency is determined by

$$\omega^2 = \frac{k}{m} \tag{6}$$

which follows from the ratio

$$-m\omega^2 + k = 0 \tag{7}$$

If we invert (4), then taking into account (7), it can be noted that it is at the natural frequency that the dynamic stiffness of the system as a whole takes zero value. In this postulate, it is assumed that the mass inertia element when considering the force loading of the system, at which the dynamic stiffness, defined as the ratio of the displacement of the protected object to the applied force, can take values equal to zero (at $b = 0$ or b 0).

Thus, from the frequency equation (5), under appropriate conditions, the general property of the system can be determined, which is formed with dynamic stiffness equal to zero and the presence of a harmonic disturbing force. The same dynamic effect is considered as resonant oscillations. In the proposed approach, dynamic stiffness is a property that extends not only to individual elements of the system, but also to the system as a whole. The latter is related to the concept of a transfer function, in which the input effect (or signal) and the output signal (in this case, the offset y) are distinguished. The given concept of the transfer function can be expanded due to the variation of representations of input and output signals, as shown in (Genkin, 1981). In the physical sense, the frequency characteristic equation (5) in the operator form is the sum of dynamic rigidities correlated with the parameters of the mass inertia element m. If this sum is zero at a certain frequency of the external force, then this means that the dynamic stiffness of the system as a whole is equal to zero (in this case, the system is represented by a material point with mass m and elastic element k at $b = 0$). If there is a wider set of typical elements, local (or partial) dynamic stiffnesses may not be equal to zero. An important circumstance is the connection between such dynamic effects as reducing the rigidity of systems as a whole to zero (or to some value) at certain frequencies, which is associated with ideas about resonant processes. Such approaches, in particular, form the basis of methods for determining natural vibration frequencies of technical objects (Genkin, 1981).

SYSTEMS WITH TWO DEGREES OF FREEDOM

Mechanical systems with two degrees of freedom are quite diverse in their physical nature and structural and technical forms, which is reflected in the variety of mathematical models and coordinate systems used. Figure 2, *a-d* show mechanical oscillatory systems of chain type (Figure 2, *a*), solid body on elastic supports (Figure 2, *b*), as well as their structural diagrams in Figure 2, *c-d*. To obtain structural mathematical models, the methods described in (Eliseev & Khomenko, 2014) are used. When considering the movements of a solid on elastic supports (Figure 2, *b*), two coordinate systems y_1, y_2 and y_0 are used, φ, which is associated with the selection of the center of mass, taking into account the moment of inertia of the solid and angular vibrations. Systems with two degrees of freedom include partial blocks (systems) and corresponding to the selected coordinate system of forms of interpartial links. In general, typical elements of mechanical oscillating systems have the same appearance as typical elements of systems with one degree of freedom, but the connections between elements, including partial systems, are very diverse. Note also that the set of typical elementary links of mechanical systems can be expanded by introducing links with transfer functions in the form of differentiating links of the second order, as well as due to other elements, the transfer functions of which correspond to the elements of the transfer functions of similar elements. In this case, a typical element with some transfer function is understood as a device that is not decomposed into separate components in the description, that is, in a sense, a typical element is considered as "indivisible." When using structural model transformation rules, typical elementary links can be used to construct more complex structures, which can consist of several related elements and have dynamic rigidity. In structural transformations, such composite structures are called quasi-springs.

Quasi-springs in structural transformations behave in the same way as ordinary typical elements, however, their dynamic rigidity generally depends on the frequency of the external force (Genkin, 1981; Eliseev, Kashuba, & Sitov, 2013).

In systems with a solid (Figure 2, *b*), the coordinates y_1, y_2 and y_0, φ are related by the relations

$$y_0 = ay_1 + by_2,\ \phi = c(y_2 - y_1),\ y_1 = y_0 - l_1\phi, \\ y_2 = y_0 + l_2\phi,\ a = \frac{l_2}{l_1 + l_2},\ b = \frac{l_1}{l_1 + l_2},\ c = \frac{1}{l_1 + l_2}. \tag{8}$$

Parameters of structural mathematical models are presented in Table 1.

Both force and kinematic disturbances from bearing surfaces *I* and *II* can be used as generalized forces in the systems of Figure 2, *a*, *b*.

Figure 2. Systems with two degrees of freedom:

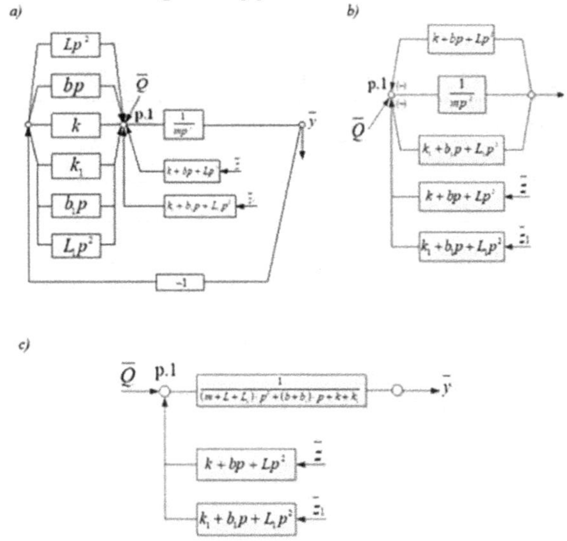

a - design diagram of chain-type system; *b* - design diagram of the system with a solid body; *c* - structural diagram of chain type in coordinates y_1, y_2; *d* - system with solid body in coordinates y_0, φ; structural diagram of the system in coordinates y_1 and y_2

1. For the structural diagram in Figure 2, *in* case of force disturbance

   $Q_1 \neq 0$, the transfer functions are determined by the expressions:

$$W_1(p) = \frac{\bar{y}_1}{\bar{Q}_1} = \frac{m_2 p^2 + k_2 + k_3}{A(p)} \tag{9}$$

$$W_2(p) = \frac{\bar{y}_2}{\bar{Q}_1} = \frac{k_2}{A(p)} \tag{10}$$

where

$$A(p) = (m_1 p^2 + k_1 + k_2)(m_2 p^2 + k_2 + k_3) - k_2^2 \tag{11}$$

Expression (11) defines the characteristic frequency equation for the mechanical oscillatory system in Figure 2, *a*. From the comparison (9) and (10), it can be seen that the transfer functions can be different depending on the choice of points of application of external disturbances and the choice of output signals giving an idea of the dynamic state of the system.

Table 1. Coefficients of structural mathematical models of systems in Figure 2, a, b

Chain type system (Figure 2, *a*)	
a_{11}	a_{12}
$m_1p^2 + k_1 + k_2$	$-k_2$
a_{21}	a_{22}
$-k_2$	$m_2p^2 + k_2 + k_3$
Generalized forces	
Q_1	Q_2
Solid system (coordinates y_1, y_2) (Figure 2, *b*)	
a_{11}	a_{12}
$(Ma^2 + Jc^2)p^2 + k_1$	$(Mab - Jc^2)p^2$
a_{21}	a_{22}
$(Mab - Jc^2)p^2$	$(Mb^2 + Jc^2)p^2 + k_2$
Generalized forces	
Q_1	Q_2

continued on following page

Table 1. Continued

Chain type system (Figure 2, *a*)	
a_{11}	a_{12}
Solid system (coordinates y_0, φ) (Figure 2, *b*)	
a_{11}	a_{12}
$Mp^2 + k_1 + k_2$	$k_2l_2 - k_1l_1$
a_{21}	a_{22}
$k_2l_2 - k_1l_1$	$Jp^2p^2 + k_1l_1^2 + k_2l_2^2$
Generalized forces	
Q_1	Q_2

When solving specific problems of dynamics, the design scheme of the system takes a certain form, for example, in the tasks of vibration protection, which is accompanied by the choice of the object of protection. Assuming that the object of protection against external influence Q_1 is related to the mass-inertial element m_1, we bring the frequency equation (11) to the form

$$(m_1p^2 + k_1 + k_2) - \frac{k_2^2}{(m_2p^2 + k_2 + k_3)} = 0 \tag{12}$$

Convert (12)

$$m_1p^2 + k_1 + \frac{k_2(m_2p^2 + k_3)}{m_2p^2 + k_2 + k_3} = 0 \tag{13}$$

The structural mathematical model of the mechanical system shown in Figure 2, *a*, can be obtained by converting the structural diagram in Figure 2, *c*, which is shown in Figure 3, *a*, *b*.

Figure 3. Variants of structural diagrams of mechanical oscillatory system chain type (Figure 2) if the y_1 coordinate is excluded: a - block diagram at exclusion of y_2 coordinate at basic model m_1, k_1; b - block diagram with allocation of the protection object; c - block diagram with generalized feedback in relation to the object of protection

a)

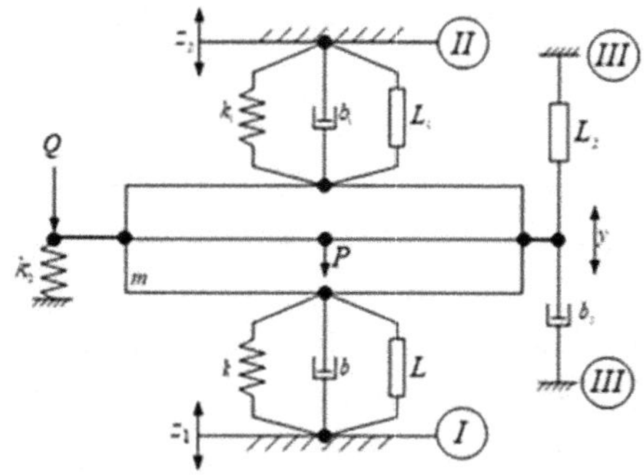

b)

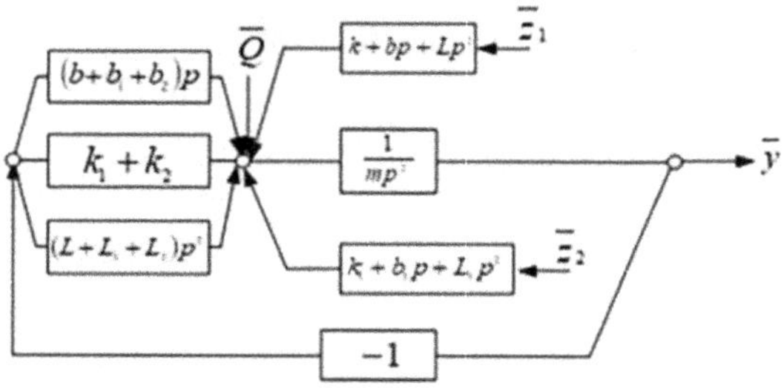

Comparison of structural diagrams in Figure 3 makes it possible to track the characteristics of negative feedback in relation to the object of protection (m). In Figure 3, *and* the negative feedback, in the physical sense, is the dynamic stiffness of the quasi-spring, which reflects the dynamic properties of a certain part of the mechanical system associated with the intermediate mass m_2.

2. Dynamic stiffness of quasi-spring

$$\bar{k}_{np}(p) = \frac{k_2(m_2p^2 + k_3)}{m_2p^2 + k_2 + k_3} \tag{14}$$

can be zero at frequency

$$\omega_1^2 = \frac{k_3}{m_2} \tag{15}$$

In this case, the system (Figure 2, *a*) operates under the action of an external force $Q_1 \neq 0$ in such a way that the original system is transformed into a system with one degree of freedom with elements m_1 and k_1. Feedback (15), in a certain sense, neutralizes the dynamic action of a part of the system. On the other hand, the dynamic stiffness of the quasi-spring at the frequency

$$\omega_2^2 = \frac{k_2 + k_3}{m_2} \tag{16}$$

takes an infinitely large value, which corresponds to the dynamic vibration damping mode. In this case, the movement along the y1 coordinate stops ($y_1 = 0$).

When moving to the structural diagram in Figure 3, *b*, the object of protection is considered in the scheme of support on two parallel springs with stiffnesses k_1 and k_2. The total dynamic stiffness of such a quasi-spring will be

$$\overline{k'}_{np}(p) = \frac{(k_1 + k_2)m_2 p^2 + k_1 k_2 + k_2 k_3 + k_1 k_3}{m_2 p^2 + k_2 + k_3} \tag{17}$$

The dynamic stiffness of the system under such conditions is zero at the frequency of

$$\omega_3^2 = \frac{k_1 k_2 + k_2 k_3 + k_1 k_3}{(k_1 + k_2)m_2} \tag{18}$$

At frequency ω_2 of (16) dynamic vibration damping mode is maintained.

Thus, when "zeroing" the dynamic stiffness of the quasi-spring, if we consider in terms of sequential release of the object of protection m_1 from additional connections, various characteristic forms of self-organization of the movement of the system elements will be observed.

3. On the basis of (12) the condition

$$m_1 p^2 + k_1 + k_2 = \frac{k_2^2}{m_2 p^2 + k_2 + k_3} \tag{19}$$

which allows you to determine on the basis of the graph-analytical approach when substituting $p^2 = -\omega^2$ the values of the natural frequencies as shown in Figure 4.

Figure 4. Schematic diagram of the solution of the graphic solution of the condition (19)

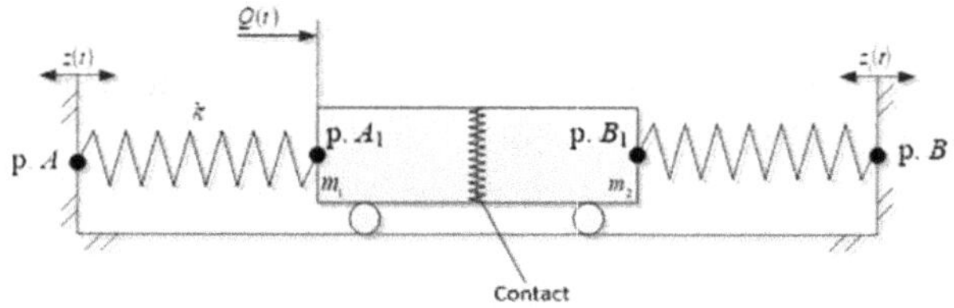

From the intersection of the relationship graphs constituting condition (19) and denoted by vols. (1) and (2), which determine the frequencies at which the dynamic stiffness of the system as a whole will be zero. Such intersection points determine the natural frequencies. Figure 4 also shows the partial frequencies and$n_1^2 = \frac{k_1 + k_2}{m_1}$,$n_2^2 = \frac{k_2 + k_3}{m_2}$, the arrangement of which corresponds to the known provisions of the theory of oscillations.

4. The physical meaning of the frequency characteristic equation in the form (12) corresponds to ideas about the summation of all forces acting on the object of protection. Moreover, each of the elements (12) essentially represents the product of the dynamic rigidity reduced to m_1 by$\bar{y}_1$, since it $\bar{y}_2$is excluded by structural transformations. In principle, this also coincides with the known provisions of the Dalembert principle.

A feature of the considered chain systems is the homogeneity of the physical nature of the partial systems. In this case (Figure 2, *a*), partial systems make rectilinear small vibrations, and the connections between partial systems are inertial.

When considering systems with a large number of degrees of freedom, the methodology for determining dynamic properties and their relationship with frequencies, the types of joint forms of movement of elements, remain the same, taking into account the complications introduced by increasing the order of the frequency characteristic equation.

TWO-DEGREE-OF-FREEDOM SYSTEMS CONTAINING SOLIDS

When considering the mechanical system in Figure 2, *b*, it can be noted that the form of partial systems depends on the choice of the coordinate system.

1. If y_1 and y_2 are selected as motion coordinates, then the structural diagram in Figure 2, *d* has partial systems with movements of the same type. The structure of partial system 2 includes the reduced mass $Ma^2 + Jc^2$, related to the coordinate y_1.

For the second partial system, we have the reduced mass $Mb^2 + Jc^2$, related to the coordinate y_2. Connections between partial systems are elastic.

Partial systems with coordinates y_1 and y_2, in fact, reflect small angular vibrations, indicated by y_1 and y_2 at respectively "stopped" movements along y_2 and y_1, which is quite acceptable at small angles of rotation relative to the center of mass.

Assuming that the external influence in the system is carried out by the harmonic force Q_1 applied at the point with coordinates y_1 ($Q_2 = 0$), we write the transfer functions

$$W_3(p)=\frac{\bar{y}_1}{\bar{Q}_1} = \frac{(Mb^2+Jc^2)p^2+k_2}{A_1(p)} \tag{20}$$

$$W_4(p)=\frac{\bar{y}_2}{\bar{Q}_1} = \frac{(Jc^2-Mab)p^2}{A_1(p)} \tag{21}$$

where

$$A_1(p)= [(Ma^2+Jc^2)p^2+k_1][(Mb^2+Jc^2)p^2+k_2]-[(Jc^2-Mab)p^2]^2. \tag{22}$$

From (22) we get the frequency equation in the coordinates y_1, y_2:

$$[(Ma^2+Jc^2)p^2+k_1][(Mb^2+Jc^2)p^2+k_2]-[(Jc^2-Mab)p^2]^2 = 0. \tag{23}$$

Figure 5, *a* - c show the variants *of* the structural diagram transformations in Figure 2, *d* when excluding the φ coordinate.

Figure 5. Options for forming feedbacks in structural diagrams using the coordinates y_1 and y_2: a - structural diagram of the system with exclusion of y_2 coordinate; b - block diagram with allocation of feedback in relation to the object with reduced mass ($Ma^2 + Jc^2$) p^2; c - block diagram with generalized feedback

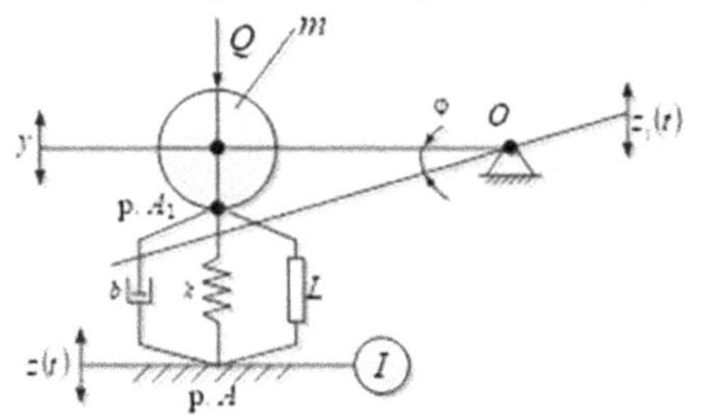

2. From the analysis of the structural diagram in Figure 5, *c*, it can be made by analogy with the previous calculations that the selected object, the dynamic state of which is estimated, is determined by the transfer function of the second-order integrating link. The object is covered by a feedback having a fractionally rational transfer function. At frequency

$$\omega_3 = \sqrt{\frac{k_2}{Mb^2 + Jc^2}} \tag{24}$$

dynamic vibration damping mode is possible in the system. In this case, the condition $y_1 = 0$ is satisfied by the force Q_1. In turn, the numerator of the feedback transfer function is a biquad equation relative to p^2, which predetermines the possibility of obtaining a zero value of the numerator of the feedback transfer function. With such a frequency of external influence, $\overline{Q}_1$the system has the parameters of the movement of the system with one degree of freedom, which manifests itself in certain forms of joint movements. Note that the feedback (Figure 5, c) determines the value of the dynamic stiffness not of the entire system as a whole, but only of its part, which takes the form of a quasi-spring.

3. To determine the frequencies of natural oscillations, we write the frequency characteristic equation of the system (23) in the form

$$(Ma^2 + Jc^2)p^2 + k_1 - \frac{[(Jc^2 - Mab)p^2]^2}{(Mb^2 + Jc^2)p^2 + k_2} = 0 \tag{25}$$

Since $p = j\,\omega$, then (25) is transformed

$$k_1 - (Ma^2 + Jc^2)\omega^2 = \frac{[(Jc^2 - Mab)]^2\omega^4}{k_2 - (Mb^2 + Jc^2)\omega^2} \tag{26}$$

Figure 6. Schematic diagram for determining natural vibration frequencies in a system with two degrees of freedom with a solid in coordinates y_1, y_2

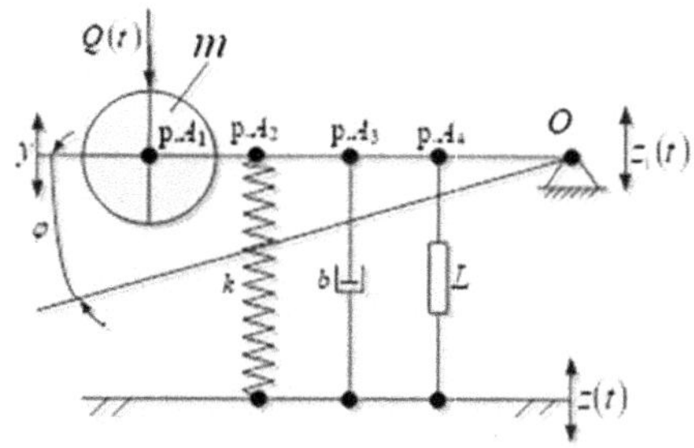

In the graphoanalytical solution, the intersection of the dependence graphs for the left and right parts of the condition (26) determines the natural vibration frequencies (Figure 6). Along the ordinate axis, Figure 6 shows the values of the dynamic stiffnesses of the corresponding fragments of the system. The intersection of the graphs determines the total dynamic stiffness of the entire system equal to zero.

The frequency characteristic equation (23) in this case is the sum of the reduced forces acting on the system as a whole when choosing a material point with a reduced mass *of* $Ma^2 + Jc^2$ as the reduction point with the coordinate y_1. Similar results can be obtained by selecting a cast point with a y_2 coordinate and a reduced mass of $Mb^2 + Jc^2$.

4. When you select the coordinate system y_0 and φ, the characteristic frequency equation becomes

$$(Mp^2 + k_1 + k_2)(Jp^2 + k_1 l_1^2 + k_2 l_2^2) - (k_1 l_1^2 - k_2 l_2^2)^2 = 0. \quad (27)$$

Convert (27)

$$Mp^2 + k_1 + k_2 - \frac{(k_1 l_1^2 - k_2 l_2^2)^2}{Jp^2 + k_1 l_1^2 + k_2 l_2^2} = 0 \quad (28)$$

Figure 7. Schematic diagram of determination of natural oscillation frequencies of the system in coordinates y_o, φ

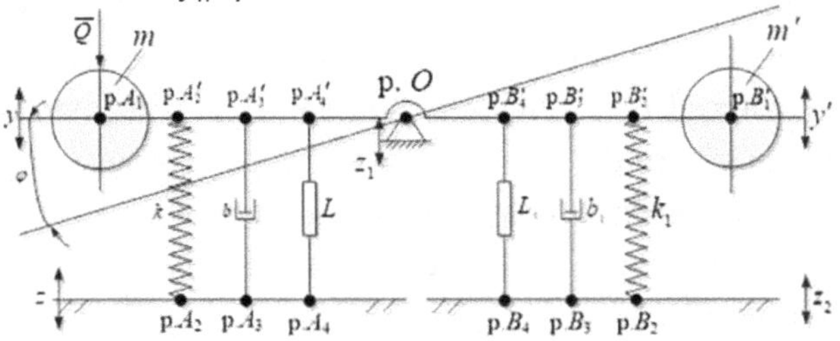

The schematic diagram of the graphoanalytical solution is shown in Figure 7 and is determined by the condition

$$k_1 + k_2 - M\omega^2 = \frac{(k_1 l_1^2 - k_2 l_2^2)^2}{k_1 l_1^2 + k_2 l_2^2 - J\omega^2} \quad (28)$$

From the analysis (26) it follows that the initial system (Figure 2, *b*) in the coordinates y_0 and φ has partial frequencies

$$n_3^2 = \frac{k_1 + k_2}{M} \tag{29}$$

$$n_4^2 = \frac{k_1 l_1^2 + k_2 l_2^2}{J} \tag{30}$$

At frequency $n_4^2$from (30) mode of dynamic damping of oscillations is possible, when at Q_1 action movement along y_2 coordinate does not occur. At the same time, there is a value of the frequency of the external force at which the feedback with respect to the object with mass M takes a zero value. In this case,

$$\omega_5^2 = \frac{(k_1 + k_2)(k_1 l_1^2 + k_2 l_2^2) - (k_1 l_1 - k_2 l_2)^2}{J(k_1 + k_2)} = \frac{k_1 k_2 (l_1 + l_2)^2}{(k_1 + k_2) J}. \tag{31}$$

At this frequency, the system moves under the influence of external perturbation as a system with one degree of freedom. In this case, the dynamic rigidity of the part of the system that forms the quasi-spring will be zero, which, in the physical interpretation, manifests itself as some form of joint movement (a form of self-organization of the joint movement of the elements of the system). In this case, the dynamic stiffness of the system as a whole is not zero (at this frequency).

The physical meaning of the frequency characteristic equation (27) when multiplied by the coordinate $\bar{y}_0$is that it is the sum of the reduced forces that arise when the system moves as a material point with mass M moving along the y_0 coordinate.

Similar results can be obtained using angular coordinates. In this case, the characteristic equation represents the sum of the moments of forces arising when moving relative to the center of reduction (i.e., relative to the center of mass - p. O)

In this case, a material solid having a moment of inertia J and a mass M has a fixed pivot point t O where the mass is concentrated. The use of the moment of inertia J in calculations presupposes an understanding of the organization of joint movements as actions taking place in a certain space, that is, taking into account the spatial metric of the location of the acting forces.

In the considered cases, systems with two degrees of freedom with a solid in different coordinate systems had different frequency characteristic equations, since the choice of a coordinate system predetermines the type of partial systems, the peculiarities of interparticle connections, which is quite clearly displayed on the structural mathematical models shown in Figure 2, *d*, *e*.The characteristic frequency equation obtained from the structural diagram in Figure 2,*d*, has the form

$$\left[(Ma^2 + Jc^2)p^2 + k_1\right]\left[(Mb^2 + Jc^2)p^2 + k_2\right] - [(Jc^2 - Mab)p^2]^2 = 0. \tag{32}$$

Expression (32) can be reduced to a form convenient for transformations:

$$
\begin{aligned}
&\left[(Ma^2 + Jc^2 - Mab)p^2 + k_1\right]\left[(Mb^2 + Jc^2 - Mab)p^2 + k_2\right] - \\
&\qquad -[(Jc^2 - Mab)p^2]^2 = 0.
\end{aligned} \tag{33}
$$

After substituting the values *a, b* and *c* from expression (8), the characteristic equation in the coordinate systems y_1, y_2 and y_0, φ takes the same form.

SYSTEMS WITH THREE DEGREES OF FREEDOM

Developing the described approach, we take as the initial mechanical oscillatory system with three degrees of freedom (Figure 8), the mathematical model of which is presented in Table 2.

Figure 8. Design scheme of the system with three degrees of freedom

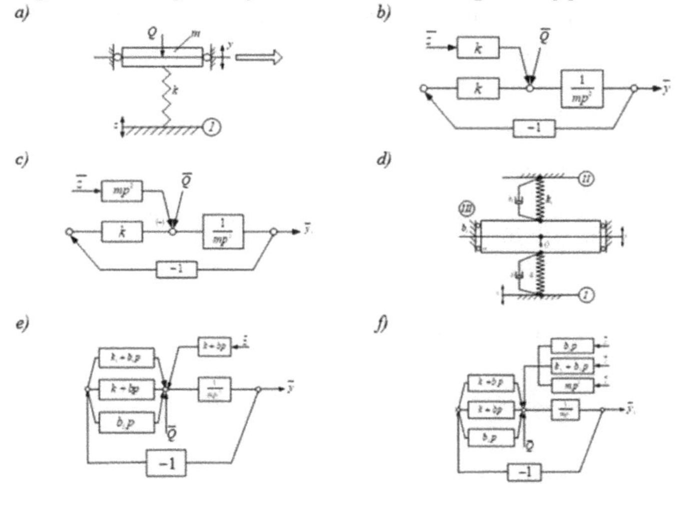

The system under consideration is closed in the sense that the connected elements of the system m_1, m_2, m_3 form a closed loop. The system has linear properties and makes small vibrations in the absence of friction forces relative to the position of stable equilibrium under the action of external harmonic force (Table 2).

Table 2 coefficients can be used to construct a structural diagram, for example, for the case of power perturbation Q_1. The structural diagram of the system (or its structural mathematical model) is shown in Figure 9.

Table 2. Coefficients of equations of motion of the system in coordinates y_1, y_2, y_3

a11	a12	a13
$m_1p^2 + k_1 + k_2 + k_5$	$-k_2$	$-k_5$
a21	a22	a23
$-k_2$	$m_2p^2 + k_2 + k_3$	$-k_3$
a31	a32	a33
$-k_5$	$-k_3$	$m_3p^2 + k_3 + k_4 + k_5$
Generalized forces		
Q_1	0	0

The transfer functions of the system and the frequency characteristic equation can be determined directly by the structural scheme.

Figure 9. Structural mathematical model (structural diagram) for mechanical oscillatory system

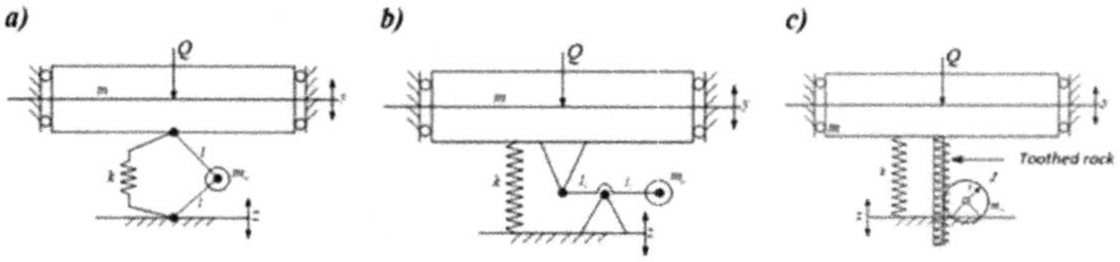

In particular, the characteristic equation in this case is

$$a_{11}a_{22}a_{33} - a_{11}a_{23}^2 - a_{22}a_{13}^2 - a_{33}a_{12}^2 + 2a_{12}a_{23}a_{31} = 0. \quad (34)$$

The expression (34) gives an idea that the system can be built from partial systems a_{11}, a_{22}, a_{33}, which enter into interactions with each other through certain connections (Figure 9).

If you select an object for monitoring the dynamic state (for example, $_{a11}$, for vibration protection), then equation (34) can be converted to the form

$$a_{11} - \frac{a_{22}a_{13}^2}{[m_2m_3]} - \frac{a_{33}a_{21}^2}{[m_2m_3]} + \frac{2a_{12}a_{23}a_{31}}{[m_2m_3]} = 0 \quad (35)$$

where

$$[m_2m_3] = a_{22}a_{33} - a_{23}^2 \quad (36)$$

As in previous considerations, the characteristic frequency equation can be interpreted as the sum of the dynamic rigidities of its individual constituents (or quasi-springs). Note that expression (36) is a characteristic equation of a system with two degrees of freedom (m_2, m_3, k_2, k_3, k_4).

This allows you to introduce partial systems of a generalized form. In this case, the structural model of the system as a whole can be built not only from ordinary partial systems with one partial systems with one degree of freedom, but also with the involvement of partial systems of a more general form with two degrees of freedom.

In the development of this approach, it can be borne in mind that combined approaches are possible, within which the interaction of two types of partial systems can be implemented. For more complex systems, partial systems with three power freedoms can be introduced, which makes it possible to create some methodological basis for considering the laws of dynamic interactions in multidimensional systems.

Thus, structural mathematical modeling, if we keep in mind its direction, is associated with the use of the apparatus of the theory of automatic control and the theory of mechanical chains, in comparison with other methods of mathematical modeling, has certain advantages. They consist in the fact that the mathematical model, due to the peculiarities of its construction, allows introducing into consideration not only the peculiarities of relations between typical elements, but also regulatory actions that reflect the influence of the structure. In general, a number of fundamental positions could be noted that could be developed autonomously, predetermining the possibility of obtaining new results, which is associated, in particular, with the ideas of expanding the set of typical elements, introducing additional connections and transforming structural models, and more.

The development of structural modeling in the theory of oscillations of mechanical linear systems is focused on the use of the apparatus of Laplace transformations, methods of frequency analysis and dynamic synthesis. The characteristic frequency equation, which is the basis for determining transfer functions, has the ability to assess the state of systems through dynamic stiffness parameters. In this sense, the characteristic equation can be considered as the sum of dynamic rigidities reduced to the corresponding point of the system, the movement of which is correlated with a certain coordinate.

The dynamic stiffness of an elastic element or some structural formation from typical elements is a generalization of the properties of not only an elastic element in the form of an idealized spring, but includes ideas about other forms of interaction, provided that the input action in the form of displacement initiates the output signal in the form of a force. Actually, on such representations, a set of typical elements is formed in linear systems, when mass inertia, dissipative, elastic and other elements are distinguished.

The concept of dynamic stiffness extends to both elementary links and structural blocks or formations from individual elements, called quasi-springs. The system as a whole also has dynamic rigidity. If the dynamic stiffness becomes zero under the action of the perturbation force, then the perturbation frequency corresponds to the resonance mode. If the dynamic rigidity of a part of the system, or some structural unit, takes a zero value, then the system implements a dynamic mode that reflects the possibility of manifesting certain forms of self-organization of joint movements of elements.

Systems with two degrees of freedom in different coordinate systems exhibit different forms of joint motions arising at corresponding frequencies. At the same time, characteristic equations in different coordinate systems are reduced to the same form, which predetermines the frequencies of natural vibrations in the form of system invariants. The characteristic equation in a detailed representation reflects the structural properties of the system, which include the spatial metric of the system, features of the forms of motion of partial systems, features of the interaction of system elements and structural formations with each other.

The construction of the system structure is diverse, which is based on the use of certain methods for converting models and highlighting the features of dynamic properties depending on the statement of the research task. The use of partial systems of a generalized form is of interest in connection with the prospects of creating a single minimum complexity methodological basis that allows creating technologies for the dynamic synthesis of systems by increasing the number of blocks attached under certain conditions.

ABOUT USING ADDITIONAL LINKS INERTIAL TYPE IN TECHNICAL SYSTEMS DYNAMICS PROBLEMS

Attention to the issues of ensuring the reliability of machines, improving their safety and dynamic quality stimulates the search and development of technical means for monitoring and controlling the dynamic state of technical systems. A number of theoretical and practical studies are reflected in the works (Belokobylsky, Eliseev, & Sitov, 2012; Eliseev, Artyunin, Ermoshenko, et al., 2013) related to the assessment of the possibilities of using elements of an unconventional type in the form of devices for transforming motion. Such elements can be implemented, in particular, on the basis of various mechanisms (screw, non-self-braking, toothed, articulated-lever, etc.) A detailed description of the dynamics features of mechanical oscillatory systems with such elements is presented in (Eliseev, Artyunin, Logunov, et al., 2013). Significant attention to the assessment of the dynamic properties of devices for transforming motion is paid in works developing the ideas of structural mathematical modeling, in which devices for transforming motion are considered at the level of typical differentiating links of the second order. The features of these structural approaches are the use of dynamic analogies in the interactions of electrical, mechanical systems, as well as automatic control systems. To date, a sufficient variety of constructive and technical solutions has been developed to predict the development of research areas. In this regard, the possibilities of using such effects from the use of APD as the introduction of additional acceleration bonds, changing the parameters of the reduced masses and rigidities, creating additional effects of dynamic damping of vibrations, "locking" the interaction of elements at certain frequencies, the formation of non-traditional types of inter-partial bonds, etc.

At the same time, a number of issues in the assessment of the dynamic properties and capabilities of MCD in the formation of the structures of mechanical vibration systems, and vibration protection systems, in particular, have not yet received the proper detail.

Possibilities of development of methodological bases for construction of mathematical models for problems of vibration protection and vibration isolation of objects of machine building are considered using structural representations of dynamic interactions of ordinary standard links and devices for motion conversion.

GENERAL PROVISIONS; FEATURES OF DYNAMIC INTERACTIONS OF SYSTEM ELEMENTS

Within the framework of structural mathematical modeling, a mechanical oscillatory system is displayed by a dynamically equivalent structural diagram of an automatic control system, which is associated with a number of operations using Laplace transforms. The methodological bases of such approaches are sufficiently detailed (Eliseev, Artyunin, Ermoshenko, et al., 2013; Eliseev, Artyunin, Logunov, et al., 2013).

Figure 10, *bf* show different forms of structural transformations of models using the example of a vibration protection system (Figure 10 *a*) with one degree of freedom. In Figure *10, a-f* the following designations are adopted*:* m - object of protection with mass m, k - spring stiffness coefficient, b - coefficient of viscous friction, L - reduced mass of the device for converting motion, *I* and *II* support surfaces. Power Q and kinematic harmonic disturbances z_1 (t) and $_z2$ (t) are considered in the system. The system has linear properties and makes small fluctuations relative to the position of static equilibrium. In Figure 10, it is assumed that p = j ω is a complex variable (the Laplace transform is performed for zero initial conditions).

Construction of the structural diagram in Figure 10, *b* requires preliminary construction of a mathematical model in the form of a differential equation. The corresponding expressions for kinetic and potential energies can be written:

$$T = \frac{1}{2}m\dot{y}^2 + \frac{1}{2}L_1(\dot{y} - \dot{z}_1)^2 + \frac{1}{2}L_2(\dot{y} - \dot{z}_2)^2 \tag{37}$$

$$\Pi = \frac{1}{2}k_1(y - z_1)^2 + \frac{1}{2}k_2(y - z_2)^2 \tag{38}$$

Figure 10. Design and structural diagrams of the system: a - circuit with full set of elementary standard links; b - block diagram with a full set of standard links; c - block diagram with UPD in the feedback circuit; d - block diagram in generalized form; f - structural diagram with an object with reduced weight; g calculation scheme with generalized elastic elements with stiffnesses k_{nr1} *and* k_{nr2}

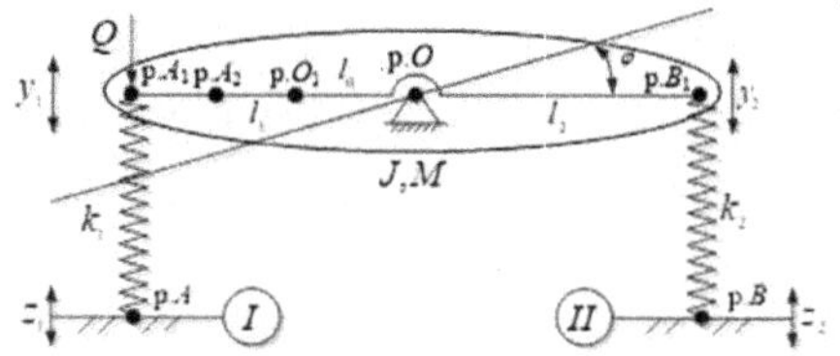

The expression for the scattering energy is taken as

$$\Phi = \frac{1}{2}b_1(\dot{y} - \dot{z}_1)^2 + \frac{1}{2}b_2(\dot{y} - \dot{z}_2)^2 \tag{39}$$

The equation of motion is obtained based on the use of the Lagrange equation of the II kind:

$$\begin{aligned} &\bar{y}(m + L_1 + L_2)p^2 + \bar{y}(b_1 + b_2)p + \bar{y}(k_1 + k_2) = \\ &= \bar{Q} + \bar{z}_1(L_1p^2 + b_1p + k_1) + \bar{z}_2(L_2p^2 + b_2p + k_2) \end{aligned} \tag{40}$$

Equation (40) is the basis for constructing a structural mathematical model. Figure 10, *a* shows the calculation scheme, which, as a result of formal transformations, is transformed into a structural scheme in Figure 10, *b*, which is a mathematical model of a different kind, but this model, as well as equation (40); Laplace transforms occur at zero initial conditions. The <-> icon indicates a Laplacian image. Note that typical elements within the structural model are interpreted as corresponding feedback loops. Since the feedback circuit has a coordinate at the input, $\bar{y}$and the transfer function of each feedback branch, in the physical sense, corresponds to the "stiffness coefficient" of the element, then when the branch exits in vol. 1 (see Figure 10, b), the feedback branch will already correspond to the communication reaction, the value of which is equal to the product of the reduced stiffness of the link by displacement.

In this case, it is obvious that typical elements with transfer functions L_1p^2, L_2p^2, b_1p, b_2p, k_1 and k_2 differ in a certain feature: there is a displacement at the input of any typical link, and a force is formed at the output. In such conditions, the rules for switching elementary links are predetermined as "as rules for parallel and serial connection of springs" (Eliseev, Volkov, & Kukharenko, 1990). Transfer functions for the system are shown in Figure 10, *and* are determined by any of the structural diagrams in Figure 10, *c-f*.

The transfer functions of the system are determined with one input signal, assuming that the other input actions will be zero. Thus

$$\underset{\bar{Q}\neq 0, \bar{z}_1=0, \bar{z}_2=0}{W_1(p)} = \frac{\bar{y}}{\bar{Q}} = \frac{1}{A(p)} \tag{41}$$

$$\underset{\bar{z}_1\neq 0, \bar{Q}=0, \bar{z}_2=0}{W_2(p)} = \frac{\bar{y}}{\bar{z}_1} = \frac{L_1p^2 + b_1p + k_1}{A(p)} \tag{42}$$

$$\underset{\bar{z}_2\neq 0, \bar{Q}=0, \bar{z}_1=0}{W_3(p)} = \frac{\bar{y}}{\bar{z}_2} = \frac{L_2p^2 + b_2p + k_2}{A(p)} \tag{43}$$

$$\underset{\bar{Q}=0, \bar{z}_1=\bar{z}_2=\bar{z}}{W_3(p)} = \frac{\bar{y}}{\bar{z}} = \frac{(L_1 + L_2)p^2 + (b_1 + b_2)p + k_1 + k_2}{A(p)} \tag{44}$$

where

$$A(p) = (m + L_1 + L_2)p^2 + (b_1 + b_2)p + k_1 + k_2 \tag{45}$$

is a characteristic frequency equation.

The dynamic properties of the vibration protection system are evaluated in Figure 10, *a* based on the frequency characteristics obtained when substituting in the expression (41) - (44) $p = j\omega$ ($j = \sqrt{-1}$), as is done in the frequency methods of system analysis (Eliseev, Kovyrshin, & Bolshakov, 2012).

PECULIARITIES OF SWITCHING OF TYPICAL ELEMENTS OF MECHANICAL OSCILLATORY SYSTEMS

From the above it follows that each of the typical elements has its own transfer function of an elastic, dissipative or mass-inertial (MCD) link. In this case, the object of protection (Figure 10, *bf*) has the transfer function of the integrating link of the second order, and the transfer functions of the system (41) - (44) are formed using the rules of convolution of elements with feedback (Kim, 2003).

Let's consider a number of cases of connecting the UDP with typical elements.

Elastic element k_1 and Lp^2 - series and parallel connections:

$$k_{\text{pr1}} = k + Lp^2 \tag{46}$$

$$k_{\text{pr2}} = \frac{k \cdot Lp^2}{k + Lp^2} \tag{47}$$

Dissipative element and UPD in parallel and serial connections have the form

$$k'_{\text{pr}} = bp + Lp^2 \tag{48}$$

$$k''_{\text{pr}} = \frac{bLp^3}{b + Lp} \tag{49}$$

Compounds L_1p^2 and L_2p^2, respectively, have the forms

$$k'''_{\text{pr}} = (L_1 + L_2)p^2 \tag{50}$$

$$k^{\text{IV}}_{\text{pr}} = \frac{L_1 L_2 p^2}{L_1 + L_2}. \tag{51}$$

Other forms of linkage of typical units in more complex configurations can be constructed in a similar manner. Such element formations form a generalized spring, and the vibration protection system can be reduced to a basic form when the initial system, for example, in Figure 10, *a*, is reduced to a simple form. In this case, the system has a protection object with mass m and one generalized spring with dynamic stiffness

$$k_{\text{pr0}} = (L_1 + L_2)p^2 + (b_1 + b_2)p + k_1 + k_2 \tag{52}$$

The dynamic stiffness k_{pr0} depends on the frequency of the external perturbation and can take a zero value at the frequency that can be found from the solution of equation (52).

In this case, k_{pr0} is presented as some complex link reflecting the properties of a quasi-spring.

Under the action of external force disturbance Q in the mechanical system (Figure 10, *a*), a forced vibration mode occurs, characterized by a transfer function

$$W_Q(p) = \frac{\bar{y}}{\bar{Q}} = \frac{1}{(m+L)p^2 + k_1 + k_2} \tag{53}$$

The frequency of resonant oscillations is determined from the expression

$$\omega^2_{res} = \frac{k_1 + k_2}{m + L} \tag{54}$$

The amplitude-frequency characteristics of the *system* under power disturbance have the usual form. In case of kinematic disturbance, for example, at $z_1 \neq 0$ ($Q = 0$, $z_2 = 0$), the transfer function of the system will be determined

$$W_{z_1}(p) = \frac{\bar{y}}{\bar{z}_1} = \frac{Lp^2 + k_1}{(m+L)p^2 + k_1 + k_2} \tag{55}$$

Under the condition $Lp^2 + k$, the system implements the dynamic vibration damping mode, but the frequency of dynamic vibration damping can be correlated with the following conditions

$$\omega^2_{dyn} < \omega^2_{res}, \omega^2_{dyn} > \omega^2_{res}, \omega^2_{dyn} = \omega^2_{res}.$$

Figure 11 shows the corresponding frequency response. From the expression (55) it follows that $\underset{p\to\infty}{W(p)} = \frac{1}{m+L}$, and with p 0 $\underset{p\to 0}{W(p)} = \frac{k_1}{k_1 + k_2}$. If the numerator and denominator (55) simultaneously become zero, $\frac{k_1}{L} = \frac{k_1 + k_2}{m + L}$ then it follows that between the values L and m there is a relationship

$$L = \frac{k_1 m}{k_2} \tag{56}$$

In case of disturbance from the side of bearing surface *II* (Figure 10, *a*)

$$\underset{z_2 \neq 0}{W(p)} = \frac{\bar{y}}{\bar{z}_2} = \frac{Lp^2 + k_2}{(m+L)p^2 + k_1 + k_2} \tag{57}$$

From (57) it follows that there may be cases$\omega^2_{2\ res} < \frac{k_1 + k_2}{L + m}$, $\omega^2_{1res} < \frac{k_1 + k_2}{L + m}$,$\omega^2_{2res} = \frac{k_1 + k_2}{L + m}$. The latter gives the relationship between L and m, with perturbation $z_2 \neq 0$, the relationship

$$m = \frac{k_1}{k_2} L \tag{58}$$

whereas with perturbation $z_1 \neq 0$ there is

$$L = \frac{k_2}{k_1} m \tag{59}$$

Figures 11, *a*, *b* show the amplitude-frequency characteristics of the system under kinematic disturbance z_1 (*t*) and z_2 (*t*), respectively.

Figure 11. Amplitude-frequency characteristics of the system:a - at excitation from the side of support surface $\bar{z}_1 \neq 0$ (dynamic quenching before resonance - 1, dynamic quenching after resonance - 2, xxx - frequencies dynamic damping and resonance coincide); b - similar frequency characteristics at kinematic disturbance $\bar{z}_1 \neq 0$ ($\bar{z}_2 = 0$, $\bar{Q} = 0$)

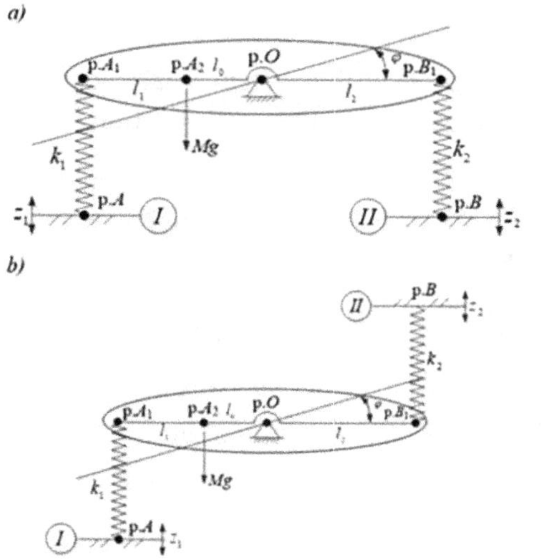

With a certain ratio of parameters, when $m = \frac{k_2}{k_1} L(z1 \neq 0)$ and at ($m = \frac{k_1}{k_2} L$, $z2 \neq 0$), the "locking" mode may occur at all frequencies. Relations between amplitudes of oscillations of protected object and support surface are determined in form at $\frac{k_1}{k_1+k_2}z1 \neq 0$ and at $\frac{k_2}{k_1+k_2}z2 \neq 0$. These conditions are fulfilled at all frequencies of the range (friction forces are not taken into account in the calculations; otherwise, the task requires other approaches).

If the critical relations between the inertial parameters m and L (expressions (58), (59)) are not fulfilled, then the locking effect is realized at high frequencies (p ∞). The limits of the amplitude ratio at p ∞ are determined by the same values as in the critical case. The systems implement, as shown in Figure 11, the modes of dynamic damping of vibrations, which can be implemented in the post-resonance and pre-resonance frequency areas of external influences. A new effect may be considered when the difference between the amplitudes of the oscillations of the object of protection is a certain value. For example, if, $z_1 = z_0 \sin\omega t$then the amount of compression of the spring k_1 is, $z_{10} - y\frac{k_1}{k_1+k_2}$which makes it possible to determine a certain margin of potential energy that the system has in the "locking" mode. Similarly, the spring k2 will be subject to deformation, for which the potential energy will be determined by the value of the amplitude of the oscillations. This $\frac{y_2 k_1}{k_1+k_2}$"reserve" of energy of elastic elements is determined by the relations between the values of m and L at the frequency of critical interactions ("locking"). This condition is met at different frequencies. When excited from the side of the support surface II, a similar pattern of interactions will be observed.

In a certain sense, the "locking" mode can be considered as a mode of "capturing" energy and storing it for the period of forced oscillations while fulfilling the necessary relationships between the system parameters. This effect can be achieved, for example, by adjusting the value of the reduced mass L, which is possible in the presence of controlled devices that create certain moments of force on the flywheel nut, if a non-self-braking screw mechanism is used as an MCD (Eliseev, Volkov, & Kukharenko, 1990).

FEATURES OF SYSTEMS WITH TWO DEGREES OF FREEDOM

Consider a chain mechanical oscillatory system with two degrees of freedom, which includes elastic elements k_1, k_2, k_3 and mass inertia elements m_1 and m_2, as shown in Figure 12.

Figure 12. Design diagram of vibration protection system with two degrees of freedom

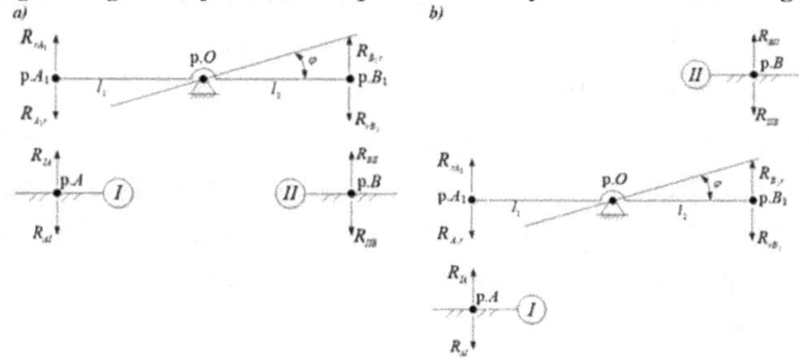

We will find expressions for kinetic and potential energies under the assumption that the system performs free oscillations and at the same time there is a relationship between the coordinates y_2 and y_1:

$$y_2/y_1 = a. \tag{60}$$

Then

$$\mathrm{T} = \frac{1}{2}m_1\dot{y}_1^2 + \frac{1}{2}m_2\dot{y}_2^2 = \frac{1}{2}\dot{y}_1^2(m_1 + a^2 m_2) \tag{61}$$

$$\Pi = \frac{1}{2}k_1 y_1^2 + \frac{1}{2}k_1(y_2 - y_1)^2 + \frac{1}{2}k_3 y_2^2 = \frac{1}{2}y_1^2(k_1 + k_2 + (k_2 + k_3)a^2 - 2k_2 a) \tag{62}$$

We introduce into consideration the frequency energy function (by analogy with the approximate energy method for determining the lower frequencies of natural vibrations), which has the form

$$\omega^2(a) = \frac{\Pi(a)}{\mathrm{T}(a)} = \frac{(k_1 + k_2 + (k_2 + k_3)a^2 - 2k_2 a)}{m_1 + a^2 m_2} \tag{63}$$

Figure 13 shows a graph of ω 2 (a) dependence at the specified parameters of the design scheme ($k_1 = k$, $k_2 = 2k$, $k_3 = 3k$, $m_1 = m_2 = m$). The graph in Figure 13 is based on the assumption that $a > 0$, that is, the form of relations between the coordinates of motion y_1 and y_2, which corresponds to the first form of principal oscillations, is taken into account. The graph has a characteristic shape and at the maximum value has two parameters ω^2, which corresponds to the fundamental frequency of natural oscillations (ω^2_{1res}), and also determines the value of the coefficient of distribution of oscillation amplitudes.

Figure 13. Plot of $\omega^2(a)$ versus parameter a

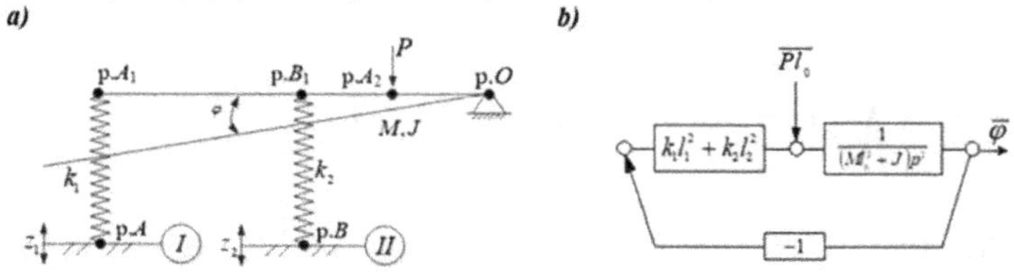

For further analysis, let's build a structural mathematical model of the system (Figure 14). The system block diagram is as shown in Figure 14. Using the presented scheme, we write down the expressions for the transfer functions (Eliseev, Reznik, & Khomenko, 2011).

Figure 14. Structural mathematical model for the system design diagram shown in Figure 10

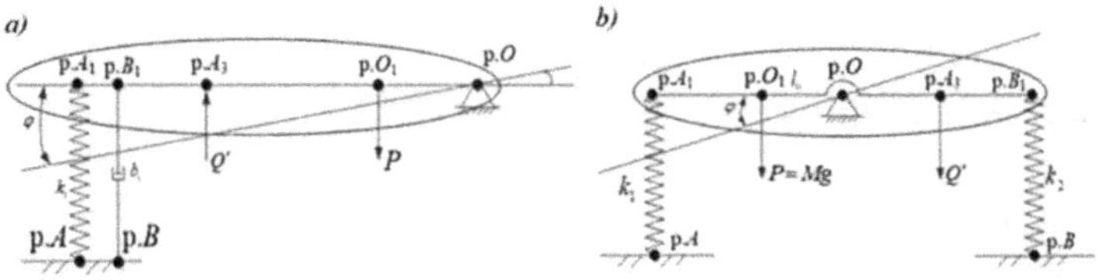

In accordance with the system diagram shown in Figure 14, we will record that

$$W'_1(p) = \frac{\bar{y}_1}{\bar{Q}_1} = \frac{m_2p^2 + k_2 + k_3}{A_0} \tag{64}$$

$$W'_2(p) = \frac{\bar{y}_2}{\bar{Q}_1} = \frac{k_2}{A_0} \tag{65}$$

$$W_{12}(p) = \frac{\bar{y}_2}{\bar{y}_1} = \frac{k_2}{m_2p^2 + k_2 + k_3} \tag{66}$$

where

$$A_0 = (m_1p^2 + k_1 + k_2)(m_2p^2 + k_2 + k_3) - k_2^2 \tag{67}$$

The expression (67) defines a characteristic frequency equation. Using (67), one can find the natural frequencies analytically, in this case

$$\omega_{1,2sob} = \frac{n_1^2}{2} + \frac{n_2^2}{2} \pm \sqrt{\frac{[m_1(k_2+k_3) - m_2(k_1+k_2)]^2 + 4m_1m_2k_2^2}{(m_1m_2)^2}} = $$
$$= \frac{n_1^2}{2} + \frac{n_2^2}{2} \pm \sqrt{(n_1^2 - n_2^2)^2\left[1 + \frac{4k_2^2}{m_1m_2(n_1^2 - n_2^2)^2}\right]}. \tag{68}$$

Here n_1, n_2 - partial frequencies

$$n_1^2 = \frac{k_1 + k_2}{m_1} \tag{69}$$

$$n_2^2 = \frac{k_2 + k_3}{m_2} \tag{70}$$

When determining natural oscillation frequencies, there are certain relationships between partial and natural oscillation frequencies.

$$\omega_{1sob}^2 < n_1 < n_2 < \omega_{2sob}^2 \tag{71}$$

In this case, the condition

$$\omega_{1sob}^2 + \omega_{2sob}^2 = n_1^2 + n_2^2 \tag{72}$$

Conditions (71), (72) are fulfilled in certain coordinate systems, which requires attention to the assessment of the type of interpartial bonds, which can be elastic and inertial. When introducing additional links, for example, when using standard elements with transfer functions of a differentiating link of the 2nd kind, inter-partial links take the form of inertial elastic ones. In such situations, inter-partial communication at certain frequencies can be “canceled,” which is associated with the appearance of specific dynamic modes of movement of the system. Combined inter-partial relationships form different values of partial frequencies, which requires separate consideration.

Since the initial system has two degrees of freedom, taking into account the features of free oscillations in the chain-type structure, the coefficient of distribution of oscillation amplitudes a will have a negative value, so we can assume that a <0. In this case, the graph of the dependence of $\omega^2(a)$ (Figure 15) takes a different form, but the presence of extremes of the frequency energy function $\omega^2(a)$ is preserved.

Figure 15. Plot of $\omega^{2}(a)$ in the area of negative values of parameter a

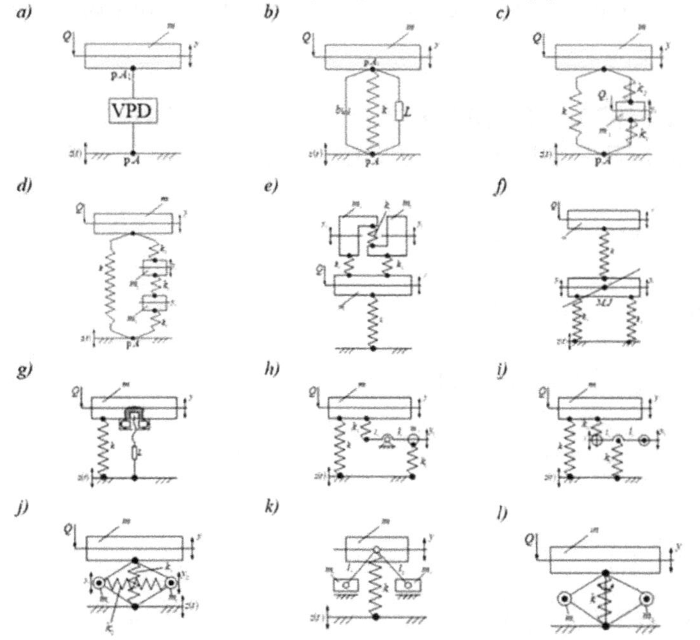

It should be noted that on schedule (Figure 15) in the field of positive values the asymptotic movement $\omega^2(a)$ is observed at limit partial system $\omega^2(a) \rightarrow \frac{k_2+k_3}{m_2} = n_2^2$, if a ∞. At $a = 0$ $\omega^2(a) = \frac{k_1+k_2}{m_1}$, which also corresponds to the partial frequency value n_1^2. In turn, when a decreases, when a negative value of a *is* ∞, the function $\omega^2(a) \rightarrow \frac{k_2+k_3}{m_2} = n_2^2$, while when $a = 0$, $\omega^2(a)$ has a value $\frac{k_1+k_2}{m_1}$. This makes it possible to introduce a generalized representation of the frequency energy function in the form of a graph in Figure 16.

Figure 16. Generalized plot of $\omega^2(a)$

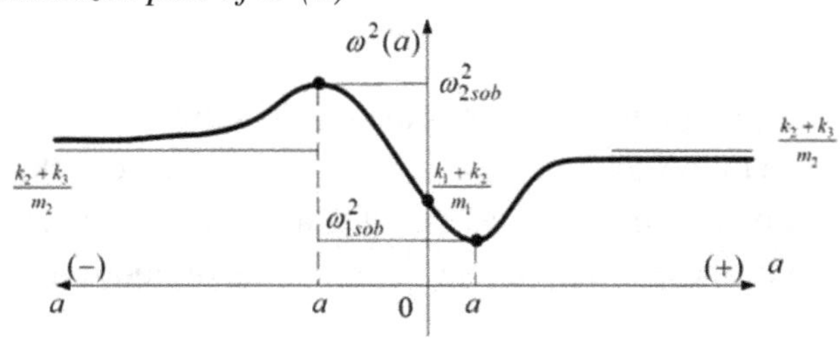

Determination of values a and $\omega^2(a)$, i.e. natural frequencies of oscillations ω^2_{1sob} and ω^2_{2sob}, is corrected in ratios determined by transfer function of interpartial links.

DETERMINATION OF NATURAL OSCILLATION FREQUENCIES BY GRAPHOANALYTICAL METHOD

From the frequency characteristic equation

$$(m_1 p^2 + k_1 + k_2)(m_2 p^2 + k_2 + k_3) - k_2^2 = 0 \tag{73}$$

follows, in particular, the condition

$$\frac{k_2}{(m_2 p^2 + k_2 + k_3)} = \frac{(m_1 p^2 + k_1 + k_2)}{k_2} \tag{74}$$

Figure 17. Schematic diagram of determination of coefficients of distribution of amplitudes of main oscillations using transfer functions of interpartial links

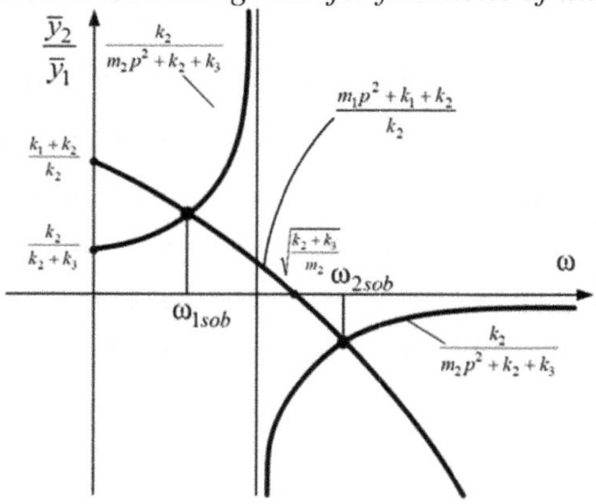

A graphical interpretation of the condition (74) is shown in Figure 17. It is taken into account that the ratio $\bar{y}_2/\bar{y}_1$ characterizes the ratio of the oscillation amplitudes in the main forms of motion. The results of numerical modeling confirmed the possibilities of the proposed ideas about the relations of coordinates of motion in the main forms.

INTRODUCTION TO MOTION CONVERSION APPARATUS SYSTEM STRUCTURE

If we introduce into the calculation scheme (Figure 12) a device for converting the movement of the APU with a reduced mass L, then this can be represented in Figure 12 as the introduction of the APU parallel to the spring k_2. In this case, the structure diagram will change, and the structural mathematical model will take the form as shown in Figure 18.

Figure 18. The structural diagram of the source system in Figure 12 with the MCD introduced parallel to elastic element k_2

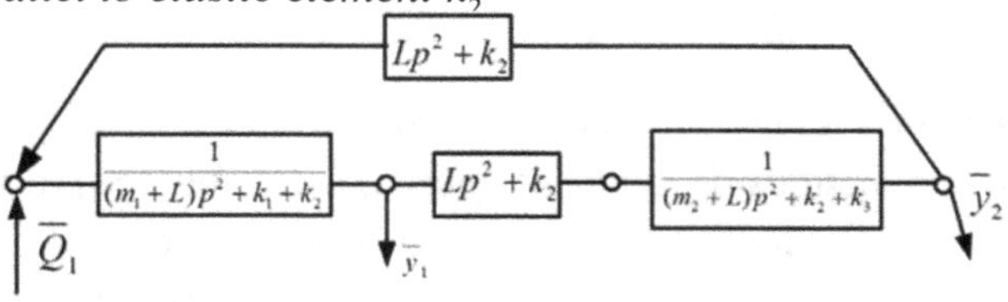

In this case, the transfer functions (64) - (66) are converted to the form

$$W'_1(p) = \frac{\bar{y}_1}{\bar{Q}_1} = \frac{(m_2 + L)p^2 + k_2 + k_3}{A'_0} \tag{75}$$

$$W'_2(p) = \frac{\bar{y}_2}{\bar{Q}_1} = \frac{Lp^2 + k_2}{A'_0} \tag{76}$$

$$W''_{12}(p) = \frac{\bar{y}_2}{\bar{y}_1} = \frac{Lp^2 + k_2}{(m_2 + L)p^2 + k_2 + k_3} \tag{77}$$

in here

$$A'_0 = [(m_1 + L_1)p^2 + k_1 + k_2)][(m_2 + L)p^2 + k_2 + k_3) - (Lp^2 + k_2)^2 \tag{78}$$

frequency characteristic equation.

In the given form (78) takes the form

$$p^4\left[(m_1 + L)(m_2 + L) - L^2\right] + p^2\left[(m_2 + L)(k_2 + k_3) + (m_2 + L)(k_1 + k_2) - 2Lk_2\right] + \\ + k_1 k_2 + k_2 k_3 + k_1 k_3 = 0. \tag{79}$$

From (79) it follows that the natural oscillation frequencies of the system will depend on the value of L. The expression for the analytical determination of natural oscillation frequencies has the form

$$\omega^2_{1,2\,sob} = \frac{m_1(k_2+k_3)+m_2(k_1+k_2)+L(k_1+k_3)}{2[m_1m_2+L(m_1+m_2)]} - \frac{1}{2}\sqrt{B} \tag{80}$$

where

$$B = \frac{\{m_1(k_2+k_3)+m_2(k_1+k_2)+L(k_1+k_3)\}^2 - L(k_1k_2+k_2k_3+k_1k_3)(m_1m_2+L(m_1+m_2))}{[m_1m_2+L(m_1+m_2)]} \tag{81}$$

If L = 0, then (81) is reduced to the expression (68). From (81) it can be seen that with interpartial connections of an elastic-inertial (or combined) form, the concept of partial frequencies acquires a more general meaning, since a different complex of dynamic interactions of constituent elements is implemented in systems. Note that the introduction of APC leads to the formation of inertial elements in the system with the so-called reduced masses, which requires the development of appropriate approaches in the tasks of dynamic analysis and synthesis.

We write down the expressions for the kinetic and potential energies of the system with MCD in the view that in the main fluctuations the relations between the coordinates also have the form (60), that is, $y_2 = ay_1$.

In this case,

$$\mathrm{T} = \frac{1}{2}m_1\dot{y}_1^2 + \frac{1}{2}m_2\dot{y}_2^2 + \frac{1}{2}L(\dot{y}_2-\dot{y}_1)^2 = \frac{1}{2}\dot{y}_1^2[m_1+L+a^2(m_2+L)-2La] \tag{82}$$

$$\Pi = \frac{1}{2}k_1y_1^2 + \frac{1}{2}k_2(y_2-y_1)^2 + \frac{1}{2}k_3y_2^2 = \frac{1}{2}y_1^2(k_1+k_2+(k_2+k_3)a^2-2k_2a) \tag{83}$$

Taking into account the introduction of UPD, the kinetic and potential energy of the system acquire a complete quadratic form, then

$$\omega^2(a) = \frac{\Pi(a)}{\mathrm{T}(a)} = \frac{k_1+k_2+(k_2+k_3)a^2-2k_2a}{m_1+L+a^2(m_2+L)-2La} \tag{84}$$

If we leave the parameters k_1, k_2, k_3, m_1, m_2 from the previous numerical experiment, then the effect *of L* can be illustrated, in a first approximation, with the number of mass ratio m and L values:,$\frac{1}{4}m$,. $\frac{2}{3}m$A plot of $5m^{\omega\,2}$(a) at - $\infty<a<\infty$ is shown in Figure 19. An increase in the reduced mass L causes the natural frequencies of the system to decrease, as shown in graphs 1 and 3, which can be compared with the graph for the system in the absence of an additional link L. However, the value *of* L can have a limit value when the movement between masses m_1 and m_2 is "locked," which corresponds to a certain relationship between L and m. In particular, in numerical modeling, this value is: $L = \frac{2}{3}m$.

Figure 19. Family of graphs $\omega^2(a)$ at different values of the reduced masses L:

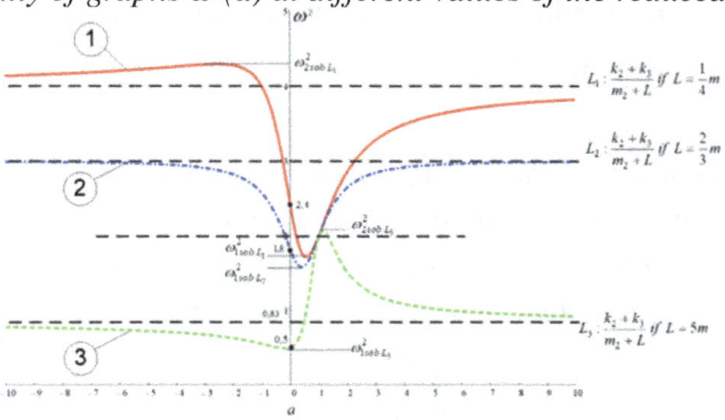

1 - at $L = \frac{1}{4}m$, 2 - at $L = \frac{2}{3}m$, 3 - at $L = 5m$

Transfer function of interpartial communication (assuming that external force is applied to the protection object m_1) has the form

$$W_{12}(p) = \frac{\bar{y}_2}{\bar{y}_1} = \frac{Lp^2 + k_2}{(m_2 + L)p^2 + k_2 + k_3} \tag{85}$$

At the frequency defined by

$$\omega^2_{\text{dyn}} = \frac{k_2}{L_2} \tag{86}$$

movement along the y_2 coordinate becomes zero. At frequency

$$\omega^2 = \frac{k_2 + k_3}{m_2 + L} \tag{87}$$

an infinitely large increase in displacement along the y_2 coordinate is possible. At the frequency determined from the ratio

$$\frac{k_2}{L} = \frac{k_2 + k_3}{m_2 + L} \tag{88}$$

follows that $\frac{\bar{y}_2}{\bar{y}_1}$ turns into a type uncertainty that $\frac{0}{0}$ can be resolved, which would amount to

$$\frac{\bar{y}_2}{\bar{y}_1} = \frac{L}{m_2 + L} \tag{89}$$

In this case, there should be a ratio between L and m_2, $k_2 m_2 + k_2 L = L k_2 + L k_3$ from where $m_2 = L\frac{k_3}{k_2}$. Graphs $\frac{\bar{y}_2}{\bar{y}_1}$ of dependence on ω are shown in Figure 20.

Figure 20. Curves of interpartial coupling ratio dependence: 1 - curve of dependence at dynamic damping in pre-resonance zone, 2 - graph of dependence at dynamic damping of oscillations by coordinate y_2 in post-resonance zone, 3 - curve of dependence in critical case when the system is "locked" over the entire frequency range

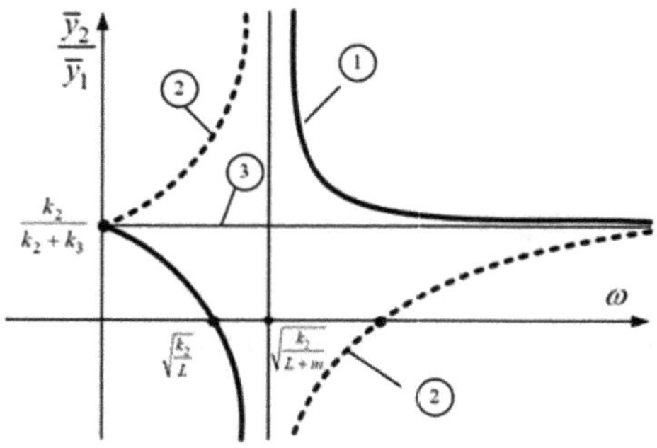

It should be noted that in the previous sections, where the system with the MCD with one degree of freedom was considered during kinematic disturbances from the support surfaces, the appearance of a critical "locking" mode of the system in the entire frequency range was also noted.

Consider the perturbation situation when $\omega^2 = \frac{k_2}{L}$. In this case, from the expression for the transfer function of the form

$$W_1(p) = \frac{\bar{y}_2}{\bar{Q}_1} = \frac{(m_2 + L)p^2 + k_2 + k_3}{[(m_1 + L)p^2 + k_1 + k_2][(m_2 + L)p^2 + k_2 + k_3] - (Lp^2 + k_2^2)} \tag{90}$$

it follows that

$$W_1(p)\Big|_{\omega^2 = \frac{k_2}{L}} = \frac{1}{m_1 p^2 + k_1} \tag{91}$$

that is, at the y_2 coordinate there is a stop ($y_2 = 0$), and at the y_1 coordinate there is a movement with the parameters of the system with one degree of freedom (m_1p^2 and k_1).

If the frequency of external exposure is

$$\omega^2 = \frac{k_2 + k_3}{m_2 + L} \tag{92}$$

that corresponds to dynamic damping of oscillations by coordinate y_1, then

$$\begin{aligned} W_2(p) = \frac{\bar{y}_2}{\bar{Q}_1} = \frac{Lp^2 + k_2}{(Lp^2 + k_2)^2} = \frac{1}{Lp^2 + k_2} = \frac{1}{\frac{-L(k_2 + k_3)}{m_2 + L} + k_2} = \\ = \frac{m_2 + L}{-Lk_2 - Lk_3 + m_2k_2 - Lk_2} = \frac{m_2 + L}{k_2m_2 - Lk_3}, \end{aligned} \tag{93}$$

which does not exclude the peculiarities of interactions at the ratios. $m_2 = \frac{k_3}{k_2}$ Note that such ratios lead to "locking" of the system in the entire frequency range.

A graphoanalytical method for determining the natural frequencies and the distribution coefficients of the amplitudes of motion in the first and second forms can also be applied using the relations obtained from the frequency equation (88), which implies that

$$\frac{Lp^2 + k_2}{(m_2 + L)p^2 + k_2 + k_3} = \frac{(m_1 + L)p^2 + k_1 + k_2}{Lp^2 + k_2} \tag{93}$$

Thus, the introduction of additional acceleration links, which are realized, in the physical sense, by devices for transforming motion (in particular, based on non-self-braking screw mechanisms), significantly affect the forms of dynamic interactions of elements of mechanical oscillatory systems.

Devices for converting motion can be introduced to construct and transform structural mathematical models of mechanical oscillatory systems, in particular, and vibration protection systems, as typical elementary links with a second-order differentiating element transfer function.

Devices for motion conversion (MCD) with elastic, dissipative, and other elements are switched based on parallel and serial spring connections, forming complex links in vibration protection systems.

Structural formations of elementary links with MCD operate as generalized springs or quasi-springs with frequency-dependent dynamic rigidity, following serial and parallel spring connection rules.

Mathematical models with MCD use structural schemes of elastic-dissipative and mass-inertial elements, with a double differentiation link transfer function connected in parallel to the elastic element.

Determining natural vibration frequencies involves a frequency energy function, defined as the ratio of potential to kinetic energies, incorporating assumed coordinate relations for main forms of motion. This method provides exact natural frequencies and analytical solutions for two-degree-of-freedom systems.

Graphoanalytical methods for natural frequency and shape coefficient determination, using int

CONNECTIVITY OF MOVEMENT OF ELEMENTS AND FORMS OF EXTERNAL INFLUENCES: MATHEMATICAL MODELS OF INTERACTIONS IN CHAIN STRUCTURES

Mechanical oscillatory systems of the chain type with two or more degrees of freedom are widely used as design circuits in various tasks of machine dynamics, including vibration protection of equipment, control of the dynamic state of technological machines and vehicles (Eliseev, Khomenko, & Kaimov, 2016; Belokobylsky & Eliseev, 2016; Belokobylsky, Eliseev, Kashuba, Nguyen, & Gypsy, 2015).

To assess the dynamic properties of mechanical systems, quite effective approaches have been developed based on analytical methods, as well as on their structural interpretations, reflecting the features of dynamic interactions of system elements with the introduction of additional bonds.

With all the variety of research and development methods that make it possible to assess the capabilities of complex systems, trends in the search and use of simplified approaches remain (Eliseev, Bolshakov, Nguyen, & Vyong, 2016). In this regard, chain mechanical systems are of great interest as model objects for studying the possibilities of evaluating dynamic effects under complex group forms of external influences. The cohesion of movements of system elements is characteristic of many technological machines that implement the technological processes of vibration transportation, vibration strengthening, etc., which is reflected in the work (Eliseev, Khomenko, & Kaimov, 2016) devoted to the dynamics of vibration benches. At the same time, such questions arise in the assessment of the properties and parameters of free oscillations of systems, problems of modal analysis and synthesis.

Methodological positions are developing in assessing the dynamic properties of mechanical oscillatory systems with three degrees of freedom while taking into account external excitations simultaneously for several inputs.

REFERENCES

Belokobylsky, S. V., & Eliseev, S. V. (2016). Change in the dynamic state of technical facilities based on the introduction of additional masses and leverage. Systems. Methods. *Technology*, *3*(31), 7–15.

Belokobylsky, S. V., Eliseev, S. V., Kashuba, V. B., Nguyen, D. H., & Gypsy, V. V. (2015). On the connections between the coordinates of movement in mechanical oscillatory systems with lever devices. Systems. Methods. *Technology*, *2*(26), 7–13.

Belokobylsky, S. V., Eliseev, S. V., & Sitov, I. S. (2012). Interaction of links through a mass-inertial element in the theory of mechanical chains. Systems. Methods. *Technology*, *2*(14), 7–15.

Belokobylsky, S. V., Eliseev, S. V., & Sitov, I. S. (2013). *Dynamics of mechanical systems. Lever and inertia-elastic links*. Polytechnic.

Druzhinsky, I. A. (1977). Mechanical chains. *Mechanical Engineering (New York, N.Y.)*.

Eliseev, S. V. (1978). *Structural theory of vibration protection systems*. Nauka.

Eliseev, S. V. (2018). Applied system analysis and structural mathematical modeling (dynamics of transport and technological machines: connectivity of movements, vibration interactions, leverage) (A. I. Artyunin, Ed.). IrGUPS.

Eliseev, S. V., Artyunin, A. I., Ermoshenko, Yu. V., et al. (2013). Methodological approaches in system analysis and mathematical modeling of mechanical oscillatory systems. Irkutsk: IrGUPS. (Deposited in VINITI 07.02.2013 No. 37).

Eliseev, S. V., Artyunin, A. I., Logunov, A. S., et al. (2013). Dynamic interactions of machine elements: design schemes and mathematical models of vibration states. Irkutsk: Deposited in VINITI 08.11.13 No. 313.

Eliseev, S. V., Bolshakov, R. S., Nguyen, D. H., & Vyong, K. Ch. (2016). Determination of natural vibration frequencies of mechanical oscillatory systems: Features of using the frequency energy function. *Bulletin of Irkutsk State Technical University*, *6*(113, PART I), 26–33.

Eliseev, S. V., Kashuba, V. B., & Sitov, I. S. (2013). Reduced stiffness of the feedback circuit. Determination of dynamic reactions in a mechanical oscillatory system. Systems. Methods. *Technology*, *2*(18), 15–22.

Eliseev, S. V., & Khomenko, A. P. (2014). *Dynamic damping of oscillations feedback concept and structural methods of mathematical modeling*.

Eliseev, S. V., Khomenko, A. P., & Kaimov, E. V. (2016). Forms of manifestation of leverage in dynamic interactions of elements of mechanical oscillatory systems. Modern technologies. Systems analysis. *Modeling*, *3*(51), 8–18.

Eliseev, S. V., Kovyrshin, S. V., & Bolshakov, R. S. (2012). Features of constructing compacts of elastic elements in mechanical oscillatory systems. Interactions with system elements and connection forms. Modern Technologies. Systems Analysis. *Modeling*, (4(36)), 61–70.

Eliseev, S. V., Kuznetsov, N. K., Bolshakov, R. S., et al. (2015). Justification and technologies of equivalent transformations of structural mathematical models of technical objects. Irkutsk: Dep. VINITI 20.10.2015 No. 179.

Eliseev, S. V., Moskovskikh, A. O., Bolshakov, R. S., & Savchenko, A. A. (2012). Possibilities of integration of methods of circuit theory and theory of automatic control in machine dynamics problems. Science and Education: Electronic Scientific and Technical Publication, 6. http://technomag.edu.ru/doc/378699.html

Eliseev, S. V., Reznik, Y. N., & Khomenko, A. P. (2011). *Mechatronic approaches in the dynamics of mechanical oscillatory systems* [Monograph]. Nauka.

Eliseev, S. V., Volkov, L. N., & Kukharenko, V. P. (1990). Dynamics of mechanical systems with additional connections. *Science*.

Genkin, M. D. (1981). Vibrations in technology. (In 6 volumes). Moscow: Mechanical Engineering. Vol. 5: Measurements and tests.

Halperin, I. N. (1968). *Automation as one-way mechanics*. Mechanical Engineering.

Khomenko, A. P., Eliseev, S. V., Artyunin, A. I., Bolshakov, R. S., Kaimov, E. V., Kinash, N. Z., & Nguyen, D. H. (2015). Structural mathematical models in problems of dynamics of mechanical oscillatory systems. Irkutsk: Dep. VINITI RAN 30.03.2015, No. 62.

Kim, P. D. (2003). Theory of automatic control. Vol. 1. Linear systems. Fizmatgiz.

Lurie, A. I. (1959). *Operational calculus and its applications to problems of mechanics*. Nauka.

Shatalov, A. S. (1962). *Structural methods in control theory and electroautomatics*. Gosenergoizdat.

ADDITIONAL READING

Contreras, N., Zhang, X., Hao, H., & Hernández, F. (2024). Application of elastic metamaterials/meta-structures in civil engineering: A review. *Composite Structures*, *327*, 117663. DOI: 10.1016/j.compstruct.2023.117663

Eliseev, A. V., Kashuba, V. B., Kinash, N. Z., & Eliseev, S. V. (2016). Features of the dynamic properties of working bodies of technological machines under inertial vibration excitation. *Bulletin of Don State Technical University*, *16*(4(87)), 63–71.

Eliseev, A. V., Kuznetsov, N. K., & Mironov, A. S. (2023). Dynamic invariant maps in assessing motion modes of mechanical oscillatory systems. MAI Proceedings, (128).

Eliseev, A. V., Pnev, A. G., Kashuba, V. B., & Sitov, I. S. (2014). Vibrational technological processes: Vibrohardening during impact periodic interactions. In Kulagin Readings: Engineering and Technologies of Production Processes. XIV International Scientific and Practical Conference: in 3 parts (pp. 281–286). Chita: FSBEI HE "Trans-Baikal State University".

Kang, X., Tang, J., Xia, G., Wei, J., Zhang, F., & Sheng, Z. (2024). Design, optimization, and application of nonlinear energy sink in energy harvesting device. *International Journal of Energy Research*, *2024*(1), 2811428. Advance online publication. DOI: 10.1155/2024/2811428

Liu, C., Zhang, X., Chang, J., Lyu, Y., Zhao, J., & Qiu, S. (2024). Programmable mechanical metamaterials: Basic concepts, types, construction strategies—a review. *Frontiers in Materials*, *11*, 1361408. Advance online publication. DOI: 10.3389/fmats.2024.1361408

Luo, J., Zhang, L., Miao, L., & Mao, Y. (2024). A new magnetic negative stiffness mechanism: A case study of fast steering mirror for quasi-zero stiffness vibration isolation. *Advances in Mechanical Engineering*, *16*(10), 16878132241288091. Advance online publication. DOI: 10.1177/16878132241288091

Tophøj, L., Grathwol, N., & Hansen, S. (2018). Effective mass of tuned mass dampers. *Vibration*, *1*(1), 192–206. DOI: 10.3390/vibration1010014

Zadpoor, A. A. (2016). Mechanical meta-materials. *Materials Horizons*, *3*(5), 371–381. DOI: 10.1039/C6MH00065G

Chapter 5

Features of Dynamic States in Mechanical Oscillatory Systems With Linkage

ABSTRACT

Peculiarities of dynamics of systems with lever mechanisms are considered, which relate to actual directions of modern dynamics of machines and theory of mechanisms and machines. Particular attention is paid to systems consisting of two solids with rotation points. Expressions are given for kinetic and potential energy, as well as equations of motion for such systems. Then we consider analogues of these systems with two degrees of freedom, which have partial blocks that make only translational movements. It is shown that these systems can be converted to equivalent schemes with material points. Methods for constructing structural and computational schemes for such systems are described, as well as expressions for transfer functions and coefficients of equations of motion. Peculiarities of dynamics of systems with lever mechanisms and their connection with systems having partial units performing only translational movements are discussed.

The issues of equivalent transformations of mechanical oscillatory systems were considered in sufficient detail in works related to the development of modern dynamics of machines and the theory of mechanisms and machines. In recent years, a number of publications have appeared in which much attention is paid to the formation of mechanical linkage systems in structures and their corresponding display using structural mathematical models.

In this regard, the study of the features of the dynamics of systems with lever mechanisms can be attributed to the current directions of modern machine dynamics. Assessment of the possibilities of structural transformations of mechanical

DOI: 10.4018/979-8-3373-0447-2.ch005

systems in which partial formations make reciprocating movements; there are still certain difficulties and requires the development of a methodological basis.

Possibilities of conversion of rotary forms of motion in their relation with structures of mechanisms having partial blocks of translational type are studied.

STRUCTURAL MATHEMATICAL MODELS OF SYSTEMS WITH CONSTITUENT ELEMENTS IN THE FORM OF SOLID BODIES HAVING ROTATION POINTS

Consider a mechanical oscillatory system consisting of two solid bodies with each fixed point of rotation (pp. O_1, O_2), in the presence of connections realized by elastic elements (Figure 1).

Figure 1. Design diagram of the system with rotary partial units (levers of the second kind)

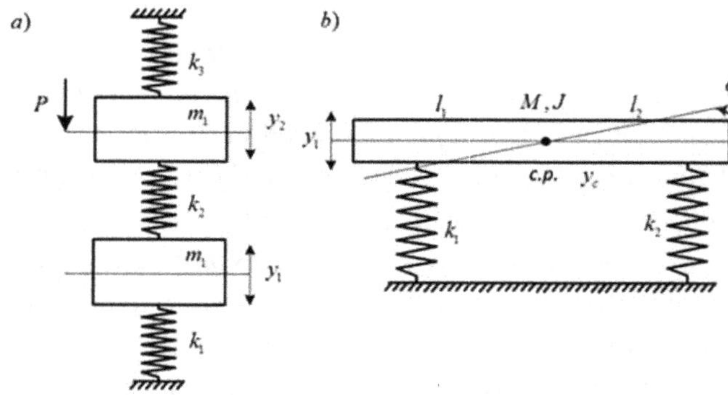

We write down the expressions for kinetic energy for kinetic and potential energies:

$$T = \frac{1}{2}J_1 \cdot \left(\dot{\phi}_1\right)^2 + \frac{1}{2}J_2 \cdot \left(\dot{\phi}_2\right)^2 \tag{1}$$

$$\Pi = \frac{1}{2}k_1 \cdot (\phi_1 l_1)^2 + \frac{1}{2}k_2 \cdot (-\phi_1 l_2)^2 + \frac{1}{2}k_3 \cdot (\phi_2 l_3 - \phi_1 l_1)^2 + \frac{1}{2}k_4 \cdot (-\phi_2 l_4 - \phi_1 l_2)^2 \tag{2}$$

After the usual transformations for Lagrange formalism, we write the equations of motion:

$$J_1\ddot{\phi}_1 + \phi_1 \cdot \left(k_1 l_1^2 + k_2 l_2^2 + k_3 l_1^2 + k_4 l_2^2\right) - \phi_2 \cdot \left(k_3 l_1 l_3 - k_4 l_2 l_4\right) = k_1 l_1 z_1 \tag{3}$$

$$J_2\ddot{\phi}_2 + \phi_2 \cdot \left(k_3 l_3^2 + k_4 l_4^2\right) - \phi_1 \cdot \left(k_3 l_1 l_3 - k_4 l_2 l_4\right) = k_2 l_2 z_2 \tag{4}$$

The system block diagram is as shown in Figure 2.

Figure 2. Structural diagram of the system in coordinates φ_1, φ_2 (rotational movements)

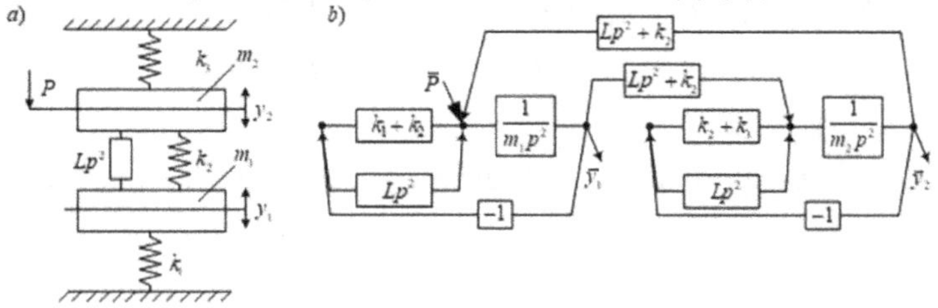

In this case, the connections between partial systems, in the physical sense, correspond to elastic interactions between solids J_1 and J_2.

If we look for analogies for the system in Figure 1 with a system with two degrees of freedom (for example, chain type), as shown in Figure 3, *a*, *b*, then it would be advisable to pay attention to the structure of partial blocks (Figure 2, *b*), in which it is reflected in the denominators of the transfer functions of the blocks of parameter k_2.

Figure 3. Calculation (a) and block diagram (b) for chain type system with translational partial blocks

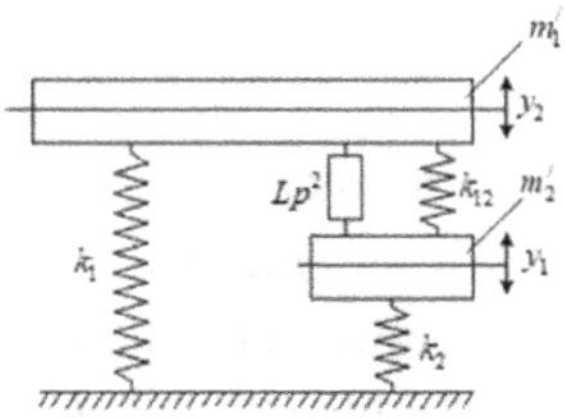

This parameter reflects the nature of the links for a particular type of system. In this case (Figure 2, *a*) at coordinates $_{y1}$ and $_{y2}$, the connection between partial systems is called elastic. The presence of k_2, repeated in the structures in Figure 2, b, predetermines the rules for transforming structural schemes obtained on the basis of Laplace transformations (Eliseev, 2018), which was used in deriving equations (3), (4) and constructing schemes (Figure 2, Figure 3, *b*). For subsequent studies, we will perform a number of transformations of equations (3), (4).

Let's change the expressions for partial systems in the diagram shown in Figure 2.

$$J_1 p^2 + l_1^2(k_1 + k_3) + l_2^2(k_2 + k_4) + (k_3 l_1 l_3 - k_4 l_2 l_4) - (k_3 l_1 l_3 - k_4 l_4) = \\ = J_1 p^2 + k_1 l_1^2 + k_2 l_2^2 + k_3 l_1 \cdot (l_1 - l_3) + k_4 l_2 \cdot (l_2 + l_4) + (k_3 l_1 l_3 - k_4 l_2 l_4). \quad (5)$$

In turn, for the second partial system:

$$J_2 p^2 + k_3 l_3^2 + k_4 l_4^2 + (k_3 l_1 l_3 - k_4 l_2 l_4) - (k_3 l_1 l_3 - k_4 l_4) = \\ = J_2 p^2 + k_3 l_3 \cdot (l_3 - l_1) + k_4 l_4 \cdot (l_2 + l_4) + (k_3 l_1 l_3 - k_4 l_2 l_4) \quad (6)$$

1. Note that the block diagram in Figure 2 is converted to a view, as shown in Figure 4.

Figure 4. Structural diagram of a system with two solid bodies having points of rotation according to the scheme of a lever of the second kind

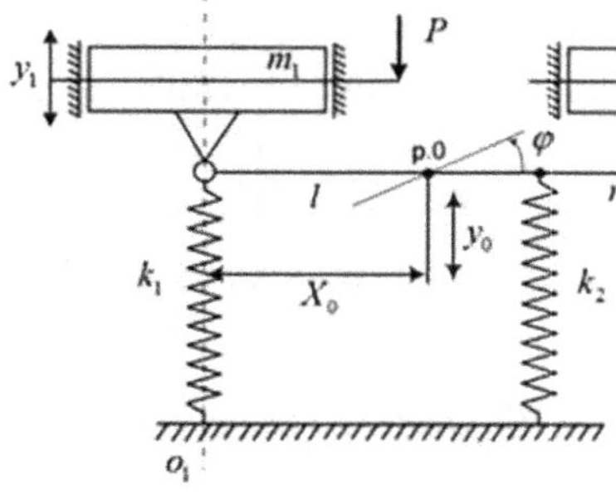

Such a block diagram displays the interactions between the angles of rotation of solids $\overline{\phi}_1$ and $\overline{\phi}_2$, just as if $\overline{\phi}_1, \overline{\phi}_2$ they were the coordinates of translational motion. In this case, the property of analogy between rotational and translational movements is clearly manifested. The external impact is taken into account in the usual way and in the structural diagram reduced to the solid $\overline{M}_{pr_1} = k_1 l_1 \overline{z}_1$ corresponds to the kinematic disturbance from the base ($z_l \neq 0$, $z_2 = 0$).

Using the input action $\overline{M}_{pr_1}$, the corresponding transfer functions can be constructed. The structural diagram in Figure 3 corresponds to the design diagram shown in Figure 5.

Figure 5. The given calculation scheme equivalent to the initial system in Figure 1

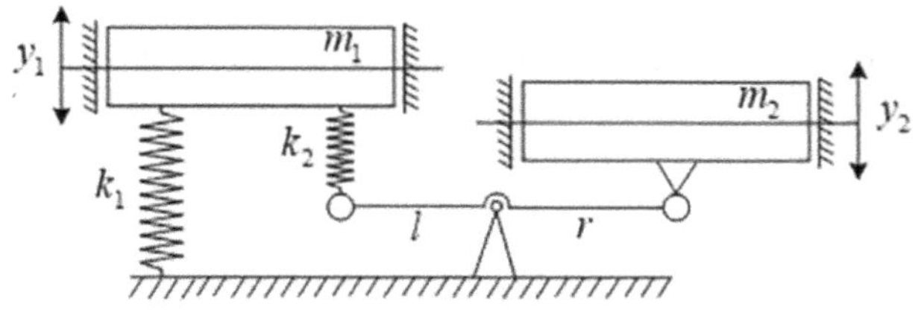

From Figure 5 it can be concluded that the system in Figure 1 is the same chain system, as shown, in particular, in Figure 3, *a*.

Note also that the calculation scheme in Figure 5 is an analogue of the calculation scheme in Figure 2, a. Using the diagram in Figure 5, it is possible to construct a structural diagram in Figure 6 in the same way as the structural diagram in Figure 6 is constructed, *b* for the design diagram in Figure 6, *a*.

The calculation diagram in Figure 5 is equivalent to the diagram in Figure 1. The differences are that the angular coordinates of φ_1 and φ_2, in this case, reflect (in a conditional plan) translational motion. That is, angular motion acts as an analogue of translational motion. At the same time, J_1 and J_2 in this scheme are analogues of masses in translational motion. The moment of force formed by the movement of the base ($z_1 \neq 0$) acts as an external perturbation (it is assumed that $z_2 = 0$, which does not reduce the commonality of consideration of dynamic states).

Elastic elements in the initial system, indicated by stiffness coefficients k_1, k_2, k_3, k_4, when switching to an analogue system, are transformed into elastic analogue elements:$k_1 l_1^2$, $k_2 l_2^2$, which, in the physical sense, determine the elastic properties of interaction in rotational motion. The main difference between the approach and the initial ideas (Figure 1) is that the system that implements the interaction of partial rotary systems is transformed into a scheme of relationships characteristic of translational movements.

2. Using the block diagram in Figure 4, we find the transfer function of the system at the output signal $\bar{\Phi}_1$and the input signal - $\bar{M}_1$- the moment of external forces created by the movement of the base z_1 ($z_2 = 0$). Thus:

$$W_1 = \frac{\bar{\Phi}_1}{\bar{M}_1} = \frac{J_2 p^2 + k_3 l_3 \cdot (l_3 - l_1) + k_4 l_4 \cdot (l_2 + l_4) + (k_3 l_1 l_3 - k_4 l_2 l_4)}{[A] \cdot [B] - (k_3 l_1 l_3 - k_4 l_2 l_4)} \tag{7}$$

where

$$A = J_1 p^2 + k_1 l_1^2 + k_2 l_2^2 + k_3 l_1 \cdot (l_3 - l_1) + k_4 l_2 \cdot (l_2 + l_4) + (k_3 l_1 l_3 - k_4 l_2 l_4) \tag{8}$$

$$B = J_2p^2 + k_3l_3 \cdot (l_3 - l_1) + k_4l_4 \cdot (l_2 + l_4) + (k_3l_1l_3 - k_4l_2l_4) \quad (9)$$

After the transformations, we obtain that:

$$W_1 = \frac{\overline{\Phi}_1}{\overline{M}_1}$$

$$= \frac{J_2p^2 + k_3l_3^2 + k_4l_4^2}{[J_1p^2 + k_1l_1^2 + k_2l_2^2 + k_3l_1^2 + k_4l_2^2] \cdot [J_2p^2 + k_3l_3^2 + k_4l_4^2] - (k_3l_1l_3 - k_4l_2l_4)^2} \quad (10)$$

If you use the calculation scheme in Figure 5, which is an analogue scheme with respect to rotational motion, then you can get the same result. Table 1 shows the coefficients of the equations of motion (3), (4), transformed to obtain the necessary ratios between analogue circuits.

Table 1. Coefficients of equations of motion by coordinates φ_1, φ_2 in the equivalent scheme of translational motion.

a_{11}	a_{12}
$J_1p^2 + k_1l_1^2 + k_2l_2^2 + k_3l_1 \cdot (l_3 - l_1) +$ $+k_4l_2 \cdot (l_2 + l_4) + (k_3l_1l_3 - k_4l_2l_4)$	$-(k_3l_1l_3 - k_4l_2l_4)$
a_{21}	a_{22}
$-(k_3l_1l_3 - k_4l_2l_4)$	$J_2p^2 + k_3l_3 \cdot (l_3 - l_1) + k_4l_4 \cdot (l_2 + l_4) +$ $+(k_3l_1l_3 - k_4l_2l_4)$
Q_1	Q_2
$M_1 = k_1l_1z_1$	$k_2l_2z_2$

Note: Q_1, Q_2 are generalized forces corresponding to generalized coordinates of φ_1, φ_2.

3. Using the calculation scheme in Figure 5 and the corresponding parameters of the coefficients of the equations (Table 1), we can obtain a similar (7) expression for determining the transfer function $W_1(p)$.

In this case, it is of interest to consider a mechanical oscillatory system consisting of two partial systems that rotate along the coordinates of φ_1 and φ_2.

Figure 6. Design diagram of an equivalent system consisting of two material points m_1 and m_3 located in points A and B

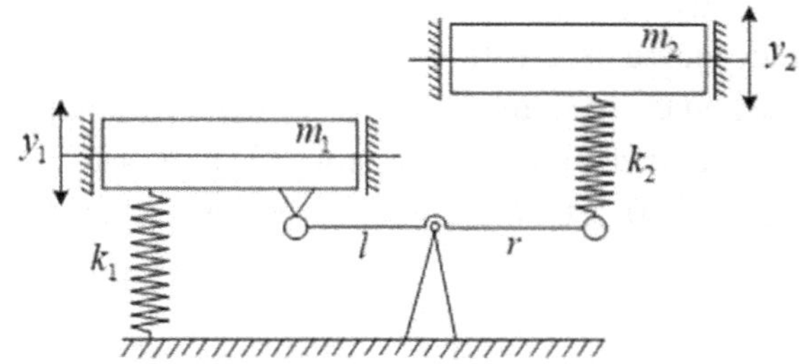

With appropriate transformations, the structural and design schemes of such an initial system become analog schemes in relation to systems that have partial blocks that make only translational movements. In this case, the torsional stiffnesses of the elastic elements correspond to the stiffnesses of ordinary springs, and the moment of inertia of solids J_1 and J_2 are analogues of solids in translational motion. If you consider the ratio $i_1 = \frac{l_2}{l_1}$ and $i_2 = \frac{l_4}{l_3}$, then the expression for the transfer function can be converted to the form:

$$W_1(p) = \frac{\bar{\Phi}_1}{\bar{M}_1}$$

$$= \frac{l_3^2 \cdot (J_3/l_3^2 + k_3 + k_4 i_2^2)}{l_1^2 \cdot l_3^2 \cdot [J_1/l_1^2 + k_1 + k_2 i_1^2 + i_{13}^2 \cdot (k_3 + k_4 i_2^2)] (J_2/l_3^2 + k_3 + k_4 i_2^2) - (k_3 - k_4 i_1 i_2)^2} \quad (11)$$

where $i_{13} = \frac{l_3}{l_1}$.
In turn:

$$W_1(p) = \frac{\bar{y}_1}{\bar{z}_1} = \frac{k_1 \cdot (J_3/l_3^2 + k_3 + k_4 i_2^2)}{[J_1/l_1^2 + k_1 + k_2 i_1^2 + i_{13}^2 \cdot (k_3 + k_4 i_2^2)] (J_2/l_3^2 + k_3 + k_4 i_2^2) - (k_3 - k_4 i_1 i_2)^2} \quad (12)$$

The expressions (11), (12) can be transformed using the relations $y_1 = \varphi_1 l_1, y_2 = \varphi_2 l_3, m_1 = \frac{J_1}{l_1^2}, m_2 = \frac{J_2}{l_3^2}$. With this approach, the mass inertia parameters are brought to the ends of the levers with the corresponding lengths l_1 and l_3:

$$W_1(p) = \frac{\bar{y}_1}{\bar{z}_1} = \frac{k_1 \cdot (m_2 p^2 + k_3 + k_4 i_2^2)}{[m_1 p^2 + k_1 + k_2 i_1^2 + i_{13}^2 \cdot (k_3 + k_4 i_2^2)] (m_2 p^2 + k_3 + k_4 i_2^2) - (k_3 - k_4 i_1 i_2)^2} \quad (13)$$

Figure 6 shows the transformed diagram, which shows that the masses m_1 and m_2 are on the same side of the points of rotation (pp. O, O_2); Note here that volumes A and B are related to masses m_1 and m_3, respectively. In turn, A_1 and B_1 points are used to connect elastic elements, creating a spatial system of interaction of elements.

The equivalent scheme in Figure 5, if used to derive the equations of motion, gives the same result as given by the expressions (3) and (4).

If we assume that $m_2 = 0$, then (13) is converted to the form:

$$W_1(p) = \frac{\bar{y}_1}{\bar{z}_1} = \frac{k_1 \cdot (k_3 + k_4 i_2^2)}{[m_1 p^2 + k_1 + k_2 i_1^2 + i_{13}^2 \cdot (k_3 + k_4 i_2^2)](k_3 + k_4 i_2^2) - (k_3 - k_4 i_1 i_2)^2} \quad (14)$$

The expression (14) can be reduced to the form:

$$W_1(p) = \frac{\bar{y}_1}{\bar{z}_1} = \frac{k_1}{[m_1 p^2 + k_1 + k_2 i_1^2 + i_{13}^2 \cdot (k_3 + k_4 i_2^2)] - \frac{(k_3 - k_4 i_1 i_2)^2}{(k_3 + k_4 i_2^2)}} \quad (15)$$

4. Using the denominator (14), you can build a calculation scheme of a single-mass system ($m_1 \neq 0$, $m_2 = 0$), which looks like shown in Figure 7.

Figure 7. Design circuit with quasi-spring

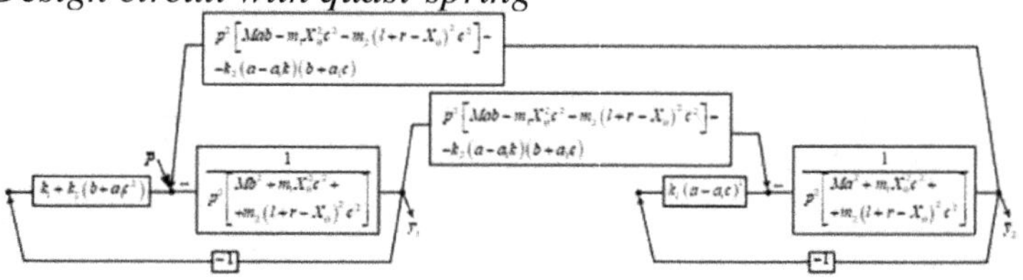

Note that at $m_2 = 0$, the elastic system turns into a structure consisting of quasi-springs: $(k_1 + k_2 i_1^2)$, $(k_3 + k_4 i_2^2) \cdot i_{13}^2$, $\frac{(k_3 - k_4 i_1 i_2)^2}{(k_3 + k_4 i_2^2)}$.

Quasi-springs are in parallel connection. In this case, the calculation scheme in Figure 7, where there is a kinetic disturbance, can be reduced to equivalent power (Figure 8).

Figure 8. Equivalent design scheme with force disturbance and three quasi-springs

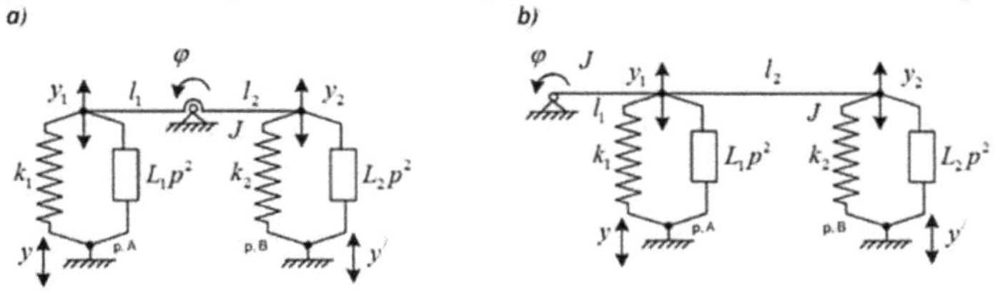

Each of the quasi-springs (meaning the expressions for reduced rigidity) reflects the features of the linkage structure, that is, the total reduced rigidity will be determined by the expression:

$$k_{pr} = k_1 + k_2 i_1^2 + i_{13}^2 \cdot (k_3 + k_4 i_2^2) - \frac{(k_3 - k_4 i_1 i_2)^2}{k_3 + k_4 i_2^2} \quad (16)$$

At $i_1 = 1$, $i_2 = 1$, $i_{13} = 1$, $k_{pr} = 4k$ (it is assumed that $k_1 = k_2 = k_3 = k_4 = k$). In expression (16), one of the quasi-spring elements has a negative sign. It should be noted that the general expression for the reduced stiffness cannot be equal to. Therefore, expression (16) can be expanded with respect to the coefficient of elasticity k_4, which gives the equation at $k_{pr} = 0$, the study of which determines the possibility of providing static stability by appropriate selection of system parameters (k_1-k_4, i_1, i_2, i_{13}).

At

$$k_3 = k_4 i_1 i_2 \quad (17)$$

The reduced stiffness takes the form:

$$k_{pr} = k_1 + k_2 i_1^2 + i_{13}^2 \cdot (k_3 + k_4 i_2^2) \quad (18)$$

It follows from (18) that it is possible to find the missing parameters by setting the value of the known ones:

$$i_1 = \frac{k_3}{k_4 i_2} \quad (19)$$

Then:

$$k_{pr} = k_1 + k_2 \cdot \frac{k_3^2}{k_4^2 \cdot i_2^2} + i_{13}^2 \cdot (k_3 + k_4 i_2^2) \cdot i_{13}^2.$$

5. Of greatest interest is the study of $k_{pr} = 0$.

$$k_{pr} = \left[k_1 + k_2 i_1^2 + i_{13}^2 \cdot \left(k_3 + k_4 i_2^2\right)\right]\left(k_3 + k_4 i_2^2\right) - \left(k_3 - k_4 i_1 i_2\right)^2 = 0 \tag{20}$$

$$k_4^2 \cdot \left[i_2^4 - i_2^2 i_1^2 i_{13}^2\right] + k_4 \cdot \left[\left(k_1 + k_2 i_1^2\right) \cdot i_2^2 + i_{13}^2 k_3 i_2^2 + 2k_3 i_1 i_2\right] + \left(k_1 + k_2 i_1^2\right) \cdot k_3 + k_3^2 i_{13}^2 - k_3^2 = 0$$

or

$$k_4^2 i_2^2\left(i_2^2 - i_1^2 i_{13}^2\right) + k_4\left[i_2^2\left(k_1 + k_2 i_1^2\right) + k_3\left(i_{13}^2 i_2^2 + 2 i_1 i_2\right)\right] + k_3^2\left(i_{13}^2 - 1\right) + k_3\left(k_1 + k_2 i_1^2\right) = 0 \tag{21}$$

Equation (21) can be used to estimate static stability.

The expression $k_3 l_1 l_3 - k_4 l_4 l_2$is, in essence, torsional or rotational stiffness in the interactions of the rotations of solids J_1 and J_2 at the coordinates of φ_1 and φ_2.

It can be noted that the respective torsional stiffnesses of the solids J_1 and J_2 considered relative to the fixed support surfaces have a rather complex appearance, which is determined by the spatial arrangement of the solids and elastic elements constituting, in general, a complex of lever links.

6. We find the transfer function by $\overline{\Phi}_2$from:$\overline{M}_1$

$$\underset{z_2=0}{W_2(p)} = \frac{\overline{\Phi}_2}{\overline{M}_1}$$

$$= \frac{k_3 l_1 l_3 - k_2 l_2 l_4}{\left[J_1 p^2 + k_1 l_1^2 + +k_2 l_2^2 + k_3 l_3^2 + k_4 l_4^2\right]\left(J_2 p^2 + k_3 l_3^2 + k_4 l_4^2\right) - \left(k_3 l_1 l_3 - k_2 l_2 l_4\right)^2} \tag{22}$$

Since, $y_2 = \varphi_2 l_3$ then:

$$\underset{J_1=0, z_2=0}{W'_2(p)} = \frac{\overline{y}_2}{\overline{z}_1} = \frac{k_1 l_1^2 l_3^2 \cdot \left(k_3 - k_2 i_1 i_2\right)}{l_1^2 l_3^2 \cdot \left[k_1 + k_2 i_2^2 + i_{13}^2 \cdot \left(k_3 + k_4 i_2^2\right)\right]\left[m_2 p^2 + k_3 + k_4 i_2^2\right] - \left(k_3 - k_4 i_1 i_2\right)^2} \tag{23}$$

Convert to:

$$W''(p) = \frac{\overline{y}_2}{\overline{z}_1} = \frac{\left(\dfrac{k_3 - k_2 i_1 i_2}{k_1 + k_2 i_2^2 + i_{13}^2 \cdot \left(k_3 + k_4 i^2\right)}\right)}{m_2 p^2 + k_3 + k_4 i_2^2 - \dfrac{\left(k_3 - k_4 i_1 i_2\right)^2}{k_1 + k_2 i_1^2 + i_{13}^2 \cdot \left(k_3 + k_4 i_2^2\right)}} \tag{24}$$

The design diagram of the system with object J_2 at $J_1 = 0$ is shown in Figure 9.

Figure 9. System design diagram at $J_1 = 0$

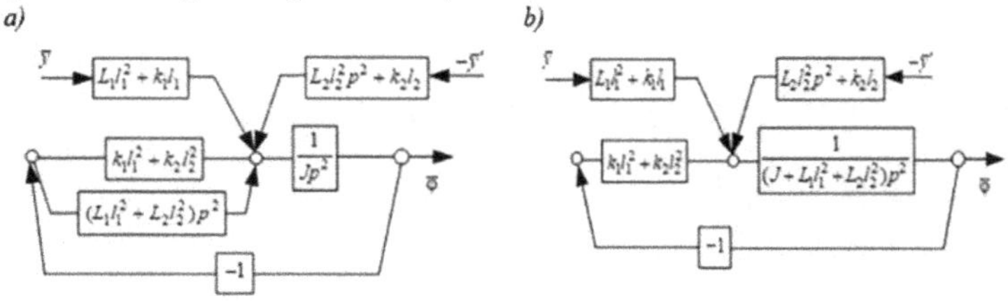

The peculiarity of the design scheme is that the mass inertia properties of the entire system at $J_1 = 0$, $J_2 \neq 0$ can be displayed by the translational movement of the element$m_2 = \frac{J_2}{l_3^2}$. In this case, the kinematic disturbance is reduced to mass m_2 (the material point attached to the movable end of lever l_3):

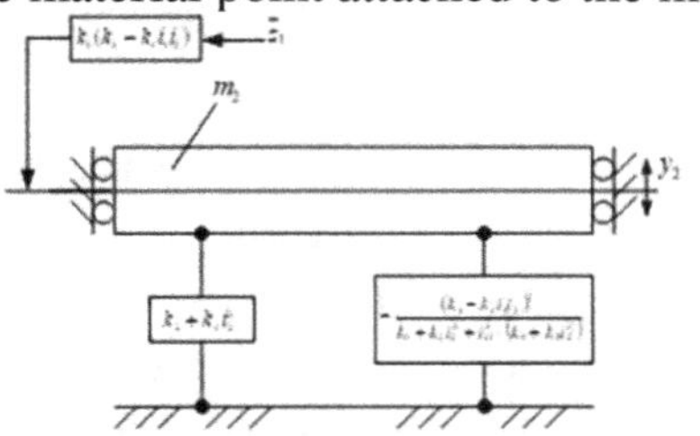

In a simpler physical form, the calculation scheme is shown in Figure 10.

Figure 10. Design diagram of the system with one degree of freedom at $J_1 = 0$

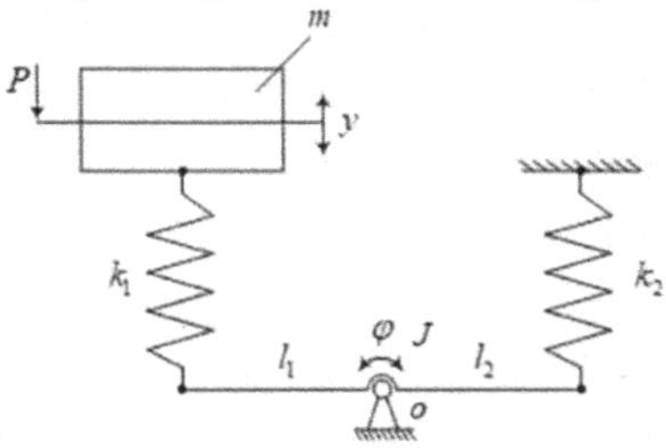

In this diagram (Figure 10), it can be assumed that the vibration protection object makes reciprocating movements around the O_2 point. In this case, a complex structure of elastic elements acts on the object.

Consider the capabilities of the system using the expression (22).

In its most general form, at $J_2 \neq 0$ ($J_1 = 0$), the reduced stiffness will be determined by the expression:

$$k_{pr} = k_3 + k_4 i_2^2 - \frac{(k_3 - k_4 i_1 i_2)^2}{k_1 + k_2 i_1^2 + i_{13}^2 \cdot (k_3 + k_4 i_2^2)} \tag{25}$$

After the transformations, the equation for determining the critical values of k_4 takes the form:

$$\begin{aligned}&k_4^2 i_2^2 \cdot \left(i_{13}^2 i_2^2 - i_1^2\right) + k_4\left[i_2^2\left(k_1 + k_2 i_1^2\right) + k_3 i_{13}^2 i_2^2 + k_3 i_1 i_2\right] + \\ &+ k_3\left(k_1 + k_2 i_1^2\right) + k_3\left(k_1 + k_2 i_1^2\right) + k_3^2 i_{13}^2 - k_3^2 = 0\end{aligned} \tag{26}$$

or:

$$k_4^2 + \frac{k_4 \cdot \left[i_2^2\left(k_1 + k_2 i_1^2\right) + k_3 i_{13}^2 i_2^2 + k_3 i_1 i_2\right]}{i_2^2 \cdot \left(i_{13}^2 i_2^2 - i_1^2\right)} + \frac{k_3\left(k_1 + k_2 i_1^2\right) + k_3\left(k_1 + k_2 i_1^2\right) + k_3^2\left(i_{13}^2 - 1\right)}{i_2^2 \cdot \left(i_{13}^2 i_2^2 - i_1^2\right)} = 0 \tag{27}$$

A critical situation occurs when the value of k_4 corresponds to zero.

The system can be unstable even in a static state if certain conditions are met.

Thus, a rotational system with two degrees of freedom, consisting of two solid bodies with rotation points, as shown in Figure 1, that is, according to the scheme of connections of levers of the second kind, can be reduced to an equivalent system in the form of a chain system with two degrees of motion of a translational type. Systems are equivalent in the sense that they are reduced to design schemes from which gear ratios can be derived by the same rules.

A method of matching similarity of design schemes is proposed, which is based on observing the presence in structural schemes of partial systems of the same terms, that is, relations corresponding to the transfer function of the inter-partial link. When aligning such relations, the rotational type system also behaves as a translational type system with corresponding analogies.

At the same time, the dynamics of systems with partial formations of the rotational type is based on more complex interactions.

It is shown that the construction of mathematical models for the problems of vibration protection and vibration isolation depends on the choice of the position of the object of protection, which can be selected in the upper stage (p. O_2) or in the lower stage (p. O_1).

Ensuring the stability of the system depends on the ratio of parameters, which requires verification by solving the second-order equation relative to the stiffness coefficients of the elastic elements of the upper stage. Corresponding analytical forms have been proposed. Note that the properties of the systems also depend on the ratio of the types of linkage.

QUASI-ELEMENTS IN MECHANICAL OSCILLATORY SYSTEMS; FEATURES OF SYSTEMS WHEN EXCLUDING DYNAMIC STATE VARIABLES

Dynamic vibration damping is one of the areas of theory and practice of vibration protection of machines, equipment and instruments. A detailed view of the properties of dynamic vibration dampers is related to tasks in which the object of protection in the form of a solid body with one degree of freedom is brought to a mode that provides a significant reduction in the vibration of the object at a fixed frequency of external harmonic perturbation. In such cases, the dynamic damper is considered as an attachable structure (in the simplest case, mass-inertial and elastic elements), adding another degree of freedom of movement to the system.

A dynamic damper can have a more complex design and have several degrees of freedom, while the object of protection is a solid with one degree of freedom.

To a lesser extent, issues of protection of objects are developed, the state of which is described by several variables. In particular, the dynamics of transport devices can be attributed to such tasks. In the simplest versions of such devices, the object of protection is a solid body on elastic supports. Its motion can be described in several coordinate systems.

Figure 11, *a*, *b* shows the design diagram of a mechanical oscillatory system with three degrees of freedom. The diagram has a fairly general view and reflects the possibilities of power effects Q_1, Q_2, Q_3 by coordinates y_1, y_2, y_3, as well as kinematic disturbances $z_1(t)$ and $z_2(t)$ from the side of support surfaces *I* and *II* (Figure 11, *a*). Figure 11, *b* shows a block diagram consisting of three partial systems. The motion of the system is described in generalized coordinates associated with a fixed basis. Such a system can be used to simulate the effects of dynamic damping in the form of an object with one or two degrees of freedom.

Figure 11. Design (a) and structural (b) schemes in the form of a mechanical oscillatory system with three degrees of freedom

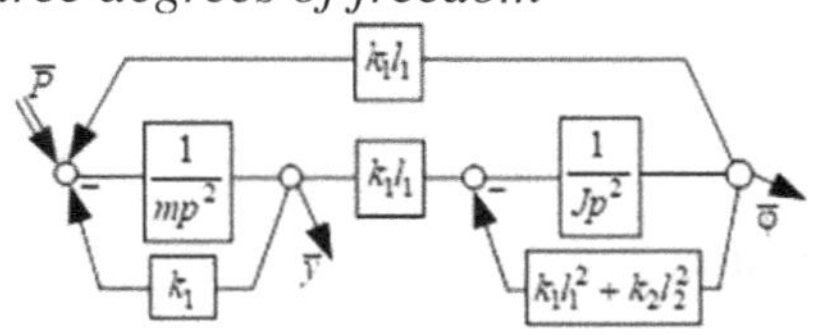

Figure 11, *a* mass inertia elements are indicated via m_1, m_2, m_3; k_1, k_2, k_3, k_4 reflect the elastic properties of the system; y_1, y_2, y_3 are associated with a fixed coordinate system. The system makes small movements; resistance forces are considered small.

External influences can be considered as kinematic (z_1 and z_2) or power harmonic disturbances applied to the respective masses as required.

The possibilities of changing the dynamic properties of vibration protection systems based on the use of additional connections and variation of structural constructions are investigated.

FEATURES OF BUILDING MATHEMATICAL MODELS

If the object of protection - the system has two degrees of freedom, then the block diagram in Figure 12, *b* can be reduced to the form, as shown in Figure 12. In this case, the coordinate y_3 as a separate observation factor is excluded.

Figure 12. Block diagram of vibration protection system with two degrees of freedom (y_1, y_2)

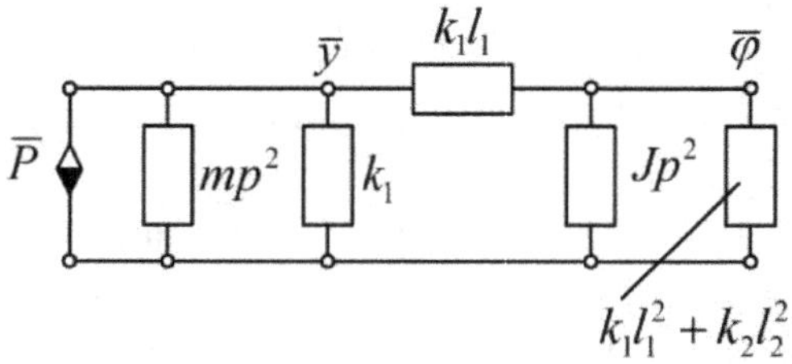

Assuming that the object has a complex structure, we find the transfer functions:

$$W_1(p) = \frac{\bar{y}_2}{\bar{z}_1} = \frac{k_1 k_2 (m_3 p^2 + k_3 + k_4)}{A_0} \tag{28}$$

$$\text{where} A_0 = (m_1 p^2 + k_1 + k_2) \cdot [(m_2 p^2 + k_2 + k_3) \cdot (m_3 p^2 + k_3 + k_4) - k_3^2] -$$

$$-k_2^2 \cdot (m_2 p^2 + k_2 + k_3) \tag{29}$$

characteristic equation of the system.

Let's convert the block diagram in Figure 12 to a view, as shown in Figure 13.

Figure 13. Structural diagram of the system with three degrees of freedom at elimination of coordinate y_2

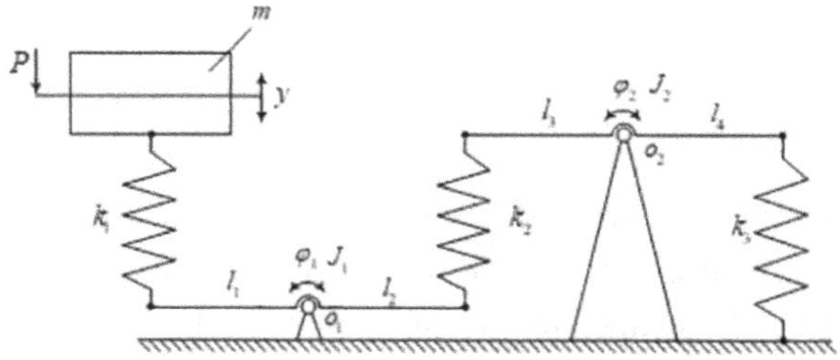

In turn, the block diagram in Figure 13 can be converted to a view as shown in Figure 14.

Note that in the structure of the circuit in Figure 14, the second partial system (coordinate y_2), unlike traditional elements, contains a generalized spring (or quasi-spring) with rigidity:

$$k_{pr}(p) = \frac{k_3(m_3p^2 + k_4)}{m_3p^2 + k_3 + k_4} \tag{30}$$

Figure 14. Structural diagram of the system as per Figure 13 with additional spring with dynamic stiffness

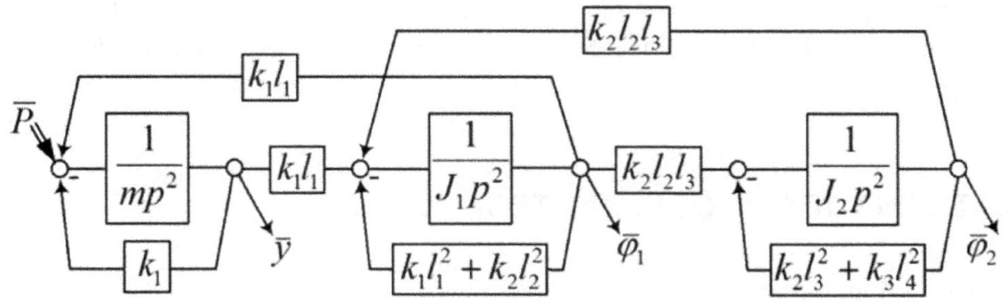

Quasi-springs have rigidity, which depends on the frequency of external influence. At frequency:

$$\omega_1^2 = \frac{k_4}{m_3} \tag{31}$$

such a spring has zero stiffness and the partial system at the coordinate of y_2 will work with the participation of the elastic element k_2. At frequency:

$$\omega_2^2 = \frac{k_3 + k_4}{m_3} \tag{32}$$

the spring constant becomes infinitely large and the influence of this partial system on the movement along the coordinate y_1 decreases. Thus, if you exclude the coordinates of y_3, the properties of the system will differ from the action of a regular system with two degrees of freedom.

Note also that at $p = 0$ ($p = j\omega$ - complex variable), the dynamic stiffness will correspond to the stiffness of a conventional spring defined by (28):

$$k'_{pr}(p) = \frac{k_3 k_4}{k_3 + k_4} \tag{33}$$

With $p \to \infty$ we get accordingly:

$$\underset{p\to\infty}{k''_{pr}(p)} = k_3 \tag{34}$$

Since the spring constant tends to take the value k_3, the system goes into interaction with a stiffer spring.

The purpose of these studies is to study the features of systems for the formation of dynamic properties when one of the coordinates is excluded, followed by a decrease in the total number of degrees of freedom and the appearance of additional connections in the system structure.

MATHEMATICAL MODELING TECHNOLOGY

The system in Figure 14 has a transfer function:

$$W(p) = \frac{\bar{y}_2}{\bar{z}_1} = k_1 k_2 / \left(m_1 p^2 + k_1 + k_2\right) \cdot \left[m_2 p^2 + k_2 + \frac{k_3\left(m_3 p^2 + k_4\right)}{m_3 p^2 + k_3 + k_4}\right] - k_2^2 \tag{35}$$

which can be reduced to the form:

$$W(p) = \frac{\bar{y}_2}{\bar{z}_1} = \frac{k_1 \cdot (m_3p^2 + k_3 + k_4) \cdot k_2}{(m_1p^2 + k_1 + k_2) \cdot [(m_2p^2 + k_2) \cdot (m_3p^2 + k_3 + k_4) + k_3(m_3p^2 + k_4)]} - \dots$$

$$\dots - k_2^2(m_3p^2 + k_3 + k_4). \tag{36}$$

It follows from (36) that at kinematic disturbance $z_1 = z_{10}sin(\omega t)$ by coordinate y_2 dynamic blanking mode is possible:

$$\omega_{dyn}^2 = \frac{k_3 + k_4}{m_3} \tag{37}$$

In turn, it follows from the frequency characteristic equation (that is, the denominator (36)) that three resonance modes can exist in the system at the coordinate y_2. Thus, excluding a coordinate does not mean that the dynamic properties of the system will change. The difference lies in the reduction in the number of observed coordinates. In this case, the elements of the mechanical oscillatory system include a generalized spring (or quasi-spring) with rigidity determined by the expression (28). With this approach, it is necessary to take into account the fact that the shape of the partial system itself changes. That is, the partial system may be a more complex formation. As such, such a partial system is a system with two degrees of freedom. So, from the structural diagram in Figure 14, you can find the transfer function of the partial system along the coordinate y_2:

$$W''(p) = \frac{(m_2p^2 + k_2) \cdot (m_3p^2 + k_3 + k_4) + k_3 \cdot (m_3p^2 + k_4)}{/m_3p^2 + k_3 + k_4} \tag{38}$$

The properties of the partial system will vary accordingly.

Also, as in simple forms, in a partial system of the extended type there is a resonance mode with a corresponding effect on the general properties of the system. At the same time, the partial system acquires two frequencies at which the expression (38) is “zeroed.” In this case, the connection with the movement along the coordinate y_1 is destroyed, which corresponds to special modes of dynamic interactions between partial systems, when the movements between the coordinates y_1 and y_2 will be unrelated.

The developed approach allows formulating the tasks of vibration protection and vibration isolation, when the object of protection can be a system with two degrees of freedom; in this case, part of the coordinates is excluded, and external forces acting on the excluded coordinates are “reduced” to the observed movements.

SYMMETRIC STRUCTURE CAPABILITIES

We believe that the mathematical model of a system with three degrees of freedom is characterized by a system of equations in the field of Laplace transformations:

$$a_{11}\bar{y}_1+a_{12}\bar{y}_2+a_{13}\bar{y}_3=\bar{Q}_1 \quad (39)$$

$$a_{21}\bar{y}_1+a_{22}\bar{y}_2+a_{23}\bar{y}_3=\bar{Q}_2 \quad (40)$$

$$a_{31}\bar{y}_1+a_{32}\bar{y}_2+a_{33}\bar{y}_3=\bar{Q}_3 \quad (41)$$

Sequential exclusion of coordinates, $\bar{y}_1$, $\bar{y}_2$ from $\bar{y}_3$ equations (30) - (41) can be made only in a certain sequence, otherwise the structural diagrams obtained on the basis of transformations (39) - (41) will not have symmetry in cross-links between partial systems. Note that, taking into account the design scheme of the initial mechanical system in the form, as shown in Figure 11, a, the coefficients of the system of equations (39) - (41) a_{11}, a_{22}, a_{33} reflect the properties of the corresponding partial systems:

$$a_{11} = m_1p^2 + k_1 + k_2$$

$$a_{22} = m_2p^2 + k_2 + k_3$$

$$a_{33} = m_3p^2 + k_3 + k_4$$

This is reflected in the structural diagram shown in Figure 11, *b*.

In order to obtain more accurate results in subsequent calculations, we will assume that the elastic relationship k_{13} is taken into account between the coordinates y_1 and y_3. In this case, the coordinates of the system of equations (39) - (41) will take the form:

$$a_{11} = m_1p^2 + k_1 + k_2 + k_{13}$$

$$a_{22} = m_2p^2 + k_2 + k_3$$

$$a_{33} = m_3p^2 + k_3 + k_4 + k_{13}$$

$$a_{12} = a_{21} = k_2$$

$$a_{23} = a_{32} = k_3$$

$a_{13} = a_{31} = k_{13}$

The type of transfer functions of partial systems in coordinates y_1 and y_3 will also change accordingly.

We assume that of (39):

$$\bar{y}_1 = \frac{1}{a_{11}}(\bar{Q}_1 - a_{12}\bar{y}_2 - a_{13}\bar{y}_3) \tag{42}$$

and substituting (42) in (40), (41), we obtain a system of two equations of motion in coordinates $\bar{y}_2$and $\bar{y}_3$:

$$\bar{y}_2\left(a_{22} - \frac{a_{12}^2}{a_{11}}\right) + \bar{y}_3\left(a_{23} - \frac{a_{13} \cdot a_{21}}{a_{11}}\right) = \bar{Q}_2 - \bar{Q}_1\frac{a_{12}}{a_{11}} \tag{43}$$

$$\bar{y}_2\left(a_{32} - \frac{a_{31}a_{12}}{a_{11}}\right) + \bar{y}_3\left(a_{33} - \frac{a_{13}^2}{a_{11}}\right) = \bar{Q}_3 - \bar{Q}_1\frac{a_{31}}{a_{11}} \tag{44}$$

Based on the system of equations (43), (44), the structural diagram shown in Figure 15 can be constructed.

Figure 15. System block diagram at $a_{13} = a_{31} \neq 0$ and exclusion$\bar{y}_1$

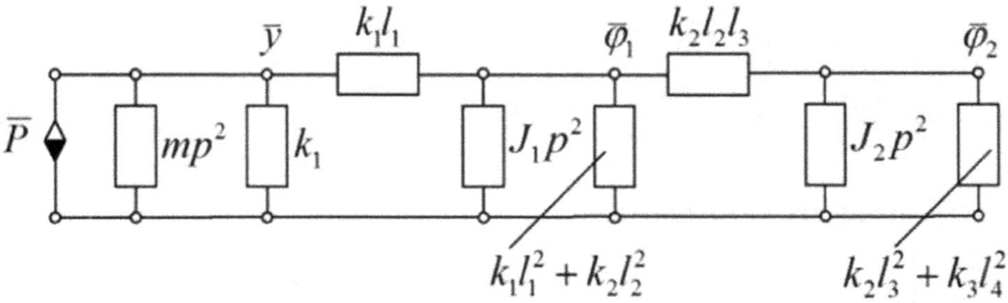

To prove that an exception $\bar{y}_1$ must be introduced only using, let us consider respectively in Figure 16, *a*, *b* structural diagrams of the system with excluded coordinates $\bar{y}_1$, but transformations made respectively not on the basis of (39), but equations (40) and (41).

Figure 16. Variants of structural schemes in coordinates y_2 and y_3:

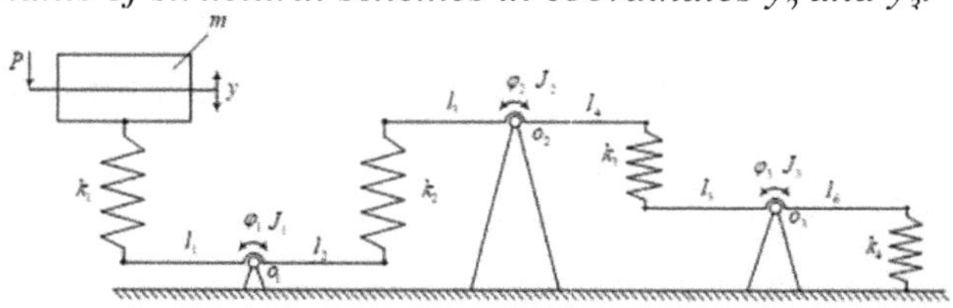

a) when obtaining y_1 from equation (40); b) when obtaining y_1 from equation (41)

From the structural diagrams in Figure 15 and Figure 16, *a*, *b* it can be seen that the transformations associated with the exception of y_1, based on the use of equations (40) and (41), give a negative result; systems lose link symmetry properties. Thus, the above assumption about the rules for choosing equations for eliminating coordinates is proved by transformations and the corresponding structural constructions in Figure 16, *a*, *b*.

In a similar way, the coordinates y_2 and y_3 can be excluded, which can be represented by systems of equations to describe movement along the coordinates y_1 and y_3 (in this case, the coordinate y_2 is excluded):

$$\bar{y}_1\left(a_{11} - \frac{a_{12}a_{21}}{a_{22}}\right) + \bar{y}_3\left(\frac{a_{13} - a_{12}a_{23}}{a_{22}}\right) = -\bar{Q}_1 - \bar{Q}_2\frac{a_{12}}{a_{22}} \tag{45}$$

$$\bar{y}_1\left(a_{31} - \frac{a_{32}a_{21}}{a_{22}}\right) + \bar{y}_3\left(a_{33} - \frac{a_{12}a_{23}}{a_{22}}\right) = \bar{Q}_3 - \bar{Q}_2\frac{a_{32}}{a_{22}} \tag{46}$$

Accordingly, the block diagram is as shown in Figure 17.

Figure 17. Structural diagram of the system in coordinates y_1 and y_3 (excluding coordinate y_2)

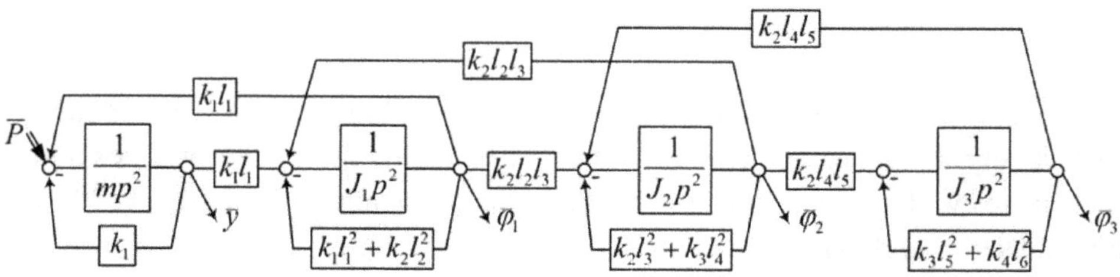

Note that when excluding the coordinate y_1, the expression (42) was used. Accordingly, equation (40) is used to exclude the coordinate y_2, which implies:

$$\bar{y}_2 = \frac{\bar{Q}_2 - a_{21} \cdot \bar{y}_1 - a_{31} \cdot \bar{y}_3}{a_{22}} \tag{47}$$

Comparison of (42) and (47) allows us to determine that the elimination of the intermediate coordinate is made using the corresponding equations that make it possible to translate the operator a_{11}, as well as a_{22} into the denominator by the relation (42), (47).

It can also be obtained by excluding the coordinates of y_3. From (41) it can be obtained that:

$$\bar{y}_3 = \bar{Q}_3 - \frac{a_{31} \cdot \bar{y}_1 - a_{32} \cdot \bar{y}_2}{a_{33}} \tag{48}$$

Substituting (48) in (39) and (40) with the corresponding transformations allows us to obtain the desired system of equations of motion.

If we exclude the coordinates of y_3, we get a mathematical model:

$$\bar{y}_1\left(a_{11} - \frac{a_{13}a_{31}}{a_{33}}\right) + \bar{y}_2\left(a_{21} - \frac{a_{13}a_{32}}{a_{33}}\right) = \bar{Q}_1 - \bar{Q}_3\frac{a_{13}}{a_{33}} \tag{49}$$

$$y_1\left(a_{31} - \frac{a_{32}a_{21}}{a_{22}}\right) + y_3\left(a_{33} - \frac{a_{12}a_{23}}{a_{22}}\right) = \bar{Q}_3 - \bar{Q}_2\frac{a_{32}}{a_{22}} \tag{50}$$

Accordingly, based on (49) and (50), a block diagram can be constructed, as shown in Figure 18.

Figure 18. Block diagram in coordinates y_1, y_2 (at excluded coordinate y_3)

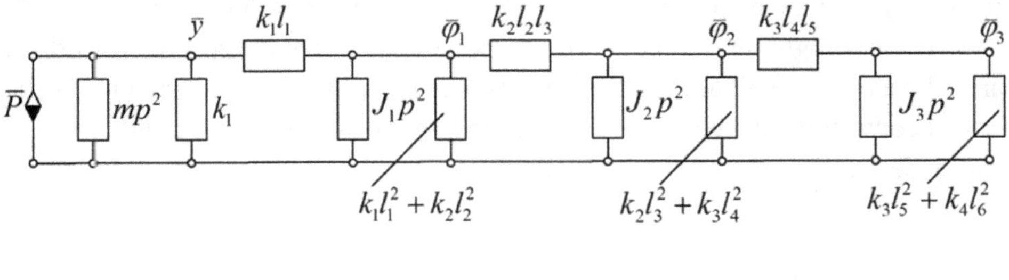

Thus, the exclusion of variables in systems with several degrees of freedom leads to the fact that a "simplified" system (by the number of degrees of freedom) retains all the properties of vibrational systems displayed by structural diagrams. At the same time, it should be noted that "simplification" leads to the introduction of new concepts that are associated with ideas about the possibility of using links of a new type. Such links are built as some blocks, structures or compacts from known typical elements in the form of mass inertia links, springs, dampers, as well as devices for transforming motion based on the rules of structural transformations based on the apparatus of the theory of automatic control (Eliseev, Khomenko, & Upyr, 2009). In the above cases, such formations were part of the structural schemes in Figure (15) - (18), and the analysis shows that complex or composite links of this kind can include other typical elements in addition to springs, ultimately forming a compact or quasi-spring. The latter names seem quite appropriate, since the formations in question are reduced springs with dynamic stiffness depending on the frequency of the excited oscillations. The concept of "quasi-spring" is associated, in particular, with the fact that the quasi-spring, if we keep in mind the possibility of being part of the blocks, behaves like an ordinary typical element (spring, damper, etc.).

The transformations above reflect some general properties of mechanical oscillatory systems, based on the notions that partial systems are defined as necessary steps in the construction of more complex structures, the development algorithm of which relies on Dalembert principles. In particular, the properties of symmetry of bonds between partial systems are associated with this. Setting such tasks can be found, in particular, in the work (Eliseev & Artyunin, 2016).

VARIANTS OF QUASI-LOADS DISPLAY IN CALCULATION DIAGRAMS

Assuming that Figure11, a shows the basic design scheme of a mechanical oscillatory system with three degrees of freedom, it can be shown that the exclusion of coordinates is associated with the introduction of new structural elements.

Developed in Figure 19, *a* the calculation scheme contains an additional element - the spring k_{13}, which connects the mass-inertial elements m_1 and m_3, which can be converted, highlighting the peculiarity of the system in that the spring k_{13} works in parallel connection (Figure 19, *c*) with the chain k_3, m_2, k_2. At the same time, such a parallel connection can be converted or rolled up into a quasi-spring with dynamic rigidity determined by the expression:

$$k_{pr_1} = k_{13} + \frac{k_2 k_3}{m_1 p^2 + k_2 + k_{13}} \tag{51}$$

The variant of the design scheme in Figure 20, *a* corresponds to the scheme in Figure 19, c. The design scheme in Figure 19, in this case will take the form as shown in Figure 20, *b*. System transformations lead to the fact that external forces change on the design scheme (they are reduced to coordinates). In addition, quasi-springs appear in the circuit. These new as parallel additional springs (with respect to k_4) between bearing surface *II* and mass m_3 (Figure 20, *b*).

Figure 19. Variants of design schemes for mechanical system with three degrees of freedom: a) design scheme of chain type; b) exclusion y_1*:* $k_3 + \frac{k_2 k_{13}}{m_1 p^2 + k_1 + k_2 + k_{13}}$*; c) exclusion* y_2*:* $k_{13} + \frac{k_3 k_2}{m_1 p^2 + \text{[illegible]}}$*; d) exclusion* y_3*:* $k_2 + \frac{k_{13} k_3}{m_3 p^2 + k_3 + k_4 + k_{13}}$

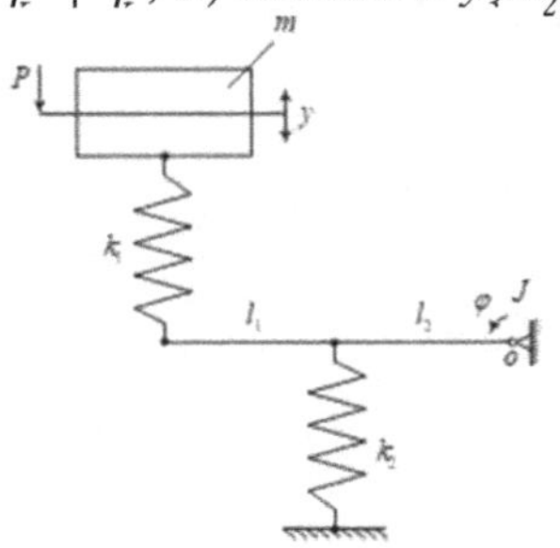

The transfer function of the quasi-spring has the form:

$$W_3(p) = \frac{m_2 p^2 \cdot k_3}{m_2 p^2 + k_2 + k_3} \tag{52}$$

A quasi-spring (Figure 20, *b*) with a transfer function also appears between the mass-inertial element m_1 and the support surface *I:*

$$W_1(p) = \frac{m_2 p^2 \cdot k_2}{m_2 p^2 + k_2 + k_3} \tag{53}$$

We also note that changes occur in the transfer function of interpartial communication (Figure 20, *b*), which takes the form:

$$W(p) = \frac{k_3 \cdot k_2}{m_2 p^2 + k_2 + k_3} \tag{54}$$

There is a one-to-one correspondence between the structural diagram of the system in Figure 20, *a* its design diagram in Figure 20, *b*. The parameters of the structural diagram in Figure 20, *a* can be obtained by direct transformations based on the rules of the theory of mechanical circuits or the structural theory of vibration protection systems (Bolshakov & Eliseev, 2013; Eliseev, Khomenko, & Upyr, 2009; Tarasik,

2004). Figure 20, *c* and *d* show structural diagrams of systems in coordinates y_2, y_3 (Figure 20, *c*) and in coordinates y_1, y_2 (Figure 20, *d*).

Figure 20. Structural diagrams of the system with design diagram as per Figure 11a: in coordinates y_1 and y_3 (a) and design diagram with quasi-springs (b); structural diagrams of systems by coordinates y_2, y_3 (c) and coordinates y_1 and y_2 (d)

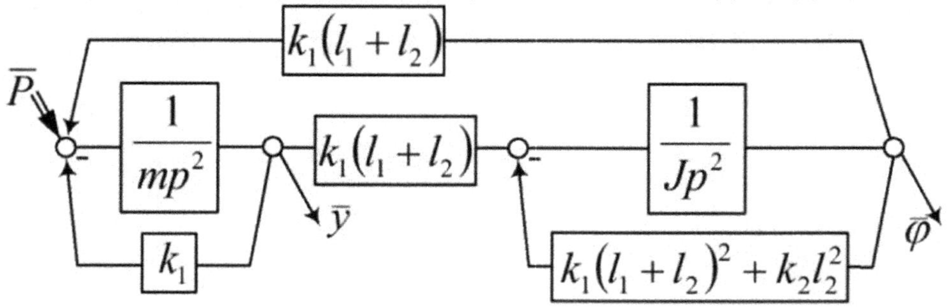

From the structural and design diagrams given in Figure 20, *a* - *d*, it can be seen that quasi-springs are included in parallel connections with typical elements k_1, k_2, k_3 and k_4. Appropriate conversions are possible. In particular, we can show the correspondence of the structural diagram in Figure 20, *a* the structural diagram in Figure 12. Both schemes reflect the properties of systems with two degrees of freedom. Characteristic of the structural diagram in Figure 12 is the presence *of* k_2 in the transfer functions of partial systems, as well as in the direct chain - element k_2. The same can be observed in the structural diagram shown in Figure 20, *a*, as well as in the calculation diagram (Figure 20, *b*), where a link with parameters is identified

$$k_{13} + \frac{k_2 k_3}{m_2 p^2 + k_2 + k_3}$$

As for partial systems, the corresponding elements are easily distinguished by simple transformations.

EVALUATION OF THE EFFECT OF ADDITIONAL ELASTICITY

When solving a number of problems related to the assessment of the influence of various types of connection of elastic elements, questions arise of taking into account the features and forms of interactions. Figure 21 shows a design scheme with three degrees of freedom, which resembles the design scheme in Figure 11, *a*, but differs in a set of elastic elements.

Figure 21. Design diagram of a system with three degrees of freedom with additional elastic elements

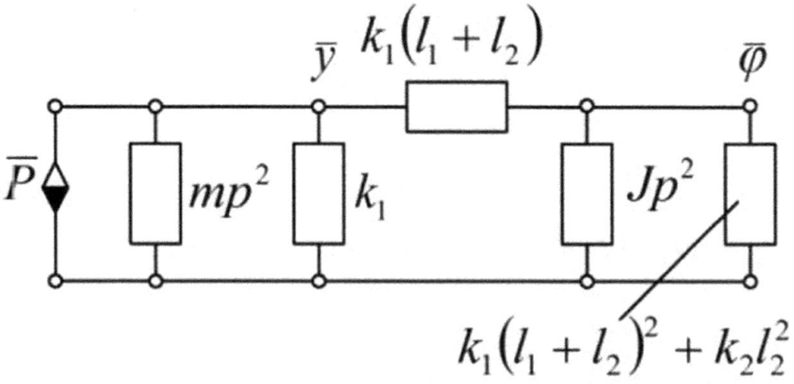

Table 2 shows the coefficients of the system of equations of motion in coordinates y_1, y_2, y_3.

It follows from Table 2 that the introduced springs k_{01}, k_{04} do not directly affect the inter-partial relationships determined by the stiffnesses of the elastic elements k_2, k_3, k_{13}.

The noted features are important when carrying out transformations of structural schemes, including when excluding variables.

Table 2. Coefficients of equations of motion in coordinates y_1, y_2, y_3

a_{11}	a_{12}	a_{13}
$m_1p^2+k_1+k_2+k_{13}$	$-k_2$	$-k_{13}$
a_{21}	a_{22}	a_{23}
$-k_2$	$m_2p^2+k_2+k_3+k_{01}$	$-k_3$
a_{31}	a_{32}	a_{33}
$-k_{13}$	$-k_3$	$m_3p^2+k_3+k_4+k_{04}$

Note: Q_1, Q_2, Q_3 - generalized forces by coordinates y_1, y_2, y_3

Consider a more complex example in the form of a computational scheme of a system with four degrees of freedom, having a wider set of elastic bonds (Figure 22).

Figure 22. Design scheme of the system with four degrees of freedom

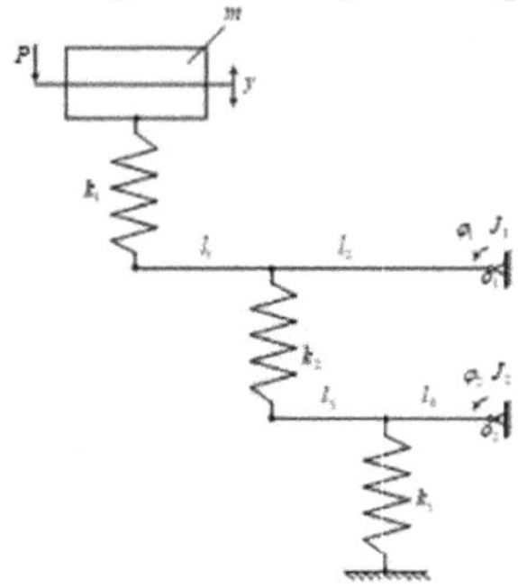

Table 3 shows the coefficients of the equation of motion (linear system) in coordinates y_1 - y_4.

Table 3. Coefficients of equations of motion in coordinates y_1 - y_4

a_{11}	a_{12}	a_{13}	a_{14}
$m_1p^2 + k_1 + k_2 + k_{13} + k_{14}$	$-k_2$	$-k_{13}$	$-k_{14}$
a_{21}	a_{22}	a_{23}	a_{24}
$-k_2$	$m_2p^2 + k_2 + k_3 + k_{42}$	$-k_3$	$-k_{42}$
a_{31}	a_{32}	a_{33}	a_{34}
$-k_{13}$	$-k_3$	$m_3p^2 + k_3 + k_4 + k_{13} + k_{35}$	$-k_4$
a_{41}	a_{42}	a_{43}	a_{44}
$-k_{14}$	$-k_{42}$	$-k_4$	$m_4p^2 + k_4 + k_5 + k_{42} + k_{41}$

It follows from Table 3 that, as in the previous case, the introduction of additional links leads to the filling of the matrix, and it does not take the form of a tape structure. That is, the mechanical system is not chained and contains non-planar connections, which makes the transformations more cumbersome. As for additional links in general, they are divided into two types: links between coordinates (k_{13}, k_{14}, k_{42}) and links between coordinates and support surfaces (k_{35}).

Note that the bonds of the first kind are included in the structure of the partial system (a_{11}, $a_{22,}$ a_{33}, a_{44}), and also occupy positions in the matrix that determine inter-partial bonds.

Links of the second type, for example, k_{35}, and in general, these can be any other links with support surfaces, then such links are included only in partial blocks (in our case, a_{33}).

The elimination of coordinates in mechanical oscillatory systems with several degrees of freedom is a formal technique for simplifying systems, which leads to a decrease in the number of variables, but does not change the mathematical model as a whole. However, the proposed approaches may be useful in several ways. The first of them is related to the possibilities of representing partial systems in the form of more complex structures, having the form of systems with two degrees of freedom. This allows us to consider the tasks of vibration protection and vibration isolation of machines and equipment in a more complex formulation, which provides control over the dynamic state of the system by two independent variables. At the same time, the possibilities of using techniques, methods and means of the structural theory of vibration protection systems remain.

The second direction for further applications is due to the fact that technologies for excluding coordinates (or variables) introduce into consideration blocks or compacts consisting of typical elements connected to each other according to certain rules; the resulting "education" can be called quasi-spring. Such a composite element behaves like an ordinary spring, bearing in mind the rules of its interaction and mating when converting structural diagrams. In fact, the quasi-spring has a reduced (or dynamic) frequency-dependent stiffness. At the same time, quasi-springs have parameters that combine them with conventional springs. These parameters appear under conditions: p 0 or p ∞.

Thirdly, the convenience of the proposed approaches is based on the fact that taking into account dissipative factors, as well as additionally introduced mass inertia devices or links, is associated with simple methods of adding introduced elements in parallel to the existing elastic links, which is detailed to a sufficient extent in the work on the structural theory of vibration protection systems.

In general, the proposed approach can be considered as the methodological basis of a new method for transforming mechanical oscillatory systems and developing the evidence base in the idea that the set of typical links can be expanded due to quasi-elements that are very diverse, but have simple principles of construction and rules for structural transformations.

COUPLING REACTIONS AS DYNAMIC STATE PARAMETERS OF OSCILLATORY MECHANICAL SYSTEMS

In solving the problems of dynamics of machines and equipment under the influence of vibration external influences, including in the tasks of vibration protection, coordinates of objects are usually used as parameters of the dynamic state, the dependence of which on the frequency is revealed in the frequency characteristics of the system. The transfer functions of mechanical vibration systems reflect the main properties of systems associated with the consideration of such dynamic effects as resonances, dynamic damping of vibrations, etc.

At the same time, the idea of the dynamic reactions of bonds arising at the points of connection of the elementary links of the system with each other, as well as at the points of contact with the supporting surfaces and the object of protection, is of great importance.

In the problems of vibration protection of technical objects within the framework of the structural theory of oscillatory systems, the reaction of bonds is interpreted as a negative feedback in the structural scheme of a dynamically equivalent automatic control system. In this case, the object of protection is interpreted in the form of a second-order integrating link, and the negative feedback, in the physical sense, corresponds to the idea of the reduced dynamic stiffness of the generalized spring (Eliseev & Artyunin, 2016).

To a lesser extent, the properties of bond reactions that occur at contact points or connections of elements of a mechanical system with each other are studied. Of particular importance in this regard is the assessment of the values of dynamic reactions at characteristic points of the system, which determine the reliability and safety of the system as a whole. In this sense, the value of the reactions of the links can be considered as parameters of the dynamic state, as well as coordinates, velocities and accelerations of the movement of the object of protection.

Peculiarities of formation of coupling reactions in linear mechanical oscillatory systems with two degrees of freedom under action of harmonic external perturbations are considered in ideas on using reactions as parameters of dynamic state of the system.

DESIGN SCHEME AND STRUCTURAL MODELS

A generalized calculation scheme for solving problems of object dynamics in systems with two degrees of freedom is shown in Figure 23. System comprises, in addition to concentrated masses m_1 and m_2, elastic elements with stiffness coefficients k_1, k_2, k_3. The system has two bearing surfaces I and II, which can perform harmonic movements z_1, z_2. The motion of the system is considered in a fixed basis using the coordinates y_1 and y_2. Harmonic forces $Q_{1,}$ Q_2 can be applied to mass inertia elements. Holding or two-way communication is assumed at the contact points (pp. *A*, *B*) (Lurie, 1968). The equation of motion of the system at kinematic disturbance ($Q_1 = 0$, $Q_2 = 0$) has the form

$$m_1 \cdot \ddot{y}_1 + y_1 \cdot (k_1 + k_2) - k_2 \cdot y_2 = k_1 \cdot z_1(t) + Q_1(t), \tag{55}$$

$$m_2 \cdot \ddot{y}_2 + y_2 \cdot (k_2 + k_3) - k_2 \cdot y_1 = k_3 \cdot z_2(t) + Q_2(t). \tag{56}$$

Figure 23. Generalized design scheme of a mechanical oscillatory system with two degrees of freedom: I, II - support surfaces; pp. A, B - contact points; Q_1, Q_2 - power disturbances; z_1, z_2 - kinematic disturbances

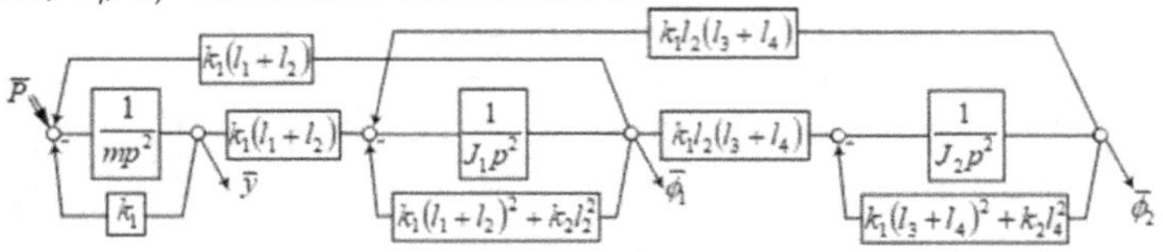

The corresponding structural diagrams in various forms of detailing representations of the possibilities of their transformations are shown in Figure 24, *a-d*. The construction of structural diagrams requires a Laplace transformation with subsequent construction of structural diagrams.

Using Laplace transforms, one can (55), (56) give the forms:

$$m_1 \cdot p^2 \cdot \bar{y}_1 + \bar{y}_1 \cdot (k_1 + k_2) - k_2 \cdot \bar{y}_2 = k_1 \cdot \bar{z}_1 + \overline{Q}_1(t), \tag{57}$$

$$m_2 \cdot p^2 \cdot \bar{y}_2 + \bar{y}_2 \cdot (k_2 + k_3) - k_2 \cdot \bar{y}_1 = k_3 \cdot \bar{z}_2 + \overline{Q}_2(t), \tag{58}$$

where $p = j\,\omega$ is a complex variable; $\bar{y}, \bar{z}, \overline{Q}$ are images *y(t), z(t)*, and *Q(t)* in the Laplace transform domain.

Figure 24. Variants representation of the structural scheme: a - structural general view corresponding to the mathematical model; b - transformed structural diagram with coordinate exclusion $\bar{y}_2$(at $z_2 = 0$, $Q_2 = 0$); c - transformed structural diagram with elimination of coordinates $\bar{y}_2$ and removal of k_2 in feedback (at $z_2 = 0$, $Q_2 = 0$); d - transformed structural diagram with formation of transfer function in the form of fractional rational expression within partial $m_1 p^2+k_1$ system (at $z_2 = 0$, $Q_2 = 0$, $Q_1 = 0$); e - transformed structural diagram with selection of protection object m_1 in the form of integrating link of the second order (at $z_2 = 0$, $Q_2 = 0$, $Q_1 = 0$)

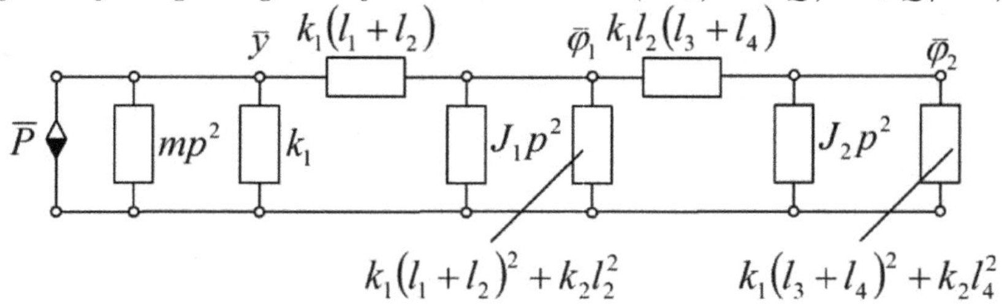

Figure 24 shows the possibilities of transforming structural diagrams with the allocation of the necessary elements or blocks of the general structural diagram (Figure 24, *a*). Determination of transfer functions for various types and combinations of external influences is possible on the basis of the superposition principle. Conventional construction techniques are associated as rules with the choice of a situation when the system defines one input and one output. For linear systems with several input effects of the same frequency, transformations with the construction of equivalent input effects are possible, which is considered in particular in (Eliseev & Artyunin, 2016).

The objective of the study is to develop a method for constructing bond reactions at characteristic points of a mechanical oscillatory system and to assess the possibilities of using bond reactions as parameters of the dynamic state of the protected object.

PECULIARITIES OF CONVERSION OF DESIGN AND STRUCTURAL DIAGRAMS OF VIBRATION SYSTEM

Let's consider the variants of transformation of design schemes, choosing different ratios of system parameters, possible changes in the position of support surfaces as shown in Figure 25, *a* - *h*.

Figure 25. Design scheme options: a - design scheme with separated support surfaces; b - design diagram with combined support surfaces; c - there are no points B and B_1 ($k_3 = 0$) in the design diagram; d - design diagram with separated support surfaces at $k_2 = 0$ and $k_3 = 0$; e - design scheme with combined support surfaces at k_2 ∞ ($y_1 = y_2$); f - design diagram with separated support surfaces at k_2 ∞ ($y_1 = y_2$); g - design diagram with separated support surfaces during k_3 ∞; h - design scheme with separated support surfaces at m_2 ∞ and $k_3 \neq 0$ or $k_3 = 0$

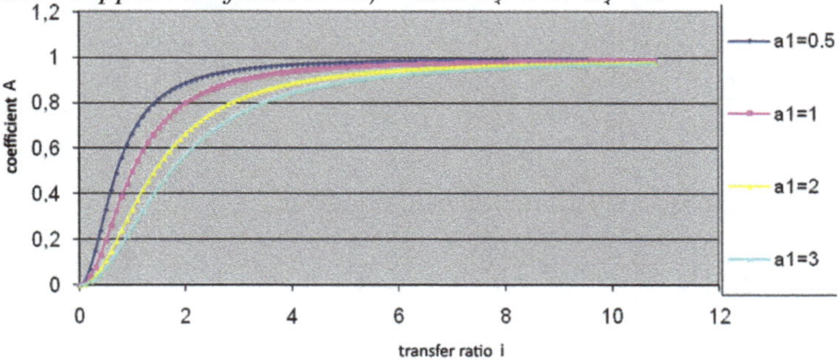

The design diagram in Figure 25, *a* corresponds to the idea of a support surface consisting of two parts, which implies the action of various kinematic disturbances $z_1(t)$, $z_2(t)$ on the object of protection m_1.

On the side of the support surface I, the bond reaction is formed by an elastic member with rigidity k_1; from the side *of* surface II, the reaction of bonds acts on the object of protection m_1, which at the point *of* A_2 is formed by a mechanical chain of series-connected elements to k_3, m_2 and k_2. In Figure 25, *b* the support surfaces *I* and *II* are combined, which corresponds to the support of the protected object on one "combined vibration isolator", which determines the overall reaction of the bonds to one support surface. At $k_3 = 0$, as shown in Figure 25, in the vibration protection system at $z_1 \neq 0$, ($Q_1 = 0$, $Q_2 = 0$), the dynamic vibration damping mode may appear on the protection object m1 in the kinematic form of external disturbance.

At $k_2 = 0$ and $k_3 = 0$, the original design scheme (Figure 25, *a*) is transformed into a system with one degree of freedom (Figure 25, *d*). When $k_2 \to \infty$, the shape of movement of two masses m_1 and m_2 with a tendency *of* $y_1 \to y_2$ (or $y_1 = y_2$) is formed, and the system as a whole in the low-frequency range is determined by the parameters of movement of the system with one degree of freedom ($m_1 + m_2$), ($k_1 + k_3$).

In Figure 25,*e* it is assumed that the movement at $m_1 + m_2$ can occur with two disturbing factors z_1 and z_2 ($Q_1 = 0$, $Q_2 = 0$). When $k_3 \to \infty$, the m_2 element becomes part of the support surface *II* (Figure 25, *g*) and the system acquires one degree of freedom.

A similar result can be obtained with $k_3 \neq 0$ and $m_2 \to \infty$ (Figure 25, *h*). If $k_3 = 0$, then in $m_2 \to \infty$ case the system will also have one degree of freedom in the vibrational form of motion.

We use the calculation diagram in Figure 23 and the block diagram in Figure 24, *a*. In this case, the transfer functions are defined by:

$$W_1(p) = \frac{\bar{y}_1}{\bar{z}_1} = \frac{k_1 \cdot (m_2 \cdot p^2 + k_2 + k_3)}{A_0} \tag{59}$$

$$W_2(p) = \frac{\bar{y}_2}{\bar{z}_1} = \frac{k_1 \cdot k_2}{A_0} \tag{60}$$

where A_0 is the frequency characteristic equation

$$A_0 = (m_1 \cdot p^2 + k_1 + k_2)(m_2 \cdot p^2 + k_2 + k_3) - k_2^2. \tag{61}$$

Kinematic disturbance $\bar{z}_1$ in this case (linear system) is replaced by equivalent force action $\bar{Q}_{eqv.1}$

$$\bar{Q}_{eqv.1} = k_1 \cdot \bar{z}_1. \tag{62}$$

then

$$W_3(p) = \frac{\bar{y}_1}{\bar{Q}_{eqv.1}} = \frac{m_2 \cdot p^2 + k_2 + k_3}{A_0} \tag{63}$$

$$W_4(p) = \frac{\bar{y}_2}{\bar{Q}_{eqv.1}} = \frac{k_2}{A_0}. \tag{64}$$

Obviously, $W_1(p) = k_1 \cdot W_3(p)$ and $W_2(p) = k_1 \cdot W_4(p)$.

In (Eliseev & Artyunin, 2016) discusses the features of constructing transfer functions in linear systems with several inputs, while there are functional connections between the input signals.

TO DEFINE DYNAMIC LINK REACTIONS AT FEATURE POINTS ON AN OBJECT

The calculation scheme in Figure 23, *a* or Figure 25 is considered, and these schemes are equivalent. All ratios are recorded in the operator form (y_1, y_2).

The reaction at point *A* is determined by the formula:

$$\overline{R}_A = k_1 \cdot \overline{y}_1 = \frac{k_1^2 \cdot \overline{z}_1 \cdot (m_2 \cdot p^2 + k_2 + k_3)}{A_0}, \tag{65}$$

or

$$\overline{R}_A = \frac{k_1 \cdot (m_2 \cdot p^2 + k_2 + k_3)}{A_0} \cdot \overline{Q}_{eqv.1}. \tag{66}$$

Let's introduce the concepts of the transfer function for the coupling reaction. An input signal is a kinematic disturbance $\overline{z}_1$ or equivalent force action $\overline{Q}_{eqv.1} = k_1 \cdot \overline{z}_1$. We get two transfer functions:

$$W'_{\overline{R}_A}(p) = \frac{\overline{R}_A}{\overline{z}_1} = \frac{k_1^2 \cdot (m_2 \cdot p^2 + k_2 + k_3)}{A_0} \tag{67}$$

$$W''_{\overline{R}_A}(p) = \frac{\overline{R}_A}{\overline{Q}_{eqv.1}} = \frac{k_1 \cdot (m_2 \cdot p^2 + k_2 + k_3)}{A_0}. \tag{68}$$

From (68), in particular, it follows that the ratio of the amplitude of oscillations of the dynamic reaction to the equivalent force $\overline{Q}_{eqv.1}$ formed by the kinematic disturbance has the form of a transfer function of the system when the input signal is considered to be a kinematic disturbance $\overline{z}_1$, and the output signal is accordingly considered to be a coordinate shift $\overline{y}_1$. This is easily verified by the block diagrams in Figure 25.

The reaction at point A_1 is formed as a result of deformation of a linear elastic element (spring) with stiffness k_1. In this case, it can be assumed that the fair statement of action is equal to counteraction

$$\overline{R}_A = -\overline{R}_{A_1}, \tag{69}$$

that is, the reactions are equal in magnitude and directed in opposite directions (in accordance with Newton's third law). However, if a mass inertia element is also placed between points A and A_1 or the mass inertia properties of the spring itself are taken into account, then such a condition will no longer be observed. Refer to the calculation diagram in Figure 25, *a*. Note that the mass inertia element m*1* supports as follows.

One leg of the support is an elastic element k_1 with attachment points A and A_1, respectively, to the support surface *I* and the mass m_1. The second branch is formed from series connection of elastic elements k_2, mass-inertial element m_2 and elastic element with rigidity $k_{3.}$ Accordingly, the branch has points of contact at point *B* with support surface *II*, and at point A_2 with mass m_1. In this case:

$$\overline{R}_{A2} \neq \overline{R}_B. \tag{70}$$

Reaction at point *B* can be found

$$\overline{R}_B = k_3 \cdot \overline{y}_2 = \frac{k_1 \cdot k_2 \cdot k_3 \cdot \overline{z}_1}{A_0}. \tag{71}$$

We get two transfer functions:

$$W'_{\overline{R}_B}(p) = \frac{\overline{R}_B}{\overline{Z}_1} = \frac{k_1 \cdot k_2 \cdot k_3}{A_0} \tag{72}$$

$$W''_{\overline{R}_B}(p) = \frac{\overline{R}_B}{\overline{Q}_{eqv.1}} = \frac{k_2 \cdot k_3}{A_0}. \tag{73}$$

The reaction at point B_1 where spring k_3 contacts mass inertia element m_2 will be determined:

$$\overline{R}_{B_1} = -\overline{R}_B. \tag{74}$$

To find the reaction at a point A_2 you must find the reduced stiffness at the A_2 point. It is at this point that the second branch (k_3, k_2, m_2) has contact with the element m_1.

The reduced stiffness will be:

$$k_{A_2PR} = \frac{k_2 \cdot (m_2 \cdot p^2 + k_3)}{m_2 \cdot p^2 + k_2 + k_3}. \tag{75}$$

The reduced stiffness reflects the dynamic stiffness of the series chain consisting of k_3, m_2p^2 and k_2. The peculiarities of this approach are presented in the monographs (Eliseev & Artyunin, 2016; Eliseev, 2018). A similar result can be obtained using the structural transformations of the original structure diagram in Figure 24, *a*. In this case, in the transformed structural diagram in Figure 24, *c* the feedback circuit with respect to the protection object with a transfer function $\frac{1}{m_1 \cdot p^2 + k_1}$, it has a transfer function, respectively.

$$W_{oc}(p) = k_{A_2PR} = \frac{k_2 \cdot (m_2 \cdot p^2 + k_3)}{m_2 \cdot p^2 + k_2 + k_3}. \tag{76}$$

The expression thus obtained is exactly the same as (73) for the reduced stiffness.

Knowing the reduced stiffness, k_{A_2PR} we find the dynamic reaction at point A_2

$$\overline{R}_{A_2} = k_{A_2PR} \cdot \overline{y}_1 = \frac{k_1 \cdot k_2 \cdot (m_2 \cdot p^2 + k_3) \cdot \overline{z}_1}{A_0}. \tag{77}$$

We introduce the transfer function for the dynamic reaction at point A_2

$$W'_{\overline{R}_{A_2}}(p) = \frac{\overline{R}_{A_2}}{\overline{z}_1} = \frac{k_1 \cdot k_2 \cdot (m_2 \cdot p^2 + k_3)}{A_0}, \tag{78}$$

$$W''_{\overline{R}_{A_2}}(p) = \frac{\overline{R}_{A_2}}{\overline{Q}_{экв.1}} = \frac{k_2 \cdot (m_2 \cdot p^2 + k_3)}{A_0}. \tag{79}$$

The dynamic reaction at the object of protection, that is, in the mass-inertial element m_1, is determined by the summation of reactions at points A_1 and A_2. Thus

$$\overline{R}_{m_1}(p) = \overline{R}_{A_1} + \overline{R}_{A_2} = \frac{k_1 \cdot \overline{z}_1 \cdot [k_1 \cdot (m_2 \cdot p^2 + k_2 + k_3) + k_2 \cdot (m_2 \cdot p^2 + k_3)]}{A_0}, \tag{80}$$

or

$$\overline{R}_{m_1}(p) = \frac{k_1 \cdot (k_1 + k_2) \cdot m_2 \cdot p^2 + k_1^2 \cdot (k_2 + k_3) + k_1 \cdot k_2 \cdot k_3}{A_0} \cdot \overline{z}_1. \tag{80'}$$

Accordingly, we obtain the transfer functions for the mass-inertial element m_1

$$W'_{\overline{R}_{m_1}}(p) = \frac{\overline{R}_{m_1}(p)}{\overline{z}_1} = \frac{k_1 \cdot (k_1 + k_2) \cdot m_2 \cdot p^2 + k_1^2 \cdot (k_2 + k_3) + k_1 \cdot k_2 \cdot k_3}{A_0}, \tag{81}$$

$$W''_{\overline{R}_{m_1}}(p) = \frac{\overline{R}_{m_1}(p)}{\overline{Q}_{экв.1}} = \frac{(k_1 + k_2) \cdot m_2 \cdot p^2 + k_1 \cdot (k_2 + k_3) + k_2 \cdot k_3}{A_0}. \tag{82}$$

Similarly, you can find the dynamic reaction that occurs on an element with a mass of m_2.

In this case, we can imagine that the mass m_2 reflects into two elastic branches at points B_1 and B_2. The dynamic response at B_1 is given by (75). In turn, the second branch is formed by a mechanical chain of series-connected elements k_1, $m_1 p^2$ and k_2, which allows, in accordance with the rules for converting mechanical chains, to find the reduced stiffness at the point B_2

$$k_{B_2PR} = \frac{k_2 \cdot (m_1 \cdot p^2 + k_1)}{m_1 \cdot p^2 + k_1 + k_2}. \tag{83}$$

From (83) the dynamic response at B_2 is determined

$$\bar{R}_{B_2} = k_{B_2PR} \cdot \bar{y}_2 = \frac{k_1 \cdot k_2^2 \cdot (m_1 \cdot p^2 + k_1) \cdot \bar{z}_1}{(m_1 \cdot p^2 + k_1 + k_2) \cdot A_0}. \tag{84}$$

Transfer functions of dynamic links at the B_2 point take the form:

$$W'_{\bar{R}_{B_2}} = \frac{\bar{R}_{B_2}}{\bar{z}_1} = \frac{k_1 \cdot k_2^2 \cdot (m_1 \cdot p^2 + k_1)}{(m_1 \cdot p^2 + k_1 + k_2) \cdot A_0}, \tag{85}$$

$$W''_{\bar{R}_{B_2}} = \frac{\bar{R}_{B_2}}{\bar{Q}_{eqv.1}} = \frac{k_2^2 \cdot (m_1 \cdot p^2 + k_1)}{(m_1 \cdot p^2 + k_1 + k_2) \cdot A_0}. \tag{86}$$

The total dynamic reaction formed on the mass inertial element m_2 will be determined as the sum of two dynamic reactions $\bar{R}_{B_1}$ and $\bar{R}_{B_2}$. Thus, we finally write that

$$\bar{R}_{m_2}(p) = \bar{R}_{B_1} + \bar{R}_{B_2} = \frac{k_1 \cdot k_2 \cdot \bar{z}_1 \cdot [(k_2 + k_3) \cdot m_1 \cdot p^2 + k_3 \cdot (k_1 + k_2) + k_1 \cdot k_2]}{(m_1 \cdot p^2 + k_1 + k_2) \cdot A_0}. \tag{87}$$

Transfer functions of dynamic links for mass-inertial element m_2 will take the form:

$$W'_{\bar{R}_{m_2}} = \frac{\bar{R}_{m_2}}{\bar{z}_1} = \frac{k_1 \cdot k_2 \cdot [(k_2 + k_3) \cdot m_1 \cdot p^2 + k_3 \cdot (k_1 + k_2) + k_1 \cdot k_2]}{(m_1 \cdot p^2 + k_1 + k_2) \cdot A_0}, \tag{88}$$

$$W''_{\bar{R}_{m_2}} = \frac{\bar{R}_{m_2}}{\bar{Q}_{eqv.1}} = \frac{k_2 \cdot [(k_2 + k_3) \cdot m_1 \cdot p^2 + k_3 \cdot (k_1 + k_2) + k_1 \cdot k_2]}{(m_1 \cdot p^2 + k_1 + k_2) \cdot A_0}. \tag{89}$$

Preliminary analysis (89) allows you to establish the properties of the amplitude-frequency response (frequency response), which are determined by the fact that:

1. dynamic reaction takes extreme values $\bar{R}_{m_2}$ three times (twice at natural oscillation frequencies or at resonances, as well as at partial frequency $\omega^2 = \frac{k_1 + k_2}{m_1}$);
2. at the oscillation frequency determined from the frequency equation of the numerator (89), a dynamic mode is formed, which could be called "zeroing of the dynamic reaction"; wherein the frequency is given by

$$\omega^2 = \frac{k_1 \cdot k_2 + k_3 \cdot (k_1 + k_2)}{m_1 \cdot (k_2 + k_3)}. \tag{90}$$

Thus, dynamic reactions of bonds can serve as parameters of the state of the mechanical oscillatory system as well as known forms of evaluation based on the use of kinematic parameters. The dynamic reaction of the bonds at the selected point of the system is defined in the operator form as the product of the displacement by the reduced dynamic stiffness and carries information about the features of resonance modes and dynamic damping of vibrations.

Structural transformation techniques are proposed to obtain dynamic reactions, which are based on the use of parameters of feedback chains formed in relation to selected mass-inertial elements; characteristically, there is a maximum coupling response effect, which in the physical sense is interpreted as an increase in the reduced dynamic stiffness at the frequency corresponding to the dynamic damping mode.

Structural mathematical modeling is a fairly developed area of research of mechanical systems, including mechanical oscillatory systems for various purposes based on the use of mathematical models, which can be built as some analogue systems in relation to the original systems, allowing the use of the principles of dynamic analogies.

The most developed technologies for representing electrical circuits, consisting of a well-defined set of typical elementary links, implement the basic laws of electromechanics, in particular, Kirchhoff's law.

The theory of electrical circuits has a fairly developed technology for constructing analog circuits or structural schemes that are built from a limited set of typical elements using certain rules for connecting elements and forming structures that reflect with sufficient accuracy the dynamic state of a real system.

In the theory of electrical circuits, several analogy systems with mechanical oscillatory systems are used. Details and features of mechanical circuits as structural mathematical models of mechanical oscillatory systems are considered in the works (Gardner & Barnes, 1949; Druzhinsky, 1977).

If inductance, capacitance and resistance are among the typical elements of electrical circuits, then in mechanical oscillatory systems the typical elements are mass-inertia links, springs and devices for energy dissipation (dampers). The relationships between the systems of parameters describing the dynamic states of the object under consideration are based on the comparison of the characteristics of electrical circuits (currents, voltage) with the corresponding characteristics of mechanical oscillatory systems, reflected by the values of movements, speeds of movement and arising force interactions.

There is a fairly clear comparability between the structural diagrams of electrical circuits and mechanical circuits in the forms of placement and connection of elements. The mathematical apparatus used, based on Laplace integral transformations, expands the boundaries of the generality of ideas about the specifics of physical processes.

In theoretical electrical engineering, this is Kirchhoff's law, in mechanical oscillatory systems, the Dalembert principle and other forms of manifestation of patterns of dynamic interactions of elements.

A characteristic feature of the analytical apparatus of circuit theory and its applications in assessing the dynamic states of electrical and mechanical circuits is that the main attention is paid to the distribution of parameters characterizing the dynamic state along the contours of the original systems, which is generally focused on the specifics of physical processes occurring in electrical circuits, electrical circuits, electrical devices, as well as in mechanical analogues of structural formations. In this regard, the theory of automatic control has, in comparison with the theory of chains, a certain feature, which consists in the introduction and development of ideas about the dynamic state of a certain object.

The technical implementation of such approaches is technological and transport machines with automatic control systems. Such systems contain a well-defined object, the dynamic state of which is maintained within certain boundaries. Automatic control systems include systems for measuring, evaluating and monitoring parameters of signal processing devices, as well as servomechanisms for various nature, which make it possible to form the desired dynamic states of the object.

The solution of such problems of machine dynamics is associated with the allocation of an object, the dynamic state of which is assessed, maintained within certain boundaries, adjusted and adjusted if necessary.

A typical example of such tasks is vibration protection of vibration isolation of equipment, apparatus, units and the environment from vibration effects.

Either mass inertia element with point mass or element in the form of solid body of limited dimensions with two or more degrees of freedom is selected as object of protection in structure of mechanical oscillatory systems with several degrees of freedom.

The dynamic state of technical objects, the physical model of which is considered in the form of a linear mechanical oscillatory system with concentrated parameters. The usual procedure for compiling a structural mathematical model involves the preliminary construction of a mathematical model based on the use of Lagrange equations of the 2nd kind. The resulting system of ordinary linear differential equations of the 2nd order undergoes the Laplace integral transformation procedure under zero initial conditions. In this case, the initial system of linear differential equations is transformed into a system of algebraic equations and is considered as a mathematical model in operator form.

The authors propose a detailed technology for converting equations in operator form to their graphical analogue, which is interpreted as a structural diagram of a dynamically equivalent automatic control system.

In essence, the task of the dynamics of a technical object, which has a set of elastic, dissipative and mass-inertial elements, is reduced to the task of forming the dynamic state of the protected object under the action of external disturbances, considered as input signals.

The structural diagram reveals dynamic connections between elements of the initial system, allows taking into account feedback connections between elements that determine the possibilities of oscillatory movements, as well as transforming structural diagrams based on the rules of the theory of automatic control.

Structural transformations in the tasks of implementing methods of structural mathematical modeling play a significant role, allowing to form appropriate models that provide rational approaches to solving the problems of analysis and dynamic synthesis of vibration systems, active vibration protection systems and other tasks in which the parameters of the dynamic state are formed according to certain conditions. A characteristic feature of the methods of structural mathematical modeling based on the analytical apparatus of the theory of automatic control is that the protection object is distinguished in the system, relative to which a system of dynamic connections is formed, realized through the connection of standard elementary links that have the transfer functions of the integrating link of the second order - for the protection object, the amplifying link - for elastic element, 1st order differentiating link - for dissipative element, 2nd order differentiating link for motion conversion device. The dynamic properties of all elementary links were discussed in detail in the previous sections.

The developed form of the method of structural mathematical modeling using structural schemes equivalent in dynamic terms, automatic control systems are unambiguously transformed into structural schemes used in the theory of mechanical circuits.

This paper discusses technical objects that make small vibrations relative to the position of static equilibrium or steady-state processes. It is assumed that the constituent elements of the mechanical oscillatory system as a physical model of a technical object contain only passive links.

Assessment of dynamic properties of active vibration protection systems is not considered in this monograph by the authors. Although in solving problems of analysis and dynamic synthesis, the basis for assessing dynamic states is mechanical oscillatory systems with elements of the passive type.

Structural mathematical modeling based on the use of the apparatus of the theory of automatic control allows us to introduce a number of new dynamic effects, features of the formation of dynamic states, taking into account the compatibility of the action of external forces, the implementation of dynamic damping modes, assessing the connectivity of movement and determining the location of a number of characteristic points, which include vibration nodes, centers of stiffness, etc.

REFERENCES

Bolshakov, R. S., & Eliseev, S. V. (2013). Methodological foundations of tasks for determining the reactions of links in mechanical oscillatory systems. Irkutsk: IrGUPS. Deposition manuscript in VINITI 29.11.2013 N°366.

Druzhinsky, I. A. (1977). *Mechanical circuits*. Mashinostroenie.

Eliseev, S. V. (2018). Applied system analysis and structural mathematical modeling (dynamics of transport and technological machines: movement connectivity, vibration interactions, lever connections). Irkutsk: IrGUPS.

Eliseev, S. V., & Artyunin, A. I. (2016). *Applied theory of oscillations in the dynamics of linear mechanical systems*. Nauka.

Eliseev, S. V., Khomenko, A. P., & Upyr, R. Yu. (2009). Mechatronics of vibration protection systems with lever connections. Contemporary Technologies. System Analysis. *Modeling*, *3*, 104–119.

Gardner, F., & Barnes, J. L. (1949). *Transient processes in linear systems with concentrated constants*. GITTL.

Lurie, A. I. (1968). *Analytical mechanics*. Nauka.

Tarasik, V. P. (2004). Mathematical modeling of technical systems: A textbook for technical universities (2nd ed., revised and expanded). Minsk: Design PRO.

ADDITIONAL READING

Chu, S. Y., & Song, Y. (2020). Study on dynamic interaction of railway pantograph-catenary including reattachment momentum impact. *Vibration*, *3*(1), 18–33. DOI: 10.3390/vibration3010003

Churinova, O. V., Zarubina, V. A., Eliseev, A. V., & Karlina, A. I. (2012). Modeling of mechanical systems with non-holding bonds. In Information and Mathematical Technologies in Science and Management (pp. 182-186).

Dumitriu, M., & Apostol, I. (2022). Influence of interference between vertical and roll vibrations on the dynamic behaviour of the railway bogie. *Vibration*, *5*(4), 659–675. DOI: 10.3390/vibration5040039

Eliseev, A. V. (2017). Diads in Mechanical Systems: Features of Dynamic Properties. Part I. *Vestnik Irkutskogo Gosudarstvennogo Tehniceskogo Universiteta*, *21*(7), 26–38. DOI: 10.21285/1814-3520-2017-7-26-38

Eliseev, A.V., Kuznetsov, N.K. (2022). Technology of Structural Mathematical Modeling of Technical Objects under Vibrational Loads: Forms of Interactions and Dynamic Invariants. iPolytech Journal, 26(3), 368-385.

Eliseev, A. V., & Kuznetsov, N. K. (2023). Structural Mathematical Modeling in Evaluation of Dynamic States of Mechanical Oscillatory Systems Based on Lever Representations. In Mathematics, Its Applications, and Mathematical Education (MPMO23). Proceedings of the VIII International Conference (pp. 90-92).

Eliseev, A. V., Kuznetsov, N. K., & Eliseev, S. V. (2021). New Approaches in the Assessment of Dynamic Properties of Oscillatory Structures: Frequency Functions and Movement Connectivity. MAI Proceedings, 120.

Eliseev, A. V., Kuznetsov, N. K., & Mironov, A. S. (2023). Maps of Dynamic Invariants in the Assessment of Modes of Motion of Mechanical Oscillatory Systems. MAI Proceedings, 128.

Eliseev, A. V., Kuznetsov, N. K., & Nikolaev, A. V. (2022). Concept of Dynamic Invariants in Evaluation of Structural Features of Mechanical Oscillatory Systems. Transport. *Mining and Construction Engineering: Science and Production*, *15*, 18–30.

Eliseev, A. V., & Mironov, A. S. (2022). Methodological approaches to evaluating the totality of dynamic states and interaction forms of elements in vibration machines for transport and technological purposes. *Journal of Advanced Research in Technical Science*, *28*, 22–25.

Eliseev, A. V., & Mironov, A. S. (2022). New approaches to evaluating dynamic damping modes of vibration machines for transport and technological purposes: Graphs of dynamic states and interaction forms of elements. *Mechatronics, Automation, and Robotics*, *9*, 23–29.

Eliseev, A. V., Pnev, A. G., Kashuba, V. B., & Sitov, I. S. (2014). Vibrational technological processes: vibrohardening in impact periodic interactions. In Kulaginsky Readings: Technology and Technology of Production Processes. XIV International Scientific and Practical Conference: in 3 volumes (pp. 281-286). Transbaikal State University.

Eliseev, A. V., Sitov, I. S., & Kuznetsov, N. K. (2022). Systemic approaches to assessing the dynamic states of technical objects under vibrational loads: Frequency functions, dynamic invariants, regularization methods. Systems. Methods. *Technologies*, *4*(56), 7–15.

Eliseev, A. V., Sitov, I. S., & Kuznetsov, N. K. (2023). Systemic approach to assessing the complete set of dynamic states of technical objects under connected vibrational loads. Systems. Methods. *Technologies*, *1*(57), 7–20.

Eliseev, S.V., Artyunin, A.I., Kaimov, E.V., Eliseev, A.V. (2014). Device for regulating the elastic-dissipative properties of the vibration protection system. Patent for utility model RU 142137 U1. Application No. 2014100299/11 from January 9, 2014.

Eliseev, S.V., Artyunin, A.I., Khomenko, A.P., Kaimov, E.V., Eliseev, A.V. (2013). Device for oscillation damping. Patent for utility model RU 136112 U1. Application No. 2013135078/11 from July 25, 2013.

Eliseev, S. V., & Eliseev, A. V. (2019). *Theory of Oscillations: Structural Mathematical Modeling in Problems of Dynamics of Technical Objects*. Springer.

Eliseev, S. V., & Eliseev, A. V. (2020). Some Applications of the Methods Structural Mathematical Modeling. Studies in Systems. *Decision and Control*, *252*, 433–521.

Eliseev, S. V., Eliseev, A. V., & Kashuba, V. B. (2013). Particularities of gap assessment in a model problem of material particle tossing with non-holding bonds. *Mechanical Engineering and Life Safety*, *1*(15), 50–56.

Eliseev, S. V., Kargapol'tsev, S. K., Bol'shakov, R. S., & Eliseev, A. V. (2020). Lever bonds: The possibilities of forming dynamic states in mechanical oscillatory systems. *Transport of the Urals*, *3*(66), 17–23. DOI: 10.20291/1815-9400-2020-3-17-23

Eliseev, S. V., Sitov, I. S., & Eliseev, A. V. (2012). Motion of a material particle with tossing in a model problem with non-holding bonds. *Mechanical Engineering and Life Safety*, *3*(13), 53–58.

Eliseev, S. V., Sitov, I. S., & Eliseev, A. V. (2012). Characteristics of the interaction of a material particle and an oscillation surface depending on the constant force with the consideration of non-holding bond. In Technology and technology: new development perspectives (pp. 20-23).

Moueddeb, M., Louf, F., Boucard, P.-A., Dadié, F., Saussine, G., & Sorrentino, D. (2022). An efficient numerical model to predict the mechanical response of a railway track in the low-frequency range. *Vibration*, *5*(2), 326–343. DOI: 10.3390/vibration5020019

Sehner, M., Nava, L., Seidl-Nigsch, M., & Loy, H. (2023). Vibration mitigation: Under-ballast mats in heavy-haul applications. *Practice Periodical on Structural Design and Construction*, *28*(4), 05023004. DOI: 10.1061/PPSCFX.SCENG-1258

Yoshino, S., Abe, K., & Koro, K. (2023). An analytic solution of mathematical expectation for bogie-track interaction problems. *Mechanical Engineering Journal*, *10*(3), 22-00300. Advance online publication. DOI: 10.1299/mej.22-00300

Zhao, Z., Gao, Y., & Li, C. (2020). Research on the vibration characteristics of a track's structure considering the viscoelastic properties of recycled composite sleepers. *Applied Sciences (Basel, Switzerland)*, *11*(1), 150. DOI: 10.3390/app11010150

Chapter 6
Dyads as Mass Structures

ABSTRACT

A methodological basis is being developed for building mathematical models of dyads - autonomous structural formations from two mass-inertial elements. A feature of the dyad is the possibility of elastic interaction of elements and their isolated movements.Possibilities of estimation of dynamic properties of dyad on mathematical models in two coordinate systems are investigated. method is proposed to determine parameters of dyad motion based on introduction of transfer function and frequency of energy function. Using different coordinate systems allows you to identify new dynamic effects triggered by new dynamic links. These relationships form inertial and inertial-elastic type dependencies. The concept of dyads and the features of their interactions can become the basis for new methods of analysis and dynamic synthesis in the theory of mechanical oscillations and nanomechanics.

INTRODUCTION: DYADS IN MECHANICAL OSCILLATORY SYSTEMS—DYNAMIC INTERACTION GENERATION MODEL

Ensuring the reliability and safety of operation of technical and transport machines operating under conditions of intensive dynamic loading requires the development of mathematical modeling methods, assessment of dynamic properties and forms of manifestation of dynamic effects at all stages of the design life cycle, creation of operation of technical objects. Significant attention is paid to the dynamics of machines, equipment and apparatus, many tasks of dynamics, in particular, the tasks of vibration protection of machines, are reflected in the works (Frolov & Furman, 1980; Eliseev, Reznik, Khomenko, & Zasyadko, 2008; Eliseev & Artyunin, 2016). Mathematical modeling methods (Clarence & de Silva, 2000; Tarasik, 2004) are widely used, based on the use of electromechanical and dynamic analogies, in which the calculation schemes of technical objects are compared with equivalent electrical

DOI: 10.4018/979-8-3373-0447-2.ch006

and mechanical circuits in dynamic terms, as well as structural diagrams of automatic control systems (Eliseev, Reznik, & Khomenko, 2011; Druzhinsky, 1977).

In the proposed report, the method of structural mathematical modeling is being developed in an application to the assessment of the dynamic properties of a structural formation in the form of a dyad, which is present in almost all chain oscillatory systems with two or more degrees of freedom with concentrated parameters.

Modeling a Dyads As a Mechanical System. Some General Provisions

1. A dyad is a structural formation consisting of two mass-inertial elements connected to each other by a spring. Dyad elements move in one line. The dyad can be considered as a mechanical circuit consisting of three series-connected elements (Figure 1).

Figure 1. Schematic diagram of a conventional dyad

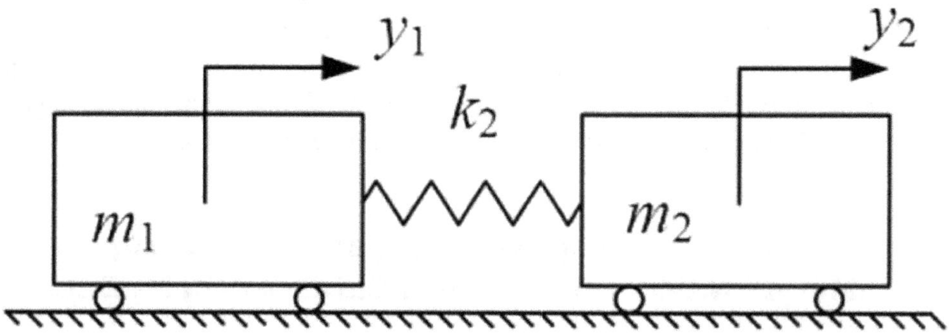

As part of the structural mathematical modeling method (Eliseev, Reznik, Khomenko, & Zasyadko, 2008; Eliseev & Artyunin, 2016; Eliseev, Reznik, & Khomenko, 2011), typical elements m_1, m_2 and k_2 can be presented in operator form; Note here that elements m_1 has transfer function $W_{m_1}(p) = m_1 p^2$,while element m_2 has $W_{m_2}(p) = m_2 p^2$;stiffness k_2. Figure 2 shows a schematic diagram of a mechanical circuit consisting of three series-connected elements, which can be called a mechanical circuit (Druzhinsky, 1977; Eliseev & Kargapoltsev, 2014).

Figure 2. Mechanical circuit diagram in operator form

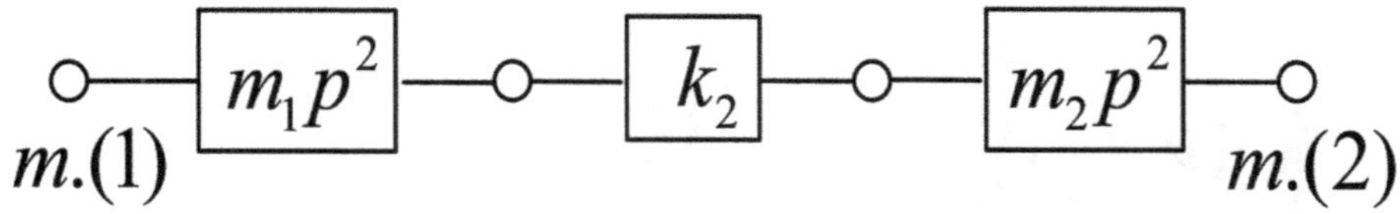

The mechanical circuit has an input of p. (1) and an output of p. (2). Using the rule of joining three elements connected in series, we obtain some structural formation that has a transfer function

$$W_0(p) = \frac{(m_1p^2)(m_2p^2)k_2}{m_1p^2k_2 + m_2p^2k_2 + m_1m_2p^4} = \frac{m_1m_2k_2p^2}{(m_1 + m_2)k_2 + m_1m_2p^2}. \quad (1)$$

The transfer function (1) has the form of a fractional rational expression. In the physical sense (1) and displays the dynamic stiffness of the quasi-spring, which depends on p^2 ($p = j\omega$ - complex variable (Eliseev, Reznik, Khomenko, & Zasyadko, 2008; Eliseev & Artyunin, 2016; Druzhinsky, 1977). Dynamic stiffness has the property that its value becomes infinitely large at frequency:

$$\omega^2_{\text{dyn}} = \frac{k_2(m_1 + m_2)}{m_1m_2} = k_2m_{\text{pr}}, \quad (2)$$

where

$$m_{\text{pr}} = \frac{m_1 + m_2}{m_1m_2} \quad (3)$$

is the reduced mass.

A dyad, thus, as a mechanical chain, can be considered as a quasi-spring having a dynamic stiffness defined by expression (1). The amplitude-frequency response of the dyad is given in Figure 3, which implies that at $\omega = 0$, $|W_0(p)| = 0$; when $\omega \to \infty$, $|W_0(p)| \to k_{2,}\omega^2 = \omega_0^2, |W_0(p)| \to \infty$.

Figure 3. Amplitude-frequency response of dyad

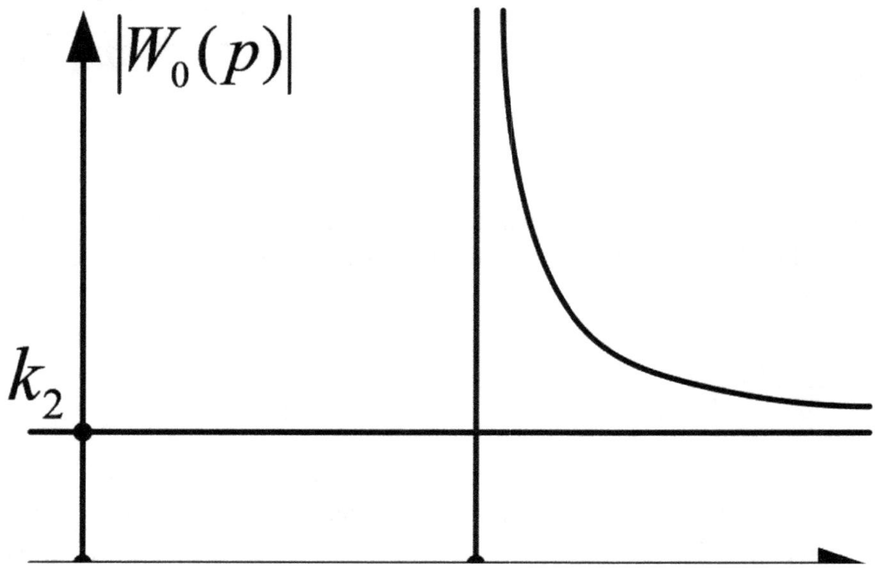

2. The structural formation (dyad), represented as a mechanical circuit in Figure 1 and Figure 2, can also be considered as a mechanical system, the position of which is determined by the coordinates y_1 and y_2 of the mass inertia elements m_1 and m_2, respectively. In this case, the kinetic and potential energies of the system will be determined by the expressions

$$T = \frac{1}{2}m_1\dot{y}_1^2 + \frac{1}{2}m_2\dot{y}_2^2, \tag{4}$$

$$\Pi = \frac{1}{2}k_2(y_2 - y_1)^2. \tag{5}$$

The equations of motion of a mechanical oscillatory system with two degrees of freedom can be constructed based on the application of Lagrange equations of the second kind (Eliseev, Reznik, Khomenko, & Zasyadko, 2008), we obtain that

$$m_1\ddot{y}_1 + k_2y_1 - k_2y_2 = 0, \tag{6}$$

$$m_2\ddot{y}_2 + k_2y_2 - k_2y_1 = 0. \tag{7}$$

Using Laplace transforms under zero initial conditions, equations (6), (7) can be transformed to the operator form

$$\bar{y}_1(m_1p^2 + k_2) - k_2\bar{y}_2 = 0, \tag{8}$$

$$\bar{y}_2(m_2p^2 + k_2) - k_2\bar{y}_1 = 0. \tag{9}$$

where $p = j\,\omega$ is a complex variable, the icon <-> above the variable means a Laplacian image. The structural mathematical model of the dyad based on equation (8), (9) can be presented in the form of a structural diagram of a dynamically equivalent automatic control system (Eliseev & Artyunin, 2016; Eliseev, Reznik, & Khomenko, 2011), as shown in Figure 4.

Figure 4. Structural mathematical model (structural diagram) of dyads as per Figure 1

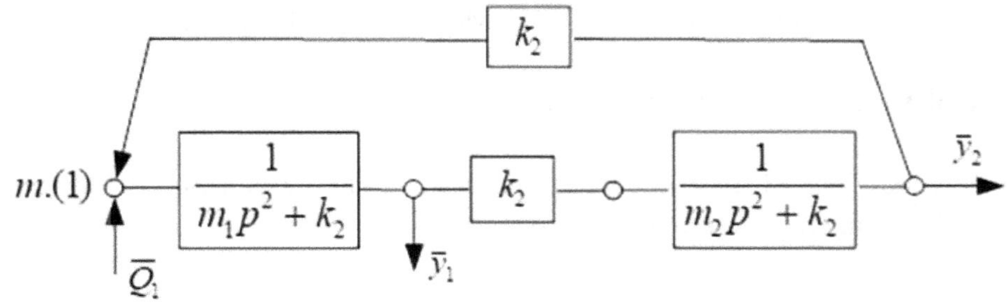

In this system, to detail the representations of its dynamic properties, we assume that the external effect has the form of a harmonic force disturbance Q_1, which is applied to volume (1) (Figure 4). In this case, the system output signals are coordinates $\bar{y}_1$and$\bar{y}_2$, which determine the positions of mass inertia elements m_1 and m_2 under the action of external disturbance.

The system in Figure 4 consists of two parallel blocks (or systems) connected by an elastic connection k_2 (Figure 5)

Figure 5. Block diagram with blocks reflecting partial properties of the system

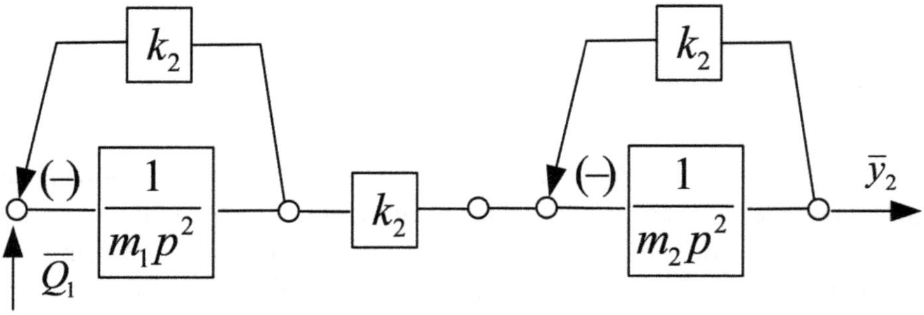

As follows from Figure 4 and Figure 5, the partial system is a mechanical oscillatory unit with one degree of freedom (Jie, Morse, & Hinkle, 1966).

Evaluation Of Dynamic Properties Of Dyad

Interaction of partial blocks occurs through elastic coupling with transfer function $W_{k_2}(p) = k_2$. Partial blocks have partial frequencies

$$n_1^2 = \frac{k_2}{m_1}, \tag{10}$$

$$n_2^2 = \frac{k_2}{m_2}. \tag{11}$$

In turn, the transfer functions of the system (Figure 4) under force disturbance $\overline{Q}_1$ have the form

$$W_1(p) = \frac{\bar{y}_1}{\overline{Q}_1} = \frac{m_2p^2 + k_2}{A(p)}, \tag{12}$$

$$W_2(p) = \frac{\bar{y}_2}{\overline{Q}_1} = \frac{k_2}{A(p)}, \tag{13}$$

where

$$A(p) = (m_1p^2 + k_2)(m_2p^2 + k_2) - k_2^2 \tag{14}$$

- is a frequency characteristic equation.

In the dyad, equation (14) takes the form

$$p^2[m_1 m_2 p^2 + k_2(m_1 + m_2)] = 0, \quad (15)$$

whence it follows that the dyad (as some system) has one cyclic coordinate, as well as one natural frequency (Eliseev, 2017a; Eliseev, 2017b)

$$\omega^2_{\text{sob}} = \frac{k_2(m_1 + m_2)}{m_1 m_2}, \quad (16)$$

Comparing expressions (16) and (2), we note that the frequency at which the dynamic stiffness of the dyad, as a mechanical chain (Figure 1), takes on an infinitely large value; this corresponds to the natural frequency of the dyad, considered in turn as some system.

1. As follows from expression (12), which determines the transfer function of the dyad as a system$W_1(p) = \frac{\bar{y}_1}{\bar{Q}_1}$, in this case there is a dynamic damping mode at the frequency

$$\omega^2_{\text{dyn1}} = \frac{k_2}{m_2}. \quad (17)$$

Frequency of dynamic damping of oscillations ω^2_{dyn1}coincides with partial frequency defined by expression (11).

2. Since the dyad has a cyclic coordinate, movement with some constant speed is possible in the system. Thus, the dyad oscillations along the coordinates y_1 and y_2 are carried out by mass inertia elements relative to the state of some steady-state movement (but not relative to the position of statistical equilibrium). Amplitude-frequency response of dyad at force disturbance Q_1 is as shown in Figure 6.

Figure 6. Amplitude-frequency characteristic of the dyad: solid line -, $W_1(p) = \frac{\bar{y}_1}{Q_1}$ dashed line -$W_2(p) = \frac{\bar{y}_2}{Q_1}$

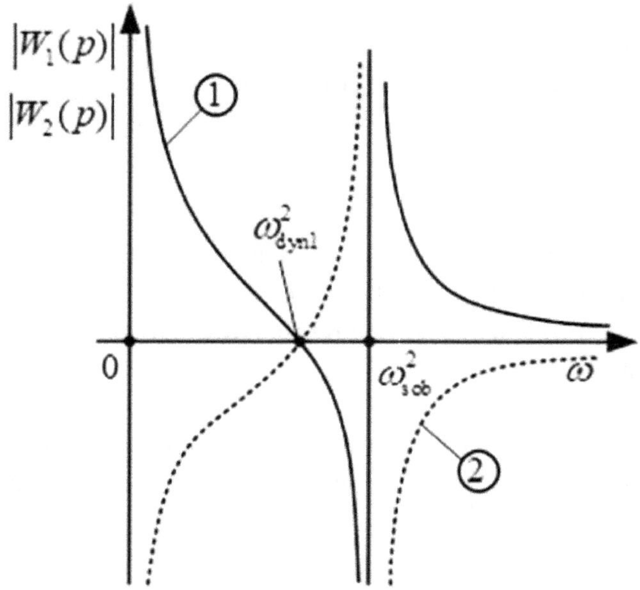

Taking into account the peculiarities of the structural scheme of the system (Figure 4), the concept of the transfer function of interpartial bonds can be introduced into consideration

$$W_{12}(p) = \frac{\bar{y}_2}{\bar{y}_1} = \frac{k_2}{m_2 p^2 + k_2}. \quad (18)$$

The ratio of oscillation amplitudes to frequency is shown in Figure 7. At the frequency of external influence, which coincides with the partial frequency, expression (11), the dependence has a $\frac{\bar{y}_2}{\bar{y}_1}(\omega)$break of the second kind, and the amplitude ratio is divided into two groups. At frequencies less than n_2, the oscillations m_1 and m_2 are in-phase; at $\omega > n_2$ - m_1 and m_2 oscillate in opposite directions.

Figure 7. Amplitude-frequency characteristic of interpartial connections

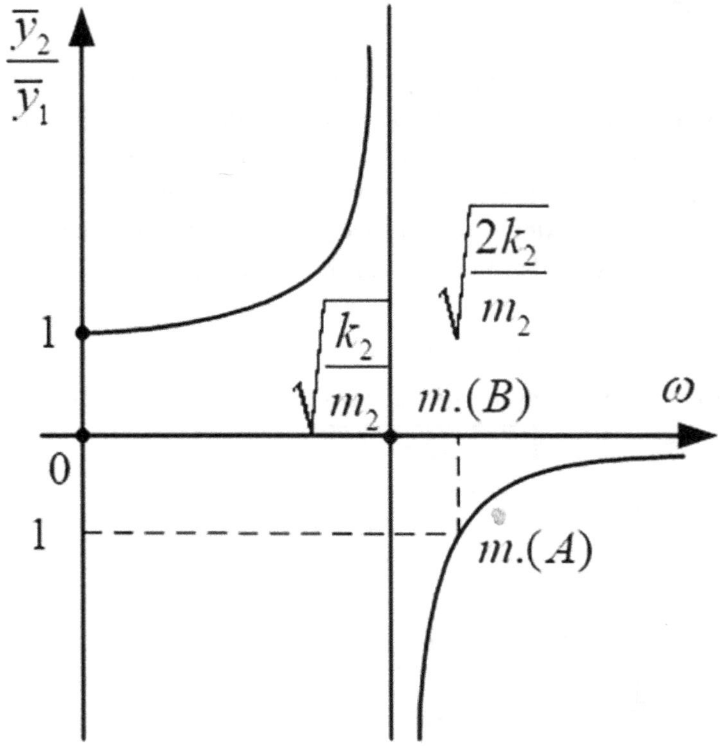

If we take that $\frac{\overline{y}_2}{\overline{y}_1} = -1$, then from this ratio, it follows, we can get an expression to determine the corresponding frequency

$$\omega_{01}^2 = \frac{2k_2}{m_2}. \tag{19}$$

Provided, the $\frac{\overline{y}_2}{\overline{y}_1} = 1$frequency cannot be implemented. From (19) it follows, in particular, that at this frequency some conditional point arises, which can be considered as the center of oscillations.

The point t (*A*) corresponds to the amplitude ratio, $\frac{\overline{y}_2}{\overline{y}_1} = -1$ a. t. (*B*) determines the frequency and position of the center of oscillation. When the frequency changes, the position of the center of vibration also changes. An analogy to such a process is the oscillation of a lever on an intermediate support. Note that the ratio of amplitudes of forced oscillations at $\overline{Q}_1 \neq 0$reflects the linkage between coordinates, which is typical for mechanical oscillatory systems. At the same time, the forms of the relationship between the coordinates "externally" have signs of rotational movements in the lever mechanisms of the first and second genera (Eliseev, Kuznetsov, & Kaimov, 2015).

3. The solution of the dyad dynamics problem can be carried out within the framework of setting the tasks of vibration protection of a certain technical device, which consists of two masses m_1 and m_2 and a connecting spring k_2. Assuming that the mass inertia element m_1 is an object whose dynamic state is evaluated under the action of an external disturbance $\overline{Q}_1$, we convert the structural diagrams in Figure 4 and Figure 5 to the form, as shown in Figure 8.

Figure 8. Block diagram of a dyad with a dedicated m1 for assessing the dynamic state

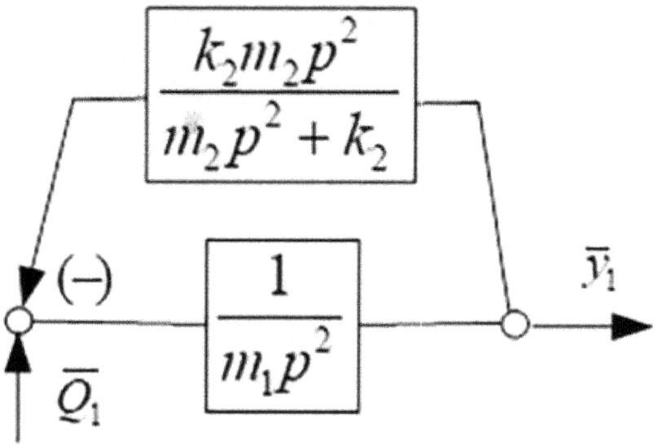

The negative feedback circuit (Figure 8) is the dynamic stiffness of the quasi-spring in relation to the protection object m_1. The problem of vibration protection of the object m_1 from $\overline{Q}_1$ is solved within a generalised basic model in form of a system with one degree of freedom. The m_1 protection object is associated with a quasi-spring that creates a dynamic communication response

$$\overline{R}_{\text{dyn}} = \overline{k}_{\text{pr}} \overline{y}_1, \tag{20}$$

where $\overline{y}_1 = W_1(p) \cdot \overline{Q}_1$ - dynamic offset by y_1 coordinate; the dynamic stiffness of the quasi-spring (Kashuba, Eliseev, & Bolshakov, 2017) is given by

$$\overline{k}_{\text{pr}} = \frac{k_2 m_2 p^2}{m_2 p^2 + k_2}, \tag{21}$$

is infinitely important, which forms a specific mode in the system. In particular

$$\overline{R}_{\text{dyn}m_1} = \frac{k_2 m_2 p^2}{m_2 p^2 + k_2} \cdot \frac{\overline{Q}_1(m_2 p^2 + k_2)}{p^2[m_1 m_2 p^2 + k_2(m_1 + m_2)]} = \frac{k_2 m_2 \overline{Q}_1}{m_1 m_2 p^2 + k_2(m_1 + m_2)}, \tag{22}$$

at which at the natural frequency of the dyad, the dynamic reaction of the bonds on element m_1 will be infinitely large ($\overline{R}_{\text{dyn}m_1} \to \infty$) (Kashuba, Eliseev, & Bolshakov, 2017).

To evaluate the dynamic state of the dyad, in this case, an amplitude-frequency characteristic can be introduced in the form of a ratio of the dynamic reaction of the bond to an external force, which can be considered as the introduction of a new parameter for the dynamic state of the dyad

$$\overline{W}_R(p) = \frac{\overline{R}_{\text{dyn}m_1}}{\overline{Q}_1} = \frac{k_2 m_2}{m_1 m_2 p^2 + k_2(m_1 + m_2)}. \tag{22}$$

Note that at $p \to \infty$, $\overline{W}_R(p) \to ;\ \frac{k_2}{m_1}$ when in the amplitude-frequency characteristic there is $\omega^2 = \omega^2_{\text{sob}}$ a gap of the second kind.

Thus, an open mechanical circuit consisting of two mass inertia elements m_1 and m_2 connected by a spring k_2 can be considered as a quasi-spring or as some structural formation of elements with dynamic rigidity depending on the frequency of dynamic periodic harmonic interactions of the elements. The dynamic stiffness of a quasi-spring can take on infinitely large values at a certain frequency. In the future, this frequency can serve as a characteristic of the properties of the system, which includes a mechanical circuit. A feature of the mechanical chain is the fact that its elements do not touch the support surfaces. A special class of open (open) mechanical chains is a chain in which the boundary elements are inertia-elastic elements for connecting inertial links. A chain of three links can be called a dyad. Such a circuit can be transformed into a mechanical oscillatory system, a mathematical model of which is displayed by a structural diagram of a dynamically equivalent automatic control system. Dyad, as a system with two degrees of freedom, under the action of harmonic force disturbance can oscillate relative to the position of steady-state motion. In this case, we note that the dyad has one cyclic coordinate. Forced oscillations of the dyad and the features of their dynamic properties can be described on the basis of the analytical apparatus of the theory of automatic control (transfer functions, amplitude-frequency characteristics, dynamic modes and states, etc.). In dyads, they appear in leverage in the distribution of vibration amplitudes at various coordinates. Characteristic parameters in the assessment of the dynamic properties of dyads are the frequency of natural vibrations and partial frequencies. With harmonic force disturbance in the dyad, the ratios of oscillation amplitudes have two forms of states. One of them corresponds to a positive ratio of oscillation amplitudes; the second - negative wearing of amplitudes by coordinates and y_1 and y_2. The transition from one form of joint movements to another is carried out at the frequencies of dynamic damping of vibrations (or at partial frequencies). The sum of the partial frequencies of the dyad is equal to the frequency of its own vibrations.

PECULIARITIES OF FORMATION OF INERTIAL-ELASTIC BONDS IN THE DYAD STRUCTURE

The creation of new technological and transport machines used in various branches of technology reflects the level of development of the theoretical foundations of modern mechanics, including the theory of oscillations (Karnovsky & Lebed, 2016). A wide range of dynamic effects arising in mechanical oscillatory systems with the introduction of additional connections is presented in works on the applied theory of oscillations (Eliseev & Artyunin, 2016) . Inertial and linkage links used to solve machine dynamics problems are discussed in (Belokobylsky, Eliseev, & Sitov, 2013). The study of the properties of structural formations with a small number of degrees of freedom capable of accumulating energy is associated with the tasks of constructing structures from typical elements (Banakh & Kempner, 2010). In recent years, attention to such tasks has been increasing in areas related to vibration technology, mechatronics and robotics (Eliseev, 2017c). A number of dynamic features of dyads, as fundamental structural formations, are reflected in the works (Eliseev & Eliseev, 2017; Eliseev, 2017a; Eliseev, 2017b; Eliseev & Eliseev, 2017; Eliseev, Orlenko, & Eliseev, 2017). At the same time, the issues of assessing the dynamic features of mechanical oscillatory systems, which are fundamental formations, require more detailed attention.

The proposed article discusses the formation of inertial-elastic bonds in a dyad considered as a fundamental structural formation of a mechanical oscillatory system.

Inertial Elastic Coupling In Dyad With Lever Mechanism. Task Statement.

A dyad formed by two mass inertia elements connected by an elastic element is considered. The elastic connection can be supplemented by a parallel connection with a motion conversion device containing a non-self-braking screw pair, which is a second-order differentiating link of the form$Lp^2 + k$. At the same time, instead of a non-self-braking screw pair, the use of a lever mechanism is proposed.

The mechanical system is formed by two mass inertia elements $m_1$and connected $m_2$by an elastic elementk_0, and performing translational movements along guides having a common axis (Figure 9). Elements $m_1$and $m_2$are connected by lever link formed by rods. At the end of one rod, at a distance $l_0$from the hinge joint, an additional mass is placedm_0. It is assumed that the system makes forced oscillations about the position of static equilibrium under the action of a harmonic force applied to $Q_1$the mass $m_1$along the guide axis.

Figure 9. Mechanical vibration system design diagram

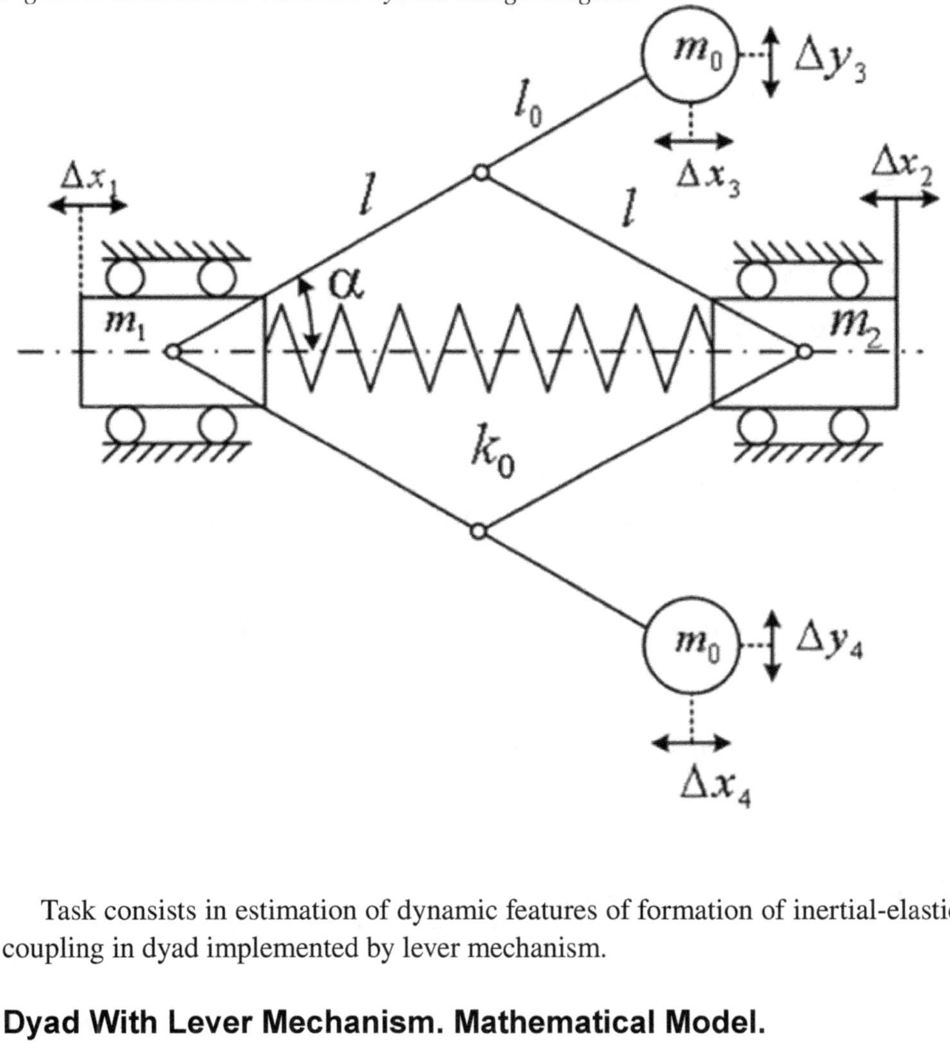

Task consists in estimation of dynamic features of formation of inertial-elastic coupling in dyad implemented by lever mechanism.

Dyad With Lever Mechanism. Mathematical Model.

We assume that the massesm_1, $m_2$and $m_0$in the position of static equilibrium occupy positions at pointsU_1, U_2,U_3,U_4. As generalized coordinates of the system, offsetsΔx_1, Δx_2massesm_1, $m_2$relative to points, are selectedU_1,U_2.

Consider the potential and kinetic energy of the system:

$$\Pi = \frac{1}{2} < K\vec{\chi},\vec{\chi} > ,T = \frac{1}{2} < M\dot{\vec{\chi}},\dot{\vec{\chi}} > , \tag{24}$$

Where

$$\vec{\chi} = \begin{bmatrix} \Delta x_1 \\ \Delta x_2 \end{bmatrix}, K = \begin{bmatrix} k_0 & -k_0 \\ -k_0 & k_0 \end{bmatrix}, M_1 = \begin{bmatrix} m_1 & 0 \\ 0 & m_2 \end{bmatrix} A = (a_{ij})_{2\times 2}$$

$$B = (b_{ij})_{2\times 2}, M = (m_{ij})_{2\times 2}, B = A^* A, M = M_1 + 2m_0 B,$$

$A^*$is the transposed matrix where

$$Aa_{11} = \frac{l - l_0}{2l}, a_{12} = \frac{l + l_0}{2l}, a_{21} = \frac{l + l_0}{2l}\frac{1}{\tan\alpha}, a_{22} = -\frac{l + l_0}{2l}\frac{1}{\tan\alpha}.$$

The mathematical model in Laplace images is:

$$(Mp^2 + K)\overline{X} = \overline{Q}$$

, (25)

where$\overline{X}$, $\overline{Q}$are Laplace images of vectors$\vec{\chi}$, $\vec{Q}$respectively.

The block diagram is shown in Figure 10. Parametersγ_0,γ_0, $\gamma_0$have values:

$$\gamma_1 = \frac{(\lambda - 1)^2}{2} + \frac{(\lambda + 1)^2}{2\tan^2\alpha}, \gamma_0 = \frac{\lambda^2 - 1}{2} + \frac{(\lambda + 1)^2}{2\tan^2\alpha}, \gamma_2 = \frac{(\lambda + 1)^2}{2} + \frac{(\lambda + 1)^2}{2\tan^2\alpha}, \lambda = \frac{l_0}{l}$$

Figure 10. Structural diagram of the mechanical oscillatory system shown in Figure 9

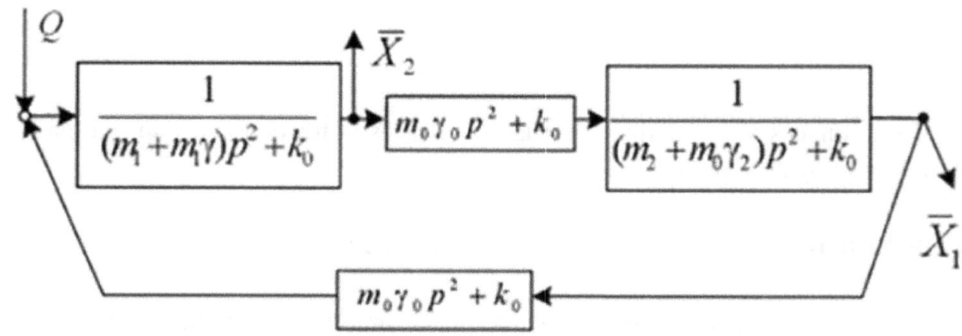

Peculiarities Of Formation Of Inertial-Elastic Coupling Characteristics

The transfer function takes the form:

$$W(p) = \frac{1}{p^2} \cdot \frac{m_0 \gamma_0 p^2 + k_0}{(m_{11} m_{22} - m_{12} m_{21}) p^2 + k_0 (m_{11} + m_{22} + m_{12} + m_{21})} \tag{26}$$

Free oscillation frequency of the system under consideration is:

$$\omega^2 = \frac{k_0}{m_{PR}} \tag{27}$$

where $m_{ПР} = \frac{m_{11} m_{22} - m_{12} m_{21}}{m_{11} + m_{22} + m_{12} + m_{21}}$ is the reduced mass.

Provided that$m_1 = m, m_2 = m, m_0 = \mu m, l_0 = \lambda l$, where μ is a dimensionless parameter characterizing the mass ratio, $\lambda$is a dimensionless parameter characterizing the length ratios, the reduced mass m_{PR}takes the form:

$$m_{PR} = \frac{m}{2} \cdot \frac{(\lambda^2 + 1)\mu + 1}{\mu + 1} + \frac{m}{2} \cdot \frac{(\lambda + 1)^2 \mu}{\tan^2 \alpha} \tag{28}$$

Figure 11 shows the dependence of reduced mass $m_{ПР}$on dimensionless parameters μandλ.

Figure 11. Reduced mass m_{PR}depending on parameters a)μ, where$\lambda = 1..10, 2$, $\alpha = \frac{6\pi}{13}$.; b)λ, where$\mu = 1..100,\ 10, m = 1$ kg.$\alpha = \frac{6\pi}{13}$.

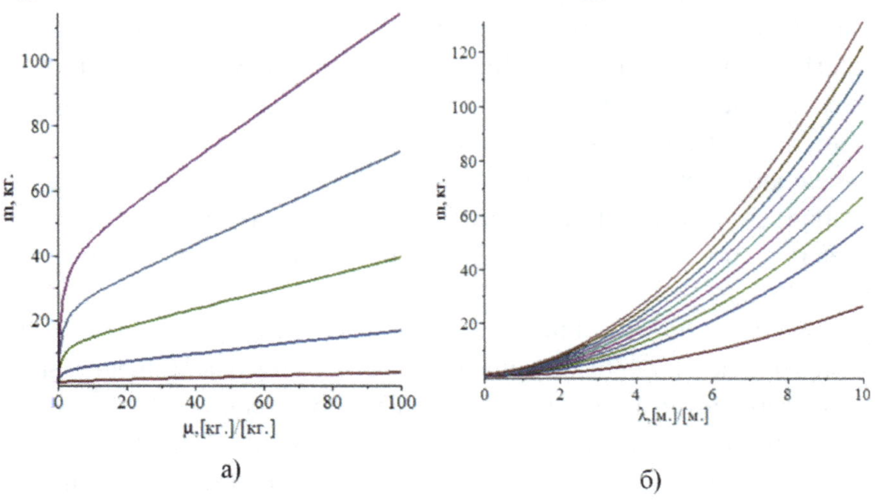

The properties of dyad, as a fundamental structural formation, are quite diverse. The dyad, considered as a flat mechanical oscillatory system isolated from the supporting surface, has characteristic dynamic properties reflecting the features of free and forced vibrations. In the framework of this approach, the inertial-elastic coupling of the mass-inertial elements of the dyad can be represented by a second-order differentiating link implemented by a flat lever mechanism. The characteristics of the link are determined based on the stiffness of the elastic element and the reduced mass depending on the characteristics of the system.

MATHEMATICAL MODELING: FEATURES OF DYNAMIC PROPERTIES OF STRUCTURAL FORMATIONS IN MECHANICAL OSCILLATORY SYSTEMS

The dynamics of mechanical objects largely determines the possibilities of effective operation under conditions of dynamic loads on elements, units and assemblies of machines. In recent years, much attention has been paid to assessing the dynamic states created by vibrations of various nature; this may be imbalances in parts and assemblies, the manifestation of external forces, etc. At the same time, a wide class of technological machines is designed and created to create vibration fields that ensure the technological processes of vibration transportation of working granular mixtures, vibration hardening, etc. (Clarence & de Silva, 2000; Eliseev, Selvinsky, & Eliseev, 2015; Eliseev, Kopylov, & Eliseev, 2017a; Eliseev, Kopylov, & Eliseev, 2017b; Kopylov & Eliseev, 2014; Eliseev, Artyunin, Nguyen, & Vuong, 2016; Eliseev & Vuong, 2016). Ensuring the reliability of operation of technological vibration machines is generally a rather difficult task (Ganiev & Kononenko, 1976; Tarasik, 2004), since the working elements of such technological objects, as a rule, are solid bodies making complex vibrational movements. A preliminary assessment of the capabilities of vibration machines in terms of the formation of the necessary vibration fields, the manifestation of specific dynamic effects associated with joint movement along several coordinates can be carried out on the basis of the use of calculation schemes in the form of mechanical vibration systems with concentrated parameters with several degrees of freedom (Orlenko, Eliseev, & Eliseev, 2018; Eliseev, 2017a).

A comparative analysis of the obtained results (Eliseev, 2017a; Eliseev, 2017b) makes it possible to identify a number of specific effects and manifestations of dynamic properties that are associated with the consideration of some structural formations from typical elements characteristic of methods of structural mathematical modeling (Eliseev, Reznik, & Khomenko, 2011; Eliseev & Artyunin, 2016; Eliseev, 2018). Such structural formations can be considered, in turn, as some forms of manifestation of the fundamental properties of mechanical oscillatory systems as a whole. Flat variants of such structural formations, called dyads, are considered in the works (Eliseev, Nikolaev, & Mironov, 2018; Eliseev, 2017c; Khomenko & Eliseev, 2017).

The proposed report addresses a number of issues related to the creation of a new methodological basis for revealing the possibilities of dyads as autonomous structural formations for compliance with the spectrum of dynamic properties for general mechanical oscillatory systems.

Characterization Of Dyads Dynamic Properties In Free Vibration. Task Statement.

Structural formation is considered in the form of a dyad, which is a mechanical oscillatory system with two degrees of freedom (Figure 12).

Figure 12. A dyad formed by the mass inertia elements m_1 and, the m_2 elastic element k_0 and the motion conversion device

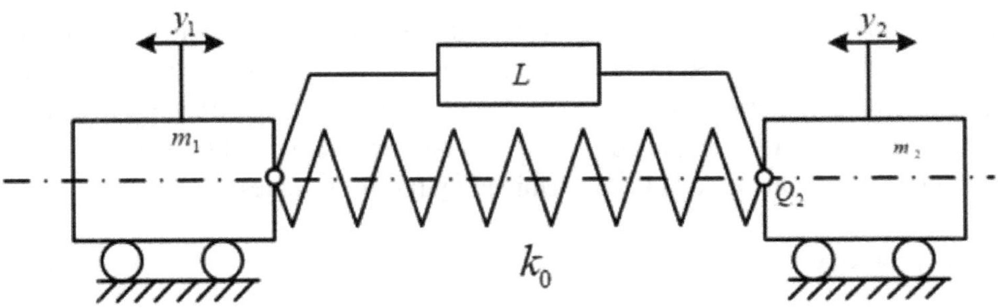

Mass inertia elements m_1 and m_2, forming a dyad, make small free translational vibrations along a common straight line. Considered dyad, masses of which are connected by elastic element with rigidity k and length l, contains in its structure additional connection implemented by device of motion conversion with mass-inertia characteristic L. No external influences. The initial displacements serve as a perturbing factor; the initial velocities are zero. Mass inertia elements of the dyad oscillate relative to static positions. In the process of free oscillations of the dyad in the structure of the mechanical oscillatory system, the emergence of characteristic points with special properties is possible. Such points include fixed points and points reflecting dynamic features associated with the formation of reactions from the elastic elements of the system.

Along with the formation of characteristic points in the structure of the dyad, which in fact are partial features of the mechanical oscillatory system, the structural features of the dyad, reflecting its metric properties, are of interest as a factor in the implementation of dynamic modes. So, on the one hand, the initial conditions determine the amplitudes of free vibrations of the elements of the dyad. On the other hand, due to the limb of the elastic element of the dyad, collisions occur provided that the amplitudes exceed critical values, which affects the shape of free oscillations. In addition to the features of the formation of the structural and dynamic properties of the dyad, the role of additional bonds represented in the structure of the dyad by the motion conversion device is of interest.

The objective of the study is to develop an approach to determining the characteristic features of the dyad depending on additional bonds that form the interaction of structural formation elements in the process of free oscillations.

Structural And Dynamic Features Of Dyad With Motion Conversion. Mathematical Model.

We consider a dyad formed by two masses, an m_1 m_2 elastic element k_0 and a motion conversion device (Figure 12). An inertial coordinate system is fixed with the origin at the point O. It is assumed that the masses m_1 m_2 in the static equilibrium position have coordinates y_{s1}, y_{s2}, while the elastic element with length l is in an undeformed state, i.e. completed $y_{s2} = y_{s1} + l$. At each moment of time, the position of the masses m_1 and m_2 dyad is determined by the coordinates Y_{T1},: Y_{T2}

$$y_{r1} = y_{s1} + y_1, y_{r2} = y_{s2} + y_2, \tag{29}$$

where y_1 and y_2 - mass displacements relative to static equilibrium positions serve as generalized coordinates of the mechanical oscillatory system.

Potential and kinetic energy in generalized coordinates have the form:

$$\Pi = \frac{1}{2}k_0(y_2 - y_1)^2, T + \frac{1}{2}m\dot{y}_1{}^2 + \frac{1}{2}m_2\dot{y}_2{}^2 + \frac{1}{2}L(\dot{y}_2 - \dot{y}_1)^2 \tag{30}$$

The Lagrange equation of the 2nd kind in images has the form:

$$\begin{pmatrix} (m_1 + L)p^2 + k_0 & -k_0 \\ -k_0 & (m_2 + L)p^2 + k_0 \end{pmatrix} \begin{pmatrix} \bar{y}_1 \\ \bar{y}_2 \end{pmatrix} = \begin{pmatrix} (m_1 + L)py_{10} \\ (m_2 + L)py_{20} \end{pmatrix} \tag{31}$$

wherey_{10}, $y_{20}$are initial displacements of mass inertia elements.

Defines The Characteristic Points Of The Dyad Based On The Motion Conversion Device.

1. The design diagram of the dyad with additional coupling is shown in Figure 12. The position of a characteristic point whose $y_{L,\lambda}$ speed is zero can be found from the condition of immobility:

$$\dot{y}_{L,\lambda} = \lambda_1 \dot{y}_1 + \lambda_2 \dot{y}_1 = 0 \tag{32}$$

where the desired point is

$$y_{L,\lambda} = \lambda_1 y_1 + \lambda_1 y_1, \lambda_1 + \lambda_2 = 1$$

In Laplace images, condition (32) takes the form:

$$\lambda_1 \bar{y}_1 + \lambda_2 \bar{y}_2 = \lambda_1 y_1(0) + \lambda_2 y_2(0) \tag{33}$$

The conditions for the compatibility of the system of equations (31) and the linear expression (33) entail the identical equality to zero of the determinant of the corresponding extended matrix of the system:

$$\begin{vmatrix} \lambda_1 p & \lambda_1 p & \lambda_1 y_{10} + \lambda_2 y_{20} \\ (m_1 + L)p^2 + k_0 & -k_0 & (m_1 + L)p y_{10} \\ -k_0 & (m_2 + L)p^2 + k_0 & (m_2 + L)p y_{20} \end{vmatrix} = 0 \tag{34}$$

The identity equality to zero of determinant (34) is equivalent to the conditions

$$k(y_{10} - y_{20})(L\lambda_1 - L\lambda_2 + \lambda_1 m_2 - \lambda_2 m_1) = 0$$

. (35)

If $y_{10} - y_{20} = 0$, then the fluctuation of the dyad is not realized. If $y_{10} - y_{20} \neq 0$, then equality is satisfied, $L\lambda_1 - L\lambda_2 + \lambda_1 m_2 - \lambda_2 m_1 = 0$ which, taking into account the condition, $\lambda_1 + \lambda_2 = 1$ determines the parameters $\lambda_1 \lambda_2$ of the fixed point

$$\lambda_1 = \frac{m_1 + L}{m_1 + m_2 + 2L}, \lambda_2 = \frac{m_2 + L}{m_1 + m_2 + 2L} \tag{36}$$

2. To determine a characteristic point that meets the conditions of zero total dynamic reaction on the part of elastic elements, an auxiliary system is considered

(Figure 13), Built on the basis of the system (Figure 12) and formed by adding mass m_3 between the elastic elements k_1, k_2.

Figure 13. Auxiliary mechanical system with motion converter L

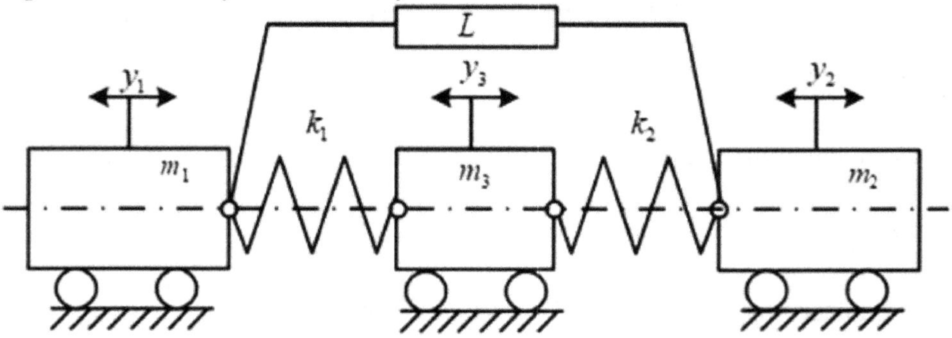

Potential and kinetic energy of the system:

$$\Pi=\frac{1}{2}k_1(y_3-y_1)^2+\frac{1}{2}k_2(y_2-y_3)^2, T=\frac{1}{2}m_1\dot{y}_1{}^2+\frac{1}{2}m_2y_2{}^2+\frac{1}{2}m_3\dot{y}_3{}^2+\frac{1}{2}L(\dot{y}_2-\dot{y}_1)^2 \quad (37)$$

The Lagrange equation of the 2nd kind has the form:

$$\begin{cases}(m_1+L)\ddot{y}_1+k_1y_1-k_1y_3 = 0\ ;\\ (m_2+L)\ddot{y}_2+k_2y_2-k_2y_3 = 0\ ;\\ m_3\ddot{y}_3-k_1y_1-k_2y_2+(k_1+k_2)\,y_3 = 0\end{cases} \quad (38)$$

The initial offsets are assumed to be given:

$$y_1(0)=y_{10}, y(0)=y_{20}, y_3(0)=0, \dot{y}_1(0)=0, \dot{y}_2(0)=0, \dot{y}_3(0)=0 \quad (39)$$

The Laplace transform of the system of differential equations (38) and the condition of equality to zero of the total dynamic reaction on the part of elastic elements $-k_1(y_3-y_1)+k_2(y_2-y_3)=0$, presented in the images, leads to the system of equations:

$$\begin{pmatrix}-k_1 & -k_2 & k_1+k_2\\ (m_1+L)p^2+k_1 & 0 & -k_1\\ 0 & (m_2+L)p^2+k_2 & -k_2\\ -k_1 & -k_2 & m_3p^2+k_1+k_2\end{pmatrix}\begin{pmatrix}\bar{y}_1\\ \bar{y}_2\\ \bar{y}_3\end{pmatrix}$$

$$
= \begin{pmatrix} 0 \\ (m_1 + L)py_{10} \\ (m_2 + L)py_{20} \\ 0 \end{pmatrix} \tag{40}
$$

The compatibility of the system (40), taking into account the equality of the initial mass displacement to m_3 zero, leads to the identity:

$$
\begin{vmatrix} -k_1 & -k_2 & 0 \\ (m_1 + L)p^2 + k_1 & 0 & (m_1 + L)py_{10} \\ 0 & (m_2 + L)p^2 + k_2 & (m_2 + L)py_{20} \\ -k_1 & -k_2 & 0 \end{vmatrix} = 0 \tag{41}
$$

The identity equality of the determinant to zero (41) is equivalent to the conditions:

$$
\begin{cases} k_1 y_{10} + k_2 y_{20} = 0 \; ; \\ (m_1 + L)y_{10} + (m_2 + L)y_{20} = 0 \end{cases} \tag{42}
$$

The conditions of existence of the solution of the system (42), taking into account, $k_1 = \frac{K}{l_1}$ make it possible to obtain conditions for the ratios of lengths of l_1, l_2 elastic elements that determine the position of the mass-inertial element characterizing the point m_3 of zero dynamic reaction:

$$
\frac{m_2 + L}{m_1 + L} = \frac{l_1}{l_2} \tag{43}
$$

Thus, the desired parameters of a characteristic point with zero total dynamic reaction on the side of elastic elements are determined by the formulas:

$$
l_{1,L} = \frac{m_2 + L}{m_1 + m_2 + 2L} l, l_{2,L} = \frac{m_1 + L}{m_1 + m_2 + 2L} l \tag{44}
$$

The found values $l_{1,L}$,$l_{2,L}$ determine the position of the characteristic point, which can be interpreted as the position of the reduced center of mass of the dyad, depending on the mass inertia parameter of the motion conversion Ldevice.

DETERMINATION OF PARAMETERS OF FREE OSCILLATION MODES OF THE DYAD

The dyad is considered taking into account the built-in device for motion conversion L (Figure 12). The structure of the solution takes the form:

$$\bar{y}_{1,L} = \frac{y_{10}p^2 + y_{00,L}\omega_{0,L}^{\;2}}{p\left(p^2 + \omega_{0,L}^{\;2}\right)}, \bar{y}_{2,L} = \frac{y_{20}p^2 + y_{00,L}\omega_{0,L}^{\;2}}{p\left(p^2 + \omega_{0,L}^{\;2}\right)} \tag{45}$$

where $m_{\Pi P,L} = \frac{(m_1 + L)(m_2 + L)}{m_1 + m_2 + 2L}$, $7_{00,L} = \frac{(m_1 + L)y_{10} + (m_2 + L)y_{20}}{m_1 + m_2 + 2L}$, $\omega_{0,L}^{\;2} = \frac{k}{m_{\Pi P,L}}$ are the given characteristics of the dyad, reflecting the concept of mass, center of mass and frequency of free oscillations.

In the time domain, the forms of movement of mass inertia elements are as follows:

$$y_{1,L} = y_{00,L} - \frac{(m_2 + L)(y_{20} - y_{10})}{m_1 + m_2 + 2L}\cos\omega_{0,L}t, y_2$$

$$= y_{00,L} + \frac{(m_1 + L)(y_{20} - y_{10})}{m_1 + m_2 + 2L}\cos\omega_{0,L}t. \tag{46}$$

Oscillation amplitudes are:

$$A_{1,L} = -\frac{m_2 + L}{m_1 + m_2 + 2L}(y_{20} - y_{10}), A_{2,L} = -\frac{m_1 + L}{m_1 + m_2 + 2L}(y_{20} - y_{10}) \tag{47}$$

The geometric characteristics of the family of dyads reflecting the dependence of the boundary positions of the mass inertia elements and the reduced center of mass on the L parameter are presented in Figure 14.

Figure 14. Geometric characteristics of the dyad depending on the mass-inertial parameter L of the motion conversion device $(m_1 \leq m_2)$

a *b*

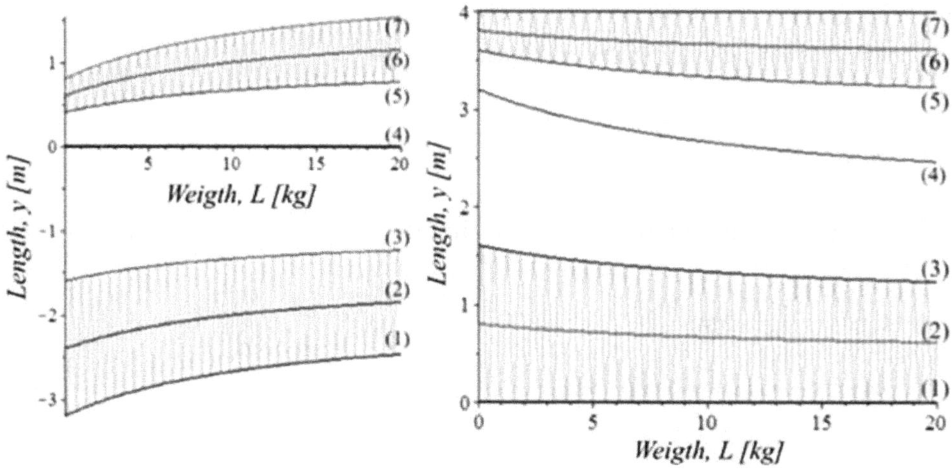

a) reduction of the dyad family to a fixed position of the reduced fixed characteristic point; b) bringing the dyad to a fixed spatial interval;

Figure 14,a shows graphs of extreme leaks of fluctuating masses of dyads (lines 1,3,5,7), centers of mass fluctuations (lines 2 and 6), the reduced cent of mass is set to the position with zero ordinate (line 4).

Figure 14,b shows the geometric characteristics of the family of dyads reduced to a fixed spatial interval: the boundaries of the oscillation of the mass-inertial elements of the dyad (lines 1,3,5,7), the centers of oscillation of the elements (lines 2 and 6, respectively), the reduced center of mass (line 4) depending on the *L* parameter.

Critical initial displacements providing unstressed movement of the dyad are the following value variants:

$$y_{20} - y_{10} = l$$

or

$$y_{20} - y_{10} = -l \quad (48)$$

This means that the initial displacements for the shock-free mode of the dyad must satisfy the conditions:

$$|y_{20} - y_{10}| \leq l \quad (49)$$

To avoid displacement relative to a fixed point, the following conditions must be met:

$$(m_1 + L)y_{10} + (m_2 + L)y_{20} \tag{50}$$

Conditions (50) can be represented as:

$$\frac{y_{20}}{y_{10}} = \frac{m_1 + L}{m_2 + L} \tag{51}$$

In such a case, the values of critical displacements, in particular, have the form:

$$\hat{y}_{10,L} = -l\frac{m_2 + L}{m_1 + m_2 + 2L}$$

,

$$\hat{y}_{20,L} = -l\frac{m_1 + L}{m_1 + m_2 + 2L} \tag{52}$$

The initial displacements (52) corresponding to (51) depend on the mass inertia characteristic of the motion conversion device L.

A formalized concept is proposed to justify the existence of characteristic points and characteristic dyad modes. Under the characteristic mode, the authors mean the mode of free oscillations of the dyad, which is performed relative to the characteristic point. This point has the following properties: a characteristic point is stationary in a given coordinate system, provided that the initial velocities are zero; the feature point coincides with the center of mass; a feature point has the property of defining the boundaries of the physical realization of a dyad. The parameter of the motion conversion device for the family of dyads in question serves as a factor determining the position of the reduced center of mass, for diads fixed in space, or a factor determining the position of the dyad in space, provided that the reduced center of mass of the dyad is fixed at some point in space.

NEW APPROACHES IN THE DYNAMICS OF OSCILLATORY SYSTEMS: BACKBONE STRUCTURAL FORMATIONS

Mechanical oscillatory systems are widely used in solving numerous problems of machine and mechanism dynamics. Mechanical oscillatory systems of various degrees of complexity act as design schemes and reflect the most significant properties of technical objects when it comes to those processes, phenomena and effects that are manifested in vibration interactions of elements in the presence of oscillatory movements characteristic of intense dynamic loads.

The creation of new machines, the development of new technological processes that use the dynamic effects of vibration interaction with working media require detailed ideas about the features of vibration processes, vibration processes that arise during the operation of machines (Panovko, 2006; Kopylov, 2011; Eliseev, Selvinsky, & Eliseev, 2015).

Although the design schemes of technical objects are sufficiently close to the assessment of characteristic features and are not linear, nevertheless, at the preliminary stages of the development of machines, the assessment of their possible dynamic properties, the determination of accident-free, safe operation conditions are carried out using mathematical models with linear properties. The dynamic properties of linear mechanical oscillatory systems in many aspects are quite well studied (Den-Hartog, 1960; Tse, Morse, & Hinkle, 1966; Panovko, 1991). A number of works have developed methods for studying the dynamics of solids with several degrees of freedom (Ganiev & Kononenko, 1976; Naidenko & Petrov, 1962; Kolovsky, 1976).

However, a number of questions need to be more detailed about the features of dynamic interactions. The possibility of solving such problems could be correlated with the emergence, development and use of so-called structural mathematical modeling methods. The essence of these methods is that structural diagrams of dynamically equivalent automatic control systems are used as mathematical models. Such approaches are based on the well-known idea that elastic elements in mechanical oscillatory systems implement so-called feedbacks, form chains of feedbacks and feedbacks and, in general, create a new subject of research, which focuses on the patterns of formation of dynamic connections, their forms of manifestation and dynamic effects. The peculiarities of the application of structural mathematical modeling methods described in sufficient detail in (Kolovsky, 1976; Eliseev, Reznik, Khomenko, & Zasyadko, 2008; Eliseev, Reznik, & Khomenko, 2011; Eliseev & Khomenko, 2014; Eliseev & Artyunin, 2016) are the identification of an object, the dynamic state of which is estimated in terms of the idea that external influences on a technical object are input signals to some systems, and the dynamic state of the object is estimated by the spectrum of transformations that acquire external influences (or signals) passing through the system. In this regard, attention is initiated to the features of the construction of systems, the dynamic state of objects of which is estimated, and the technical means that together with the object form the system.

Evaluation of the dynamic properties of various systems with two or more degrees of freedom shows that there are some structural formations in the system of vibrational circuits, the dynamic properties of which can be considered in a certain sense as fundamental. In turn, various variants of mechanical oscillatory systems reflecting the features of real objects or real technical systems are certain varieties generated or created on the basis of initial fundamental formations (Rocard, 1949; Banakh & Kempner, 2010; Eliseev, 2017a; Eliseev, 2017b; Eliseev, Orlenko, & Eliseev, 2017; Eliseev, Orlenko, & Nguyen, 2017; Eliseev, 2017c; Eliseev & Eliseev, 2017; Khomenko & Eliseev, 2017; Eliseev, Nikolaev, & Mironov, 2018). Such approaches currently seem quite interesting, quite productive, if we bear in mind the questions arising in the applied problems of nano- and meso- mechanics, robotics and mechatronics.

The proposed report presents some of the results of research work carried out to detail the ideas about the dynamic properties of dyads as structural formations consisting of two mass-inertia elements connected by an elastic element, having well-defined local dynamic properties and constituting structural formation in systems with two degrees of freedom.

DYAD AS A STRUCTURAL BASIS FOR SYSTEMS WITH MULTIPLE DEGREE OF FREEDOM. TASK STATEMENT.

The connection of several partial blocks having one degree of freedom usually represents the basis for building a structural mathematical model of systems with several degrees of freedom. Approaches reflecting the technology of obtaining structural diagrams are detailed in (Eliseev, Reznik, & Khomenko, 2011; Khomenko, Eliseev, & Yermoshenko, 2012; Emelyanov & Korovin, 1997; Lenck, 1978). At the same time, attempts have been made to use partial systems with two degrees of freedom in structural schemes with three or more degrees of freedom. The features of such ideas about interactions of this kind are discussed in more detail in (Tse, Morse, & Hinkle, 1966; Eliseev, Reznik, Khomenko, & Zasyadko, 2008). A dyad, which is a structural formation with a set of stable dynamic properties and the ability to act as a system-forming principle when constructing mechanical oscillatory systems of various types, can be chosen as an indicative form of detailing ideas about partial systems (Frolov & Furman, 1985). Figure 15 shows a schematic diagram of the variants of dyads formed by interiors and elastic typical elements, as well as using a device for converting motion.

Figure 15. Types of dyads

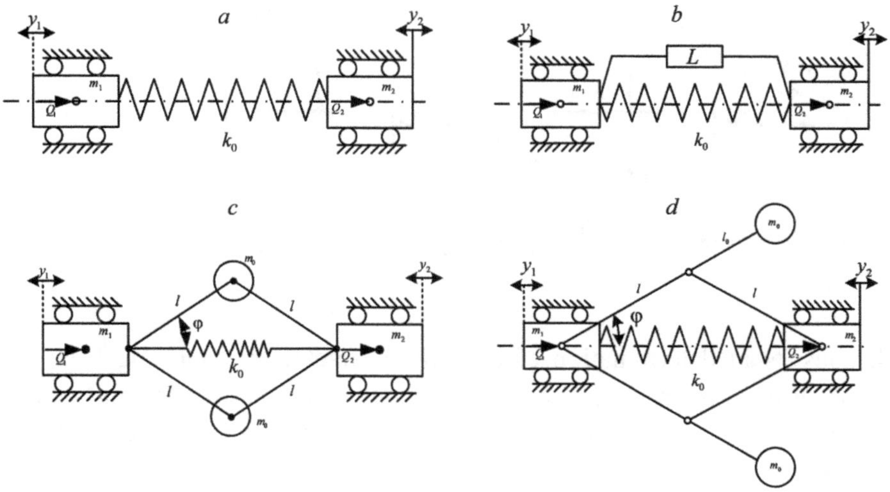

a) - a dyad formed by two masses and an elastic element, b) - a dyad formed by a motion conversion device, c) - a dyad formed by lever links, d) a dyad formed by lever links with spaced mass-inertial elements

As a basis for comparison, in relation to linear chain mechanical systems having several degrees of freedom and performing rectilinear oscillatory movements, the dyad can be represented in the form of two bodies connected to each other by an elastic element performing rectilinear oscillations, as shown in Figure 16. External harmonic forces $Q_{1s} = Q_1 \sin(\omega t)$and $Q_{2s} = Q_2 \sin(\omega t)$are applied directly to mass-inertial elements $m_1$andm_2; resistance forces are supposed to be small; system has linear properties. To describe motion, generalized coordinates are used, $y_1 y_2$denoting displacements relative to the positions of static equilibrium in some inertial frame of reference.

Figure 16. Schematic diagram of the dyad

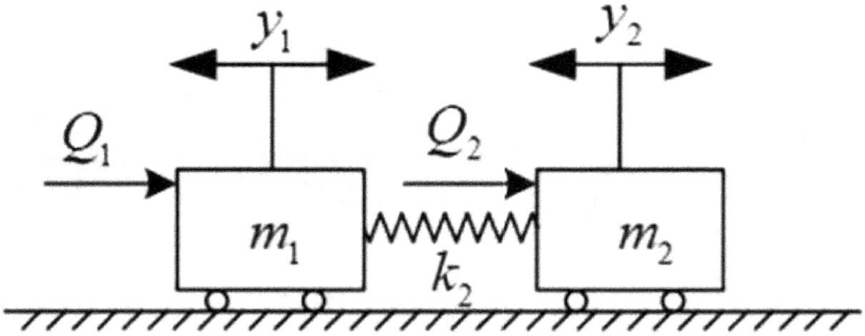

Task consists in determination of dynamic characteristics forming base for comparison of dyad formed by concentrated masses and elastic element, taking into account initial conditions and action of external harmonic disturbances in fixed coordinate system.

DYNAMIC AND STRUCTURAL ANALYSIS OF DYADS IN MECHANICAL SYSTEM. MATHEMATICAL MODEL

The structural mathematical model is built on the basis of an algebraic system obtained from the Laplace transformation of Lagrange equations of the 2nd kind. Initial conditions are assumed to be zero. In particular, if there is only one external force disturbance, the system in the images has the form:

$$\begin{cases} (m_1 p^2 + k_2)\bar{y}_1 - k_2\bar{y}_2 = \bar{q}_1, \\ -k_2\bar{y}_1 + (m_2 p^2 + k_2)\bar{y}_2 = 0, \end{cases}$$

(53) (54)

where is a $p = j\omega$ complex variable. $j = \sqrt{-1}$ The icon <-> above the variable indicates a Laplacian transform (Eliseev, Reznik, Khomenko, & Zasyadko, 2008; Eliseev & Artyunin, 2016; Belokobylsky, Eliseev, & Sitov, 2013)

Based on (53) and (54), a block diagram of a dynamically equivalent automatic control system is constructed (Figure 17).

Figure 17. The structural diagram of the dyad (according to Figure16) in coordinates $\bar{y}_1, \bar{y}_2$

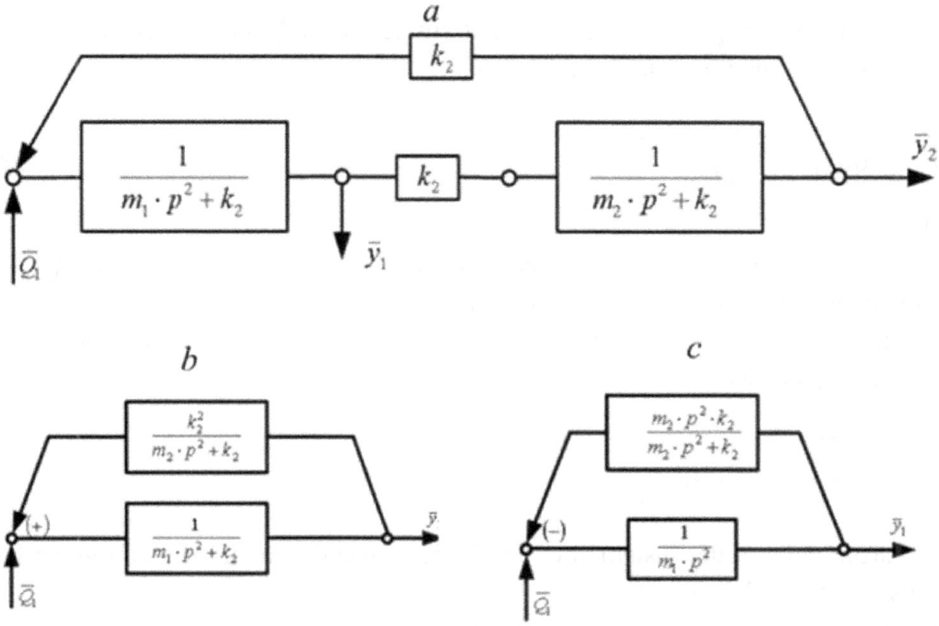

a - general diagram; b - structural diagram with coordinate exclusion $\bar{y}_2$; c - block diagram with selection of the object and reduced $m_1$spring

Structural diagrams (Figure 17) reflect the dynamic properties of the dyad taking into account the selection of the object covered by the feedback loop.

Transfer Functions

To evaluate the properties of the dyad, transfer functions are used by coordinates$\bar{y}_1$, $\bar{y}_2$with force disturbance with the condition$\bar{Q}_2 = 0$:

$$W_1 = \frac{\bar{y}_1}{\bar{Q}_1} = \frac{m_2 p^2 + k_2}{A(p)} \tag{55}$$

$$W_2 = \frac{\bar{y}_2}{\bar{Q}_1} = \frac{k_2}{A(p)} \tag{56}$$

where $A(p) = p^2(m_2 m_1 p^2 + k_2(m_1 + m_2))$is the characteristic equation of the system.

The transfer function of interpartial bonds is as follows:

$$W_{21} = \frac{\overline{y}_2}{\overline{y}_1} = \frac{k_2}{m_2 p^2 + k_2} \tag{57}$$

The natural frequencies ω_1, ω_2 are determined from the characteristic equation of the system and are equal, respectively:

$$\omega_1 = 0, \omega_2 = \sqrt{\frac{k_1}{M_{\Pi}}} \tag{58}$$

At the frequency of external impact, the dynamic damping mode $n_2 = \sqrt{\frac{k_2}{m_2}}$ is implemented according to the first coordinate y_1.

The system (Figure17, *a*) in coordinates $\overline{y}_1 \overline{y}_2$ consists of two partial blocks with partial frequencies:

$$n_1 = \sqrt{\frac{k_2}{m_1}}, n_2 = \sqrt{\frac{k_2}{m_2}} \tag{59}$$

Partial units of the system have an elastic connection realized by a link with a transfer function $W(p) = k_2$

The transfer functions are determined from the structural diagrams in Figure 17 and have the form:

$$\frac{\overline{y}_1}{\overline{Q}_1}(p) = \frac{m_2 p^2 + k_2}{A(p)} \tag{60}$$

$$\frac{\overline{y}_2}{\overline{Q}_1}(p) = \frac{k_2}{A(p)} \tag{61}$$

$$\frac{\overline{y}_2}{\overline{y}_1}(p) = \frac{k_2}{m_2 p^2 + k_2}$$

, (62)

where $A(p) = (m_2 p^2 + k_2)(m_1 p^2 + k_2) - k_2^2$.

Taking into account the transfer functions (55), (56), (57), the amplitude-frequency characteristics,, can be constructed $\frac{\overline{y}_1}{\overline{Q}_1}(j\omega) . \frac{\overline{y}_2}{\overline{Q}_1}(j\omega) \frac{\overline{y}_2}{\overline{y}_1}(j\omega)$

AMPLITUDE-FREQUENCY CHARACTERISTICS

Figure 18 shows the amplitude-frequency response curve $\frac{\overline{y}_1}{\overline{q}_1}(\omega)$.

Figure 18. Amplitude-frequency characteristic

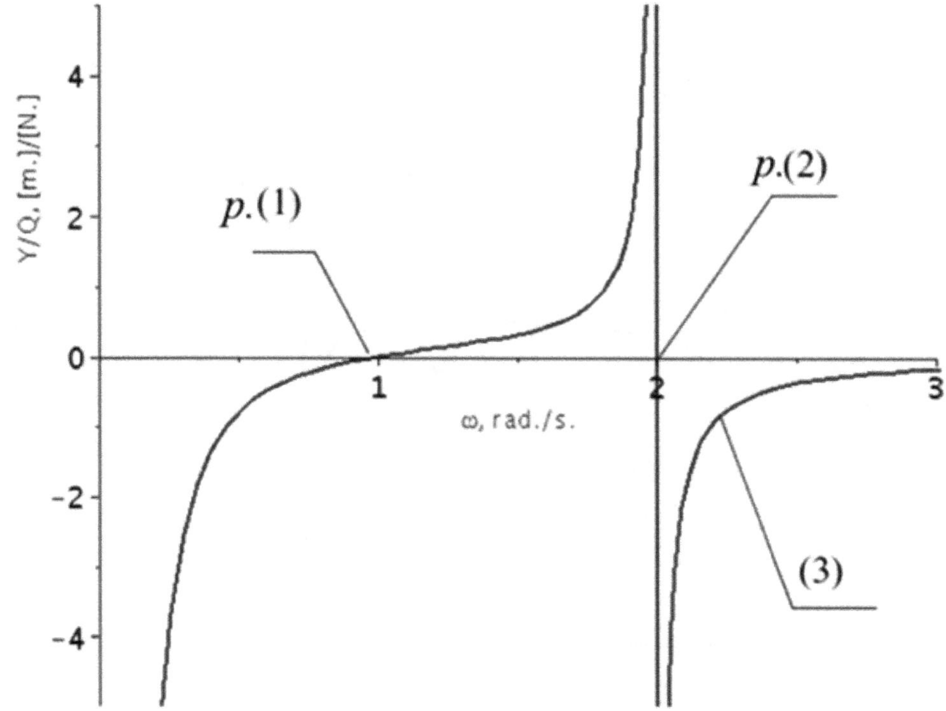

$\frac{\bar{y}_1}{q_1}(\omega)$point (1) - frequency of dynamic cancellation of the first coordinate; point (2) - natural frequency, curve (3) - amplitude-frequency characteristic plot.

Figure 19 shows the amplitude-frequency response curve. $\frac{\bar{y}_2}{q_1}(\omega)$Frequency of external ωforce disturbance is plotted along abscissa axis. At points and, the function $\omega = 0 \omega = \omega_2$suffers a break of the second kind. At point (2) in Figure 19, the resonance mode is implemented in the system. Amplitude and (ω_1, ω_2)forced $y_2$force amplitude fluctuation are in antiphase at the interval. After passing through the resonance (p. 2, Figure 19$\omega = \omega_2$), the phases of the oscillation of the coordinate and the oscillation $\bar{y}_2$of the forcing force q_1 are "synchronized". Diagram(1, Figure19) has no zeros, then there are no dynamic blanking modes in the system.

Figure 19. Amplitude-frequency response

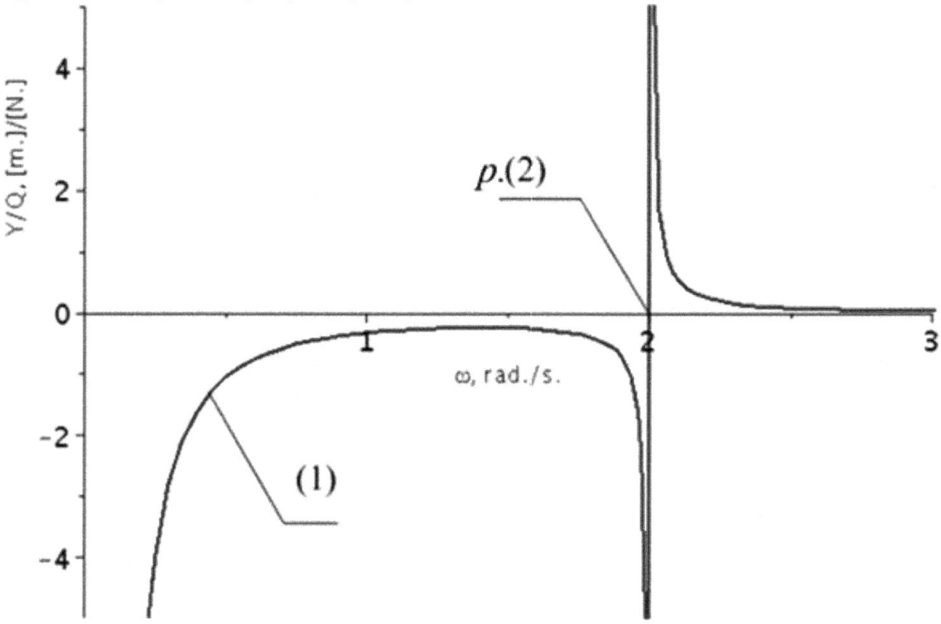

$\frac{\overline{y}_2}{q_1}(\omega)$curve (1) is a plot of the amplitude-frequency response; point (2) - natural frequency.

The change in the sign of the amplitude-frequency characteristic when crossing the point (2) shows a change in the mode of relative oscillation of the coordinate $\overline{y}_2$and the $m_1$force applied to the mass, $\overline{Q}_1$in particular, there is a change in the mode of oscillation in antiphase to the mode of in-phase oscillation

Figure 20 shows the frequency characteristic of interpartial connections, which gives an idea of the forms of connectivity of motion.

Figure 20. Amplitude-frequency characteristic: $\frac{\bar{y}_2}{\bar{y}_1}(\omega)$curve (1) - graph of the inter-partial frequency function with a break of the second kind at the point (2); point (2) - frequency of dynamic damping of the coordinatey,

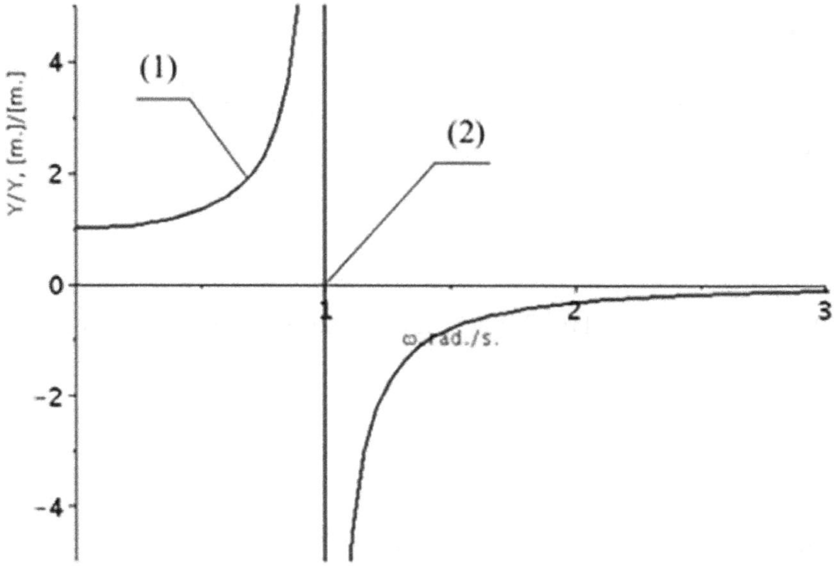

In particular, in the first frequency range from $\omega = 0$to point (2) in Figure 20, representing the interval of coordinate (ω_1, ω_{01})movement, $y_1 y_2$it occurs in phase (the frequency characteristic is positive). After passing the dynamic blanking frequency along the coordinatey_1, which is the break point of the second kind for the partial frequency function, movement along the coordinates $y_1 y_2$occurs in antiphase (the frequency function is negative at the interval(ω_{01}, ∞)). More detailed information on the use of transfer functions of interpartial bonds can be found in (Eliseev & Lonchikh, 2012; Eliseev, Trofimov, & Kaimov, 2014).

FREQUENCY-ENERGY FUNCTION DEPENDING ON THE COORDINATE SYSTEM

1. The coordinate system $y_1$and are considered. y_2Kinetic and potential energies are determined respectively:

$$T = \frac{1}{2}m_1\dot{y}_1^2 + \frac{1}{2}m_2\dot{y}_2^2 \tag{63}$$

$$\Pi = \frac{1}{2}k_2(y_2 - y_1)^2 \tag{64}$$

It is assumed that the system makes its own fluctuations$y_2 = A_2 \sin\omega t$. The amplitude $y_1 = A_1 \sin\omega t$ratio is $\alpha = A_2/A_1$. It can be shown that the maximum of kinetic and potential energy coincide:

$$T_{\max} = \Pi_{\max} \tag{65}$$

Thus, the frequency-energy function, which is an analogue of the Rayleigh ratio, takes the form:

$$\omega^2 = \frac{k_2(\alpha - 1)^2}{m_1 + \alpha^2 m_2} \tag{66}$$

To determine the natural frequencies, the first derivative of the frequency-energy function (66) is found:

$$\frac{\partial(\omega^2)}{\partial\alpha} = \frac{2k_2(\alpha - 1)(m_1 + \alpha m_2)}{(m_1 + \alpha^2 m_2)^2} \tag{67}$$

The extreme points invalidating the first derivative have the values:

$$\alpha_1 = 1, \alpha_2 = -\frac{m_1}{m_2} \tag{68}$$

Note that the following conditions are met for the dyad:

$$\omega^2(\alpha_1) = 0, \omega^2(\alpha_2) = k_2\frac{m_1 + m_2}{m_1 m_2}, \omega^2(0) = \frac{k_2}{m_1}, \lim_{\alpha\to\infty}\omega^2(\alpha) = \frac{k_2}{m_2}, \omega^2 = \frac{k_2(\alpha - 1)^2}{m_1 + \alpha^2 m_2} \tag{69}$$

Figure 21 shows a graph of the frequency energy function. Along the abscissa axis, the shape coefficient is plottedα. The graph in the coordinate system y_1, y_2 has a characteristic shape that determines the possibility of manifesting extreme properties at two frequencies, in this case, $\omega^2(\alpha_1)$,$\omega^2(\alpha_2)$. The graph of the frequency-energy function (1, Figure 21) at $\alpha \to \infty$ has an asymptote (2, Figure 21), the ordinate of which is equal to the frequency of dynamic blankingω_{01}.

Figure 21. Energy function in the coordinate system y₁, y₂

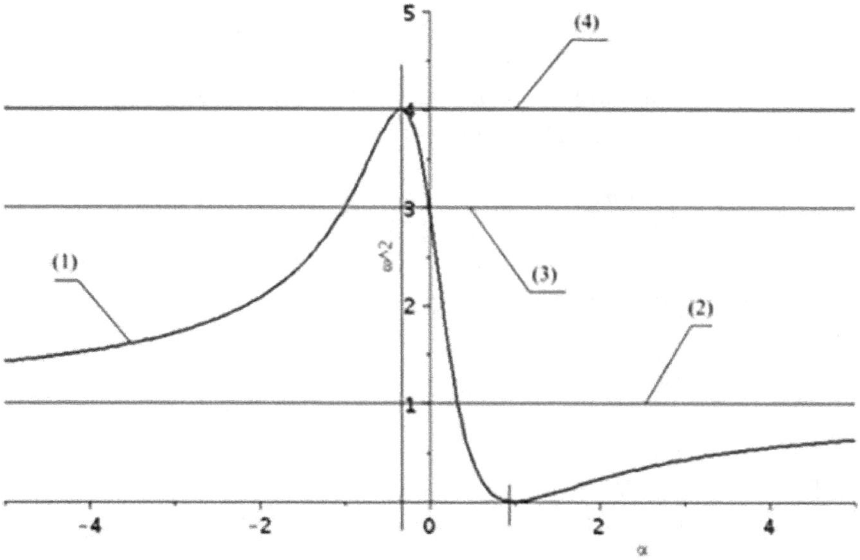

Curve (1) is a graph of the frequency-energy function, line (2) is the asymptote of the frequency-energy function at, $\alpha \rightarrow \infty$ equal to the frequency of dynamic blanking, line ω_{01}(3) intersects the frequency-energy function in the frequency value of the dynamic value, (4) ω_{02} is the maximum.

The intersection of the graph of the energy function of the ordinate axis determines the second frequency of dynamic dampingω_{02}, with the value represented by the line (3, Figure 21). The maximum and minimum of the energy function determines the natural frequencies.

2. A coordinate system showing the motion of the center of mass of $v_1 v_2$the dyad and the relative displacement of mass inertia elements is considered.

Coordinates (y_1, y_2) and (v_1, v_2) are interconnected by the following relationships:

$$\begin{cases} v_1 = \dfrac{m_1 y_1 + m_2 y_2}{m_1 + m_2}, \\ v_2 = y_2 - y_1, \end{cases} \tag{70}$$

where $v_1$is the center of mass of the dyad, $v_2$is the difference in the positions of the concentrated masses.

Transition to the coordinate system, the type of frequency energy function v_1 $v_2$changes. To construct frequency-energy functions in the coordinate system, the kinetic and potential energy $v_1 v_2$are considered, respectively:

$$T = \frac{1}{2}(m_1 + m_2)\dot{v}_1^2 + \frac{1}{2}\frac{m_1 m_2}{m_1 + m_2}\dot{v}_2^2 \tag{71}$$

$$\Pi = \frac{1}{2}k_2 v_2{}^2 \tag{72}$$

It is assumed that the system makes its own fluctuations$v_2 = B_2 \sin\omega t$,. The amplitude $v_1 = B_1 \sin\omega t$ratio is$\beta = B_2/B_1$. From the relation $T_{max} = \Pi_{max}$presented in the variables, the relation $\beta B_1$is obtained:

$$\frac{1}{2}(m_1 + m_2)\omega^2 B_1^2 + \frac{1}{2}\frac{m_1 m_2}{m_1 + m_2}\omega^2 \beta^2 B_1^2 = \frac{1}{2}k_2 \beta^2 B_1^2 \tag{73}$$

The resulting expression can be converted to:

$$\frac{1}{\omega_0{}^2 \beta^2} + \frac{1}{\omega_2{}^2} = \frac{1}{\omega^2} \tag{74}$$

Where $\omega_0{}^2 = \dfrac{k_2}{m_1 + m_2}$

Thus, the frequency-energy function has the form (Figure 22):

$$\omega^2 = \frac{\beta^2 \omega_0{}^2 \omega_2{}^2}{\beta^2 \omega_0{}^2 + \omega_2{}^2} \tag{75}$$

The first derivative is:

$$\frac{\partial(\omega^2)}{\partial\beta} = \frac{2\beta \omega_2{}^4 \omega_0{}^2}{(\beta^2 \omega_0{}^2 + \omega_2{}^2)^2} \tag{76}$$

The function increases $\omega^2(\beta)$monotonically on the positive half-axis$\beta \geq 0$. If the shape factor, $\beta_1 = 0$the function ω^2reaches a local minimum of zero.

Figure 22. Frequency-energy function in the coordinate system v_1, v_2

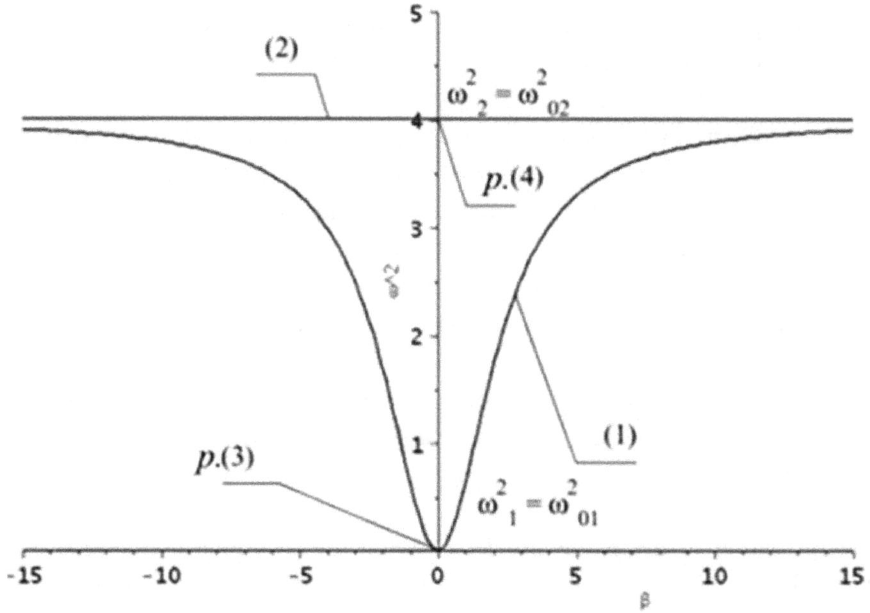

curve (1) - graph of frequency-energy function, line (2) with ordinate - asymptote of $\omega^2{}_2 = \omega_{02}{}^2$graph 1 at; t. ($\beta \to +\infty$3) - the value of the frequency-energy function is equal to$\omega^2{}_1 = \omega_{01}{}^2$

If the shape coefficient tends to infinity$\beta \to \infty$, then the function$\omega^2 \to \omega_2{}^2$. Figure 22 shows the frequency-energy function for a coordinate system in which the eigenvalues coincide with the partial frequencies:

$$\omega^2{}_1 = \omega_{01}{}^2, \omega^2{}_2 = \omega_{02}{}^2 \tag{77}$$

Curve (1) in Figure 22 is a variant of the frequency-energy function, which has a number of features associated with the coincidence of natural frequencies of oscillation and partial frequencies. The minimum function is achieved in vol. (3) in Figure 22 at $\beta = 0$ and is equal to$\omega^2{}_1$, coinciding with the square of the frequency of dynamic damping of vibrations$\omega_{01}{}^2$. The maximum of the function degenerates into the limit value at $\beta \to +\infty$, since the function is monotonic on the interval $(0, +\infty)$ and bounded by the square of the second natural frequency$\omega^2{}_2$. The line (2) in Figure 22, parallel to the abscissa axis, intersects the ordinate axis in vol. (4) and is the asymptote of the graph (1) at $\beta \to +\infty$.

Thus, the local maximum determining the eigenvalue of the system degenerates into an asymptotic value (Khomenko & Eliseev, 2016).

The dyad reflects the properties that an oscillatory system would have under isolation conditions.

The concept of dyad can be seen as a generalization of ideas about partial systems used to construct mathematical models of mechanical oscillatory systems with several degrees of freedom.

Unlike ordinary partial systems, the dyad is a structure with two degrees of freedom, has two mass-inertia elements connected by a linear spring. The dyad, in its original representation, is not associated with datum surfaces, that is, it has no external connections.

Methods of constructing structural mathematical models reflecting basic dynamic properties and capabilities of determining values of partial and natural frequencies of oscillations are proposed.

The dynamic properties of dyads form some fundamental basis for the construction of mechanical oscillatory systems for a specific purpose, which can be used in the use of dynamic effects arising in combinations of translational, rotational and helical modes of motion.

The formation of a method for determining the dynamic characteristics of a dyad based on the frequency-energy method is justified. A physical concept of the connectivity of movements between elements of the dyad and determination of the forms of the ratio of the amplitude of oscillations depending on the dynamic characteristics of the system is proposed. Analytical and graphoanalytical approaches are proposed to determine the main dynamic parameters of the system based on the use of frequency characteristics obtained from the structural schemes of the dyad.

Analytical relations determining dynamic properties of dyad by harmonic external disturbances are obtained. The possibilities and forms of interaction of the dyad with the imposition of additional bonds are determined.

DYNAMIC PROPERTIES OF DYADS WITH HETEROGENEOUS PARTIAL SYSTEMS

In recent years, considerable attention has been paid to the dynamics of mechanical oscillatory systems with several degrees of freedom in the presence of connections in their structure, realized by the introduction of additional elements, mechanisms and devices for transforming motion (de Silva, 2000; Eliseev, Reznik, & Khomenko, 2011; Karnovsky & Lebed, 2016; Eliseev & Artyunin, 2016). New approaches (Banakh & Kempner, 2010; Banakh, 2011) and methods of constructing mathematical models, based, among other things, on the analytical apparatus of the theory of chains and the theory of automatic control (Kolovsky, 1976; Druzhinsky, 1977; Galperin, 1964; Emelyanov & Korovin, 1997; Genkin & Ryaboy, 1988; Khomenko, Eliseev, & Yermoshenko, 2012), have developed. A detailed understanding of the dynamic properties of systems using the concepts of structural mathematical models, transfer functions and their features has been reflected in the works of recent years devoted to studies of the connectivity of movements, dynamic rigidities and the given mass-inertial characteristics of structural formations (Tarasik, 2004; Belokobylsky, Eliseev, & Kashuba, 2015; Khomenko & Eliseev, 2016; Khomenko & Eliseev, 2016; Belokobylsky, Eliseev, & Kashuba, 2016). Expanding ideas about the forms of dynamic interactions of elements of oscillatory systems, manifestations of self-organization of movements initiated interest in studying the features of establishing certain and characteristic connections between the coordinates of movement in systems with several degrees of freedom (Khomenko & Eliseev, 2016; Eliseev, Artunin, & Bolshakov, 2015; Eliseev, Parshuta, Kaimov, & Kinash, 2015; Eliseev, Parshuta, Kaimov, & Kinash, 2014). In this regard, one can quite definitely note the essential importance of the forms of transformation of mutual motions of elements of mechanical oscillatory systems under various forms of external harmonic disturbances.

The proposed article develops methodological positions in the evaluation of interaction of elements of oscillatory systems on the basis of ideas about dyads as certain structures consisting of two mass-inertial links connected to each other by an elastic element, taking into account different forms of movements at the level of inter-partial connections and isolation from contacts with support surfaces.

STRUCTURAL AND DYNAMIC FEATURES OF DYADS WITH MULTIPLE MOTION FORMS

Assuming that the dyad is a structural formation of two mass inertia elements connected to each other by an elastic element, we note that partial systems can make the same or different movements in shape. The simplest options are related to the consideration of two solids (material points) making rectilinear movements. In this case, the dyad has a center of mass, and its movement can be considered in various coordinate systems, including in a fixed basis. An analogue of such a dyad can be considered a structural formation of two inertial elements capable of rotating around the same axis, taking into account mutual vibrations with elastic connections in relative movements (Baty, Djanelidze, & Kelzon, 1966; Guskov, Panovko, & Shokhin, 2012). At the same time, dyads are possible in which one mass-inertial element makes a rectilinear movement, and the second rotates relative to an axis perpendicular to the axis of movement of the first element (while the axes intersect each other). Systems of this kind are considered in works where an oscillating solid body has an attached pendulum (Sommerfeld, 2001; Den Hartog, 1960). Finally, the dyad can also be considered in such a version of partial interactions that are characteristic of non-self-braking screw mechanisms: one of the elements of the dyad (screw) implements rectilinear vibrations, and the second (nut - flywheel) performs torsional vibrations relative to the screw and forms screw movements: the elements of the dyad are interconnected by a spring (Eliseev, Lukyanov, Reznik, & Khomenko, 2006; Eliseev et al., 2015) . In this case, the axis of rotation and the direction of rectilinear motion coincide. Thus, the dyad, in its simplest forms, can be represented (at the preliminary stage of research) by four physical simplest options and forms of combinations of movements that can be used in problems of the dynamics of mechanical oscillatory systems.

A solid body, for example, as an object of vibration protection on elastic supports with stiffnesses k_1 and k_2, mass M and moment of inertia J, can be represented by a schematic diagram of Figure 23, *a*. In turn, Figure 23, *b* and Figure 23, *c* show the structural mathematical models of the object, respectively, in the coordinate systems y_1, y_2 and y_0, φ, where y_0 is the coordinate of the center of gravity, φ is the angle of rotation relative to the center of gravity.

Figure 23. Schematic diagram. a) vibration protection object: b) - structural mathematical model in coordinates y_1, y_2. c) block diagram in coordinates y_0, φ

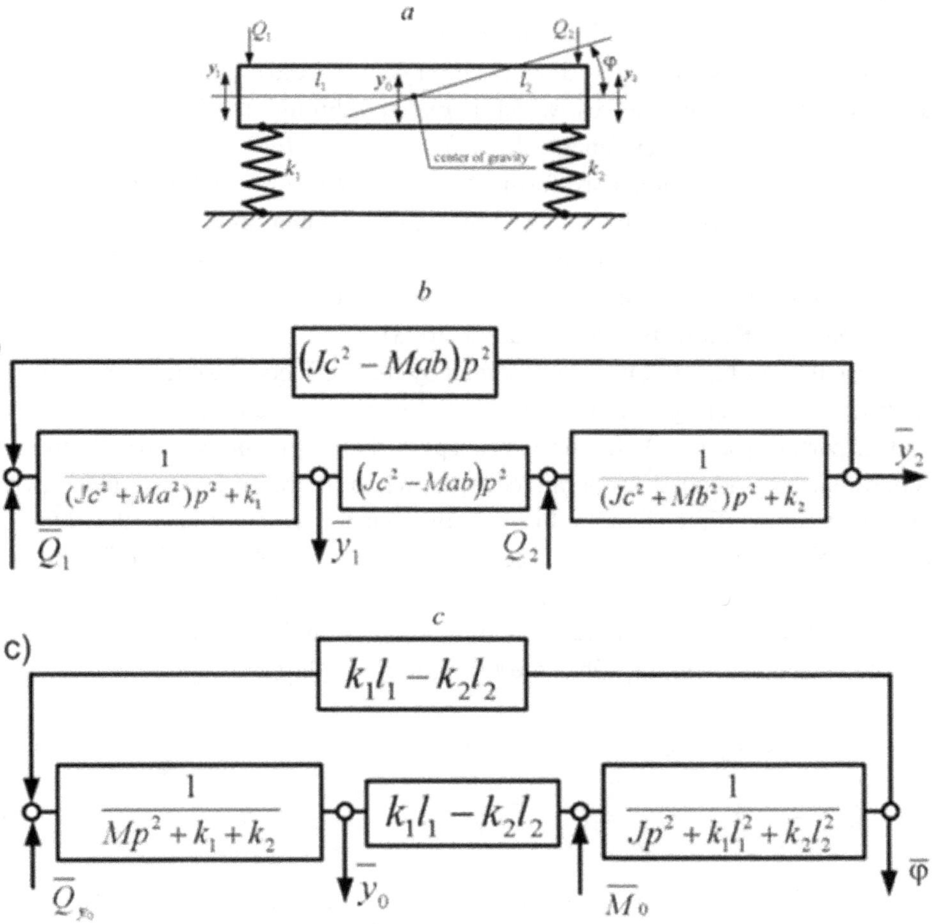

Construction of structural mathematical models involves preliminary obtaining mathematical models in the form of a system of ordinary differential equations of the 2nd order with constant coefficients under zero initial conditions, followed by the construction of structural schemes of dynamically equivalent automatic control systems: detailed descriptions and examples of such approaches are given in (Eliseev & Artyunin, 2016; Eliseev, Reznik, Khomenko, & Zasyadko, 2008). In Figure 23, *b* and Figure 23, *c* it is assumed that $p = j\omega$ is a complex variable ($j = \sqrt{-1}$); the <-> icon above the variable corresponds to its Laplace image.

When deriving equations of motion and building models, the relationships between the coordinates are taken into account: $y_0 = ay_1 + by_2$; $\varphi = c \cdot (y_2 - y_1)$; $y_1 = y_0 -$

$l_1\varphi$; $y_2 = y_0 + l_2\varphi$; $a = l_2/(l_1 + l_2)$; $b = l_1/(l_1 + l_2)$;, $c = 1/(l_1 + l_2)$where l_1, l_2- distances to the center of gravity from attachment points of elastic elements k_1 and k_2 (Figure 23, *a*).

From Figure 23, *b* - *c* it follows that the structural mathematical model consists of two partial blocks. In the coordinate system y_1, y_2, partial systems have an inertial connection between themselves, which is realized through the link of interpartial connection with a transfer function$W_{par}(p) = (Jc^2 - Mab)p^2$. In the coordinate system y_0, φ partial systems are formed in a different form and their interpartial connections are displayed by a link with a transfer function $W'_{par}(p) = k_1 l_1 - k_2 l_2$. Inter-artillery bonds are elastic; their physical meaning corresponds to stiffness of elastic elements in angular vibrations of the object (Eliseev & Artyunin, 2016; Eliseev, Reznik, Khomenko, & Zasyadko, 2008).

External disturbance can be represented by force effects $\overline{Q}_1$and$\overline{Q}_2$, as shown in the structural diagram given in Figure 23, *b*. Input actions create excitation by two inputs to partial systems. As for the external perturbations in the system in Figure 23, *c* (coordinates y_0 and φ), the generalized forces will be represented as $\overline{Q}_{y_0}$ and $\overline{M}_0$with the corresponding dimension. In this case, the transformation of generalized forces must comply with the condition of equality of the work of generalized forces on virtual displacements in both coordinate systems (Lurie, 1961).

The relationship between the applied forces Q_1, Q_2andQ_{y_0}, moment M_0in the coordinate systems $y_1, y_2$and y_0, ϕhas the form (at$Q_1 = Q$, $Q_2 = Q$):

$$\overline{Q}_1 = a\overline{Q}_{y_0} - c\overline{M}_0, \quad \overline{Q}_2 = b\overline{Q}_{y_0} + c\overline{M}_0 \tag{78}$$

$$\overline{M}_0 = l_2\overline{Q}_2 - l_1\overline{Q}_1, \quad \overline{Q}_{y_0} = \overline{Q}_1 + \overline{Q}_2 \tag{79}$$

Transfer functions of the system in coordinates y_1, y_2 at $Q_1 \neq 0$ ($Q_2 = 0$)

$$W_1(p) = \frac{\overline{y}_1}{\overline{Q}_1} = \frac{(Mb^2 + Jc^2)p^2 + k_2}{A(p)} \tag{80}$$

$$W_2(p) = \frac{\overline{y}_2}{\overline{Q}_1} = \frac{(Jc^2 - Mab)p^2}{A(p)} \tag{81}$$

$$W_{12}(p) = \frac{\overline{y}_2}{\overline{y}_1} = \frac{(Jc^2 - Mab)p^2}{(Mb^2 + Jc^2)p^2 + k_2} \tag{82}$$

where

$$A(p) = ((Ma^2 + Jc^2)p^2 + k_1)\cdot((Mb^2 + Jc^2)p^2 + k_2) - ((Jc^2 - Mab)p^2)^2 \tag{83}$$

characteristic equation of the system.

The objective of the study is to study the features of constructing mathematical models of dyads and the possibilities of assessing their dynamic properties within the framework of the concept that assumes that the dynamic properties of systems having connections with support surfaces are determined on the basis of the fundamental properties of structural formations called dyads.

Dynamic Properties Of Dyads

The mechanical system in Figure 23, *a* can be represented by structural mathematical models in different coordinate systems (Figure 23, *b*; Figure 23, *c*). Obtaining dyads can be realized by zeroing the stiffness of the elements k_1 and k_2, which are associated with the support surfaces. For the case of the physical representation of the original system according to Figure 23, *a* (that is, in the form of a solid body with mass M and moment of inertia *J*) in the coordinate system y_1, y_2, the dyad will take the form as shown in Figure 24, *a* -*b*. In structural diagrams, it is assumed that only one external force acts $\overline{Q}_1$ (Figure 24, *a*) when the y_2 coordinate is excluded as a result of transformations (it is assumed that $\overline{Q}_2 = 0$). Mass inertia properties of the protected object (Figure24, *a*) are reflected by the link with the transfer function of the second-order $W(p) = 1/(Ma^2 + Jc^2)p^2$ integrating link. Figure 24, *b* shows a block diagram of the same system, but excluding the y_1 coordinate. In this case, the reduction of all external forces is realized on the mass inertial element along the y_2 coordinate. The external force Q_1 is also transformed when transferred to the coordinate y_2 (Figure 24, *b*), taking into account that $\overline{Q}_2 = 0$.

Figure 24. Structural (a) mathematical model of dyad with excluded coordinate $\overline{y}_2$ at external harmonic force $\overline{Q}_1$; structural (b) mathematical model of a dyad with an excluded coordinate $\overline{y}_1$

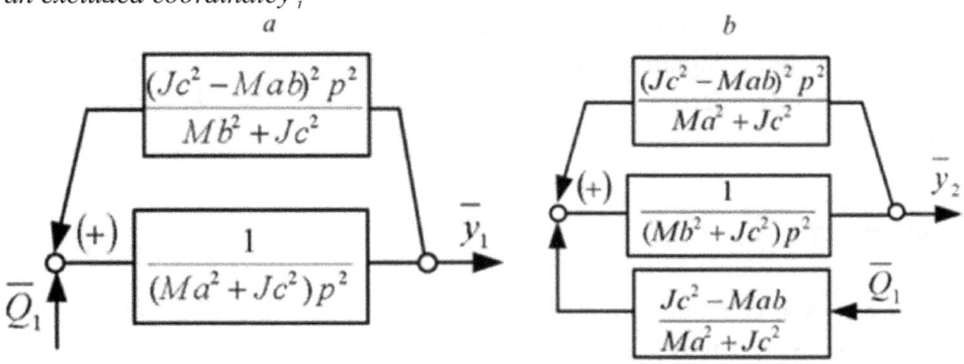

Accordingly, we write the transfer functions of the dyad using the expressions (80) - (82):

$$W'_1(p) = \frac{\bar{y}_1}{\bar{Q}_1} = \frac{(Mb^2 + Jc^2)p^2}{A'(p)} \tag{84}$$

$$W'_2(p) = \frac{\bar{y}_2}{\bar{Q}_1} = \frac{(Jc^2 - Mab)p^2}{A'(p)} \tag{85}$$

$$W'_{12}(p) = \frac{\bar{y}_2}{\bar{y}_1} = \frac{(Jc^2 - Mab)p^2}{Mb^2 + Jc^2} \tag{86}$$

Here:

$$A'(p) = ((Ma^2 + Jc^2)p^2) \cdot ((Mb^2 + Jc^2)p^2) - ((Jc^2 - Mab)p^2)^2 = MJc^2p^4 \tag{87}$$

characteristic frequency equation which is simplified compared to form (83).

Note that the transfer functions of the dyad are significantly different from the transfer functions of the system shown in Figure 23, *b*:

$$W''_1(p) = \frac{\bar{y}_1}{\bar{Q}_1} = \frac{(Mb^2 + Jc^2)}{MJc^2p^2} \tag{88}$$

$$W''_2(p) = \frac{\bar{y}_2}{\bar{Q}_1} = \frac{(Jc^2 - Mab)}{MJc^2p^2} \tag{89}$$

Consideration of the transfer functions (87) to (89) shows that they do not reflect the oscillatory properties of the dyad, that is, a mechanical oscillatory system consisting of a solid body making a flat movement cannot be represented as a dyad due to the lack of elastic elements. The presence of elastic elements is necessary for the accumulation of potential energy, since the oscillatory process ensures the exchange of energy, which during interactions passes from one form to another, leaving the value of the total mechanical energy constant.

In the structural diagram of the dyad shown in Figure 24, *a*, the link reflecting the mass-inertial properties of the object is covered by a feedback having a transfer function:

$$W_{os}(p) = \frac{(Jc^2 - Mab)p^2}{Mb^2 + Jc^2} \tag{90}$$

Taking into account this circumstance, convert the structural diagram to Figure 24, *a*, then after a number of simplifications we get the transfer function of the protected object as an integrating link of the 2nd order:

$$W_{os}(p) = \frac{1}{M_{np}p^2} \tag{91}$$

where

$$M_{pr} = \frac{Mb^2Jc^2}{Mb^2 + Jc^2} \tag{92}$$

Thus, a solid body, as an object of protection in an elastic mechanical system (Figure 23, *a*), is finally transformed into a mass-inertial element, which will have a reduced mass M_{pr} and move under the action of force $\overline{Q}_1$ along the y1 coordinate. But such a movement will not reflect the property of the dyad, since its transformation did not occur.

When choosing the coordinate system y_0, φ, the same approach is used, but at the same time for the structural diagram in Figure 23, *c* we obtain generalized forces in the form:

$$\overline{Q}_{y_0} = \overline{Q}_1 + \overline{Q}_2, \overline{Q}_{\phi} = l_2\overline{Q}_2 - l_1\overline{Q}_1 = \overline{M}_0$$

. (93)

In this case, assuming $\overline{Q} = \overline{Q}_1 = \overline{Q}_2$ we find that $\overline{Q}_{y_0} = 2\overline{Q}$, $\overline{M}_0 = \overline{Q}(l_2 - l_1)$.

Partial systems of the structural diagram in Figure 23, *c* receive perturbation at two inputs, then the transfer functions can be constructed taking into account the superposition:

$$W_{01} = \frac{\overline{y}_0}{\overline{Q}} = \frac{2(Jc^2p^2 + k_1l_1^2 + k_2l_2^2) + (l_2 - l_1)(k_1l_1 - k_2l_2)}{A''(p)} \tag{94}$$

$$W_{02} = \frac{\overline{\Phi}_0}{\overline{Q}} = \frac{2(l_2 - l_1)(Mp^2 + k_1 + k_2) + (k_1l_1 - k_2l_2)}{A''(p)} \tag{95}$$

$$W_{012} = \frac{\overline{\Phi}_0}{\overline{Q}} = \frac{2(l_2 - l_1)(Mp^2 + k_1 + k_2) + (k_1l_1 - k_2l_2)}{2(Jc^2p^2 + k_1l_1^2 + k_2l_2^2) + (l_2 - l_1)(k_1l_1 - k_2l_2)} \tag{96}$$

where

$$A''(p) = (Mp^2 + k_1 + k_2)(Jc^2p^2 + k_1l_1^2 + k_2l_2^2) - (k_1l_1 - k_2l_2)^2 \tag{97}$$

- is the characteristic frequency equation in the coordinate system $\overline{y}_0, \overline{\Phi}$.

Figure 25 shows, as an example, the structural diagram of the initial system (Figure 23, *c*) with the exception of the φ coordinate (Figure 25, *a*) and the y_0 coordinate (Figure 25, *b*).

Figure 25. Structural diagram of the system as per Figure23 coordinates y_o, φ

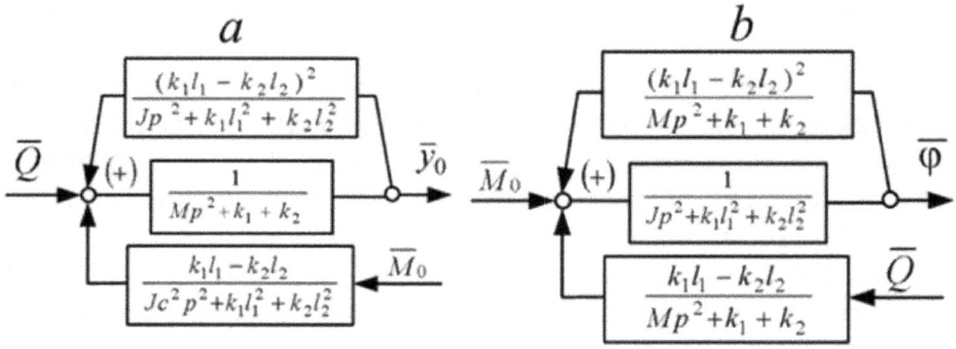

a) at exclusion of coordinate φ, b) at exclusion of coordinate y_0

In the structural diagrams, external disturbances $\overline{Q}_0$,$\overline{M}_0$are reduced to the corresponding input of partial systems. When switching to the dyad, it is necessary to accept that $k_1 = 0$ and $k_2 = 0$, then the structural diagram in Figure 25 *a-b* is transformed, which is presented in Figure 26 *a-b* by coordinates y_0, φ, respectively.

Figure 26. Structural diagram for dyads obtained from structural diagrams in Figure25, a-b by coordinate y_o (a) and φ (b)

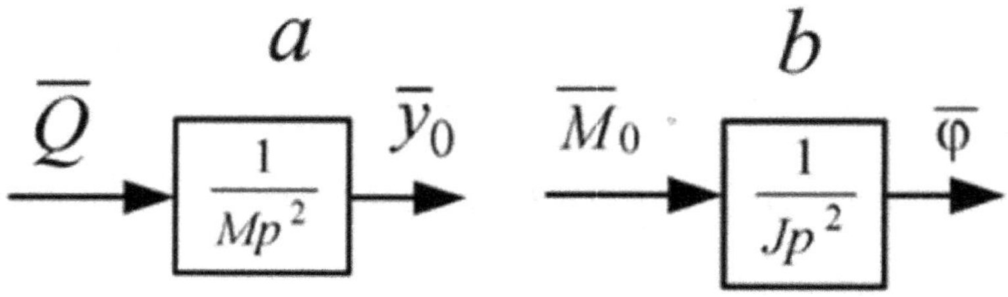

Thus, a dyad in the coordinate system of a solid according to the scheme in Figure 23, *a* is not formed in coordinates$\overline{y}_0$, $\overline{\phi}$just as in coordinates$\overline{y}_1$, $\overline{y}_2$which was shown above.

A methodological basis has been developed for constructing mathematical models of dyads in systems of a combined type, formed from two partial systems that reflect the features of translational and rotational movements.

Dyad is considered as an "autonomous" structural formation of two mass-inertial elements with the possibility of elastic interactions with each other. A combined type dyad has two degrees of freedom. A feature of the dyad is the isolation of its movements from interaction with support surfaces and other elements not included in the dyad.

It is shown that a solid body on elastic supports as a mechanical oscillatory system with two degrees of freedom, performing flat oscillatory movements under the action of external harmonic forces, cannot be reduced to the form of a dyad due to the impossibility of organizing oscillatory movements in the connections between partial systems.

Disclosed is a method of constructing mathematical models of dyads for mechanical oscillatory systems in which there is a connection between two mass-inertial elements, realised by a fifth-class kinematic rotary pair.

The possibilities of evaluating the dynamic properties of a dyad on mathematical models operating in two coordinate systems have been investigated.

Disclosed is a method of determining parameters of connectivity of movements in a dyad between coordinates based on introduction of a transfer function of inter-partial connections and frequency of the energy function. The latter is the ratio of potential and kinetic energies, the extremes of which coincide with the values of the natural frequency of the dyad. The approach allows expansion into systems in which the dyad acquires connections with supporting surfaces.

A feature of the dyad is the possibility of energy accumulation in elastic deformations of the elements during relative movements of the mass-inertial elements, which predetermines the motion parameters of the mechanical oscillatory system obtained after connecting the dyad to support points (or supports).

The use of various coordinate systems opens up the possibility of identifying new dynamic effects that are initiated by new dynamic links, and, unlike links in chain systems having elastic interpartial interactions, are able to form motion dependencies of an inertial and inertial elastic type.

It is assumed that the introduction of the concept of dyads and the features of their interactions in the formation of a mechanical oscillatory system can become the basis of new methods of analysis and dynamic synthesis not only in the theory of mechanical vibrations, but also in nanomechanics.

REFERENCES

Banakh, L., & Kempner, M. (2010). *Vibrations of mechanical systems with regular structure*. Springer. DOI: 10.1007/978-3-642-03126-7

Banakh, L. Y. (2011). Vibrations of mechanical systems with self-similar structure. *Vestnik of Nizhny Novgorod Lobachevsky State University*, *4*(2), 52–53.

Baty, M. I., Djanelidze, G. Yu., & Kelzon, A. S. (1966). *Theoretical mechanics in examples and problems* (Vol. 2). Dynamics. Science.

Belokobylsky, S. V., Eliseev, S. V., & Kashuba, V. B. (2015). Impedance approaches as a form of evaluating dynamic properties of mechanical oscillatory systems in structural mathematical modeling. Systems. Methods. *Technologies*, *4*(28), 7–15.

Belokobylsky, S. V., Eliseev, S. V., & Kashuba, V. B. (2016). Elastic elements: Dynamic properties, generalized approach possibilities, and quasi-springs. Systems. Methods. *Technologies*, *2*(30), 7–17.

Belokobylsky, S. V., Eliseev, S. V., & Sitov, I. S. (2013). *Dynamics of mechanical systems: Lever and inertial-elastic connections*. Polytechnica.

Belokobylsky, S. V., Eliseev, S. V., & Sitov, I. S. (2013). *Dynamics of mechanical systems. Lever and inertial-elastic connections*. Polytechnika.

Clarence, W., & de Silva. (2000). Vibration. Fundamentals and Practice. Boca Raton, London, New York, Washington, D.C.: CRC Press. 957 p.

de Silva, C. W. (2000). *Vibration: Fundamentals and practice*. CRC Press.

Den-Hartog, D. P. (1960). *Mechanical vibrations*. Fizmatgiz.

Druzhinsky, I. A. (1977). *Mechanical chains*. Mashinostroenie.

Eliseev, A. V. (2017). Diads in mechanical systems: Features of dynamic properties. Part I. *Vestnik Irkutskogo Gosudarstvennogo Tehniceskogo Universiteta*, *21*(7), 26–38. DOI: 10.21285/1814-3520-2017-7-26-38

Eliseev, A. V. (2017). Diads in mechanical systems: Features of dynamic properties. Part II. *Vestnik Irkutskogo Gosudarstvennogo Tehniceskogo Universiteta*, *21*(8), 22–37. DOI: 10.21285/1814-3520-2017-8-22-37

Eliseev, A. V. (2017). Dynamic properties of a dyad with heterogeneous partial systems. *Vestnik Irkutskogo Gosudarstvennogo Tehniceskogo Universiteta*, *21*(5), 32–53. DOI: 10.21285/1814-3520-2017-5-32-53

Eliseev, A. V. (2017). Dynamic properties of a diad with heterogeneous partial systems. *Vestnik Irkutskogo Gosudarstvennogo Tehniceskogo Universiteta, 21*(5), 32–53. DOI: 10.21285/1814-3520-2017-5-32-53

Eliseev, A. V., Artyunin, A. I., Nguyen, D. H., & Vuong, K. C. (2016). Control of the vibration field of a technological machine based on changes in equivalent inertial properties. In Proceedings of the Conference "Vibration Technologies, Mechatronics, and Controlled Machines" (pp. 343). Kursk: Southwest State University.

Eliseev, A. V., & Eliseev, S. V. (2017). The dyad as the basis for the formation of mechanical oscillatory systems. Systems. Methods. *Technologies*, *36*(4), 25–38.

Eliseev, A. V., & Eliseev, S. V. (2017). Diad as a basis for forming mechanical oscillatory systems. Systems. Methods. *Technologies*, *4*(36), 25–38.

Eliseev, A. V., Kopylov, Y. R., & Eliseev, S. V. (2017). Possibilities of dynamic correction of vibration fields of working elements of technological machines. In Proceedings of the International Conference on Recent Advances in Engineering, Technology and Applied Sciences (pp. 65–71). Detroit: Scientific Public Organization "Professional Science." 336 p.

Eliseev, A. V., Kopylov, Y. R., & Eliseev, S. V. (2017). Possible methods and means of assessing and controlling the vibration state of technological machines. In Proceedings of the Conference "Vibro-Wave Processes in High-Tech Parts Processing Technology" (pp. 20–22). Rostov-on-Don: DSTU. 240 p.

Eliseev, A. V., Nikolaev, A. V., & Mironov, S. V. (2018). Diads in mechanical oscillatory systems: Model of dynamic interaction formation. In Problems of Mechanics of Modern Machines: Proceedings of the VII International Scientific Conference (Vol. 1, pp. 33-40). Ulan-Ude: VSGUTU.

Eliseev, A. V., Selivinsky, V. V., & Eliseev, S. V. (2015). *Dynamics of vibrational interactions of elements of technological systems with consideration of non-retaining connections: Monograph*. Nauka.

Eliseev, A. V., Selvinsky, V. V., & Eliseev, S. V. (2015). *Dynamic interactions of elements in technological systems considering non-retentive connections*. Nauka.

Eliseev, A. V., & Vuong, K. C. (2016). Some possibilities for controlling one-dimensional vibration fields of technological machines. Modern Technologies. Systems Analysis. *Modeling*, *49*(1), 33–41.

Eliseev, S. V. (2018). *Applied systemic analysis and structural mathematical modeling (dynamics of transport and technological machines: movement connectivity, vibrational interactions, lever connections) (Monograph)*. Irkutsk State University of Railway Transport.

Eliseev, S. V., Artunin, A. I., & Bolshakov, R. S. (2015). Mechanical chains in structural diagrams of vibration protection systems: Methodology for determining dynamic reactions. *Bulletin of the All-Russian Scientific Research and Design Institute of Electric Locomotive Engineering*, *1*(69), 93–111.

Eliseev, S. V., & Artyunin, A. I. (2016). *Applied theory of oscillations in the dynamics of linear mechanical systems*. Nauka.

Eliseev, S. V., & Artyunin, A. I. (2016). *Applied theory of vibrations in the dynamics of linear mechanical systems*. Nauka.

Eliseev, S. V., Eliseev, A. V., & Mironov, A. S. (2018). Approaches to evaluating the features of vibration fields during the interaction of a working medium with the working body of technological machines. In Proceedings of the IV International Scientific and Practical Conference "Open Innovation" (pp. 318). Penza: MCSNS "Science and Education."

Eliseev, S. V., Eliseev, A. V., & Mironov, A. S. (2018). Possibility of changing the vibratory field of working bodies of technological machines. Journal of Advanced Research in Technical Science, Issue 9–1, 50–56.

Eliseev, S. V., & Kargapoltsev, S. K. (2014). Representation of mechanical element connection features in the theory of mechanical chains. In L. V. Massel (Ed.), Information and mathematical technologies in science and management: Proceedings of the XIX Baikal All-Russian Conference (pp. 100–108).

Eliseev, S. V., & Khomenko, A. P. (2014). *Dynamic damping of vibrations: Feedback concept and structural methods of mathematical modeling*. Nauka.

Eliseev, S. V., Kuznetsov, N. K., & Kaimov, E. V. (2015). On the theory of lever connections in the dynamics of mechanical oscillatory systems. *Bulletin of the Irkutsk State Technical University*, *12*(107), 30–40.

Eliseev, S. V., & Lonchikh, P. A. (2012). Evaluation of interaction forms between partial systems in mechanical chains. Possible simplifications. *Vestnik Irkutsk State Technical University*, *6*(65), 17–21.

Eliseev, S. V., Loncikh, P. A., Kashuba, V. B., Parshuta, E. A., Kaimov, E. V., & Nguyen, D. H. (2015). New approaches in the dynamics of mechanical elastic-inertial systems with lever connections. In Science and education in the life of modern society: Collection of scientific papers from the International Scientific and Practical Conference (Vol. 12, pp. 75–85).

Eliseev, S. V., Lukyanov, A. V., Reznik, Y. N., & Khomenko, A. P. (2006). *Dynamics of mechanical systems with additional ties*. Irkutsk State University Publishing.

Eliseev, S. V., Orlenko, A. I., & Eliseev, A. V. (2017). Structural formations in mechanical oscillatory systems: Dyads, their properties, and the possibilities for changing dynamic states. *Transport of the Urals*, *3*(54), 56–63. DOI: 10.20291/1815-9400-2017-3-56-63

Eliseev, S. V., Orlenko, A. I., & Eliseev, A. V. (2017). Structural formations in mechanical oscillatory systems: Diads, their properties, and the potential for changing dynamic states. *Transport of the Urals*, *3*(54), 56–63. DOI: 10.20291/1815-9400-2017-3-56-63

Eliseev, S. V., Orlenko, A. I., & Nguyen, D. H. (2017). Devices for motion transformation in the structure of a diad in a mechanical oscillatory system. *Vestnik of Don State Technical University*, *17*(3), 46–59. DOI: 10.23947/1992-5980-2017-17-3-46-59

Eliseev, S. V., Parshuta, E. A., Kaimov, E. V., & Kinash, N. Z. (2014). Mechanisms in the structure of vibration protection systems: Mathematical models, dynamic property assessment (Part I). *Bulletin of VSGUTU*, *6*(51), 37–44.

Eliseev, S. V., Parshuta, E. A., Kaimov, E. V., & Kinash, N. Z. (2015). Mechanisms in the structure of vibration protection systems: Mathematical models, dynamic property assessment (Part II). *Bulletin of VSGUTU*, *1*, 52–60.

Eliseev, S. V., Reznik, Y. N., & Khomenko, A. P. (2011). *Mechatronic approaches in the dynamics of mechanical oscillatory systems*. Nauka.

Eliseev, S. V., Reznik, Y. N., & Khomenko, A. P. (2011). *Mechatronic approaches in the dynamics of mechanical oscillatory systems*. Nauka.

Eliseev, S. V., Reznik, Y. N., Khomenko, A. P., & Zasyadko, A. A. (2008). *Dynamic synthesis in generalized problems of vibration protection and vibration isolation of technical objects*. Irkutsk State University Publishing House.

Eliseev, S. V., Reznik, Y. N., Khomenko, A. P., & Zasyadko, A. A. (2008). *Dynamic synthesis in generalized vibration protection and vibration isolation problems of technical objects*. Irkutsk State University Publishing.

Eliseev, S. V., Reznik, Y. N., Khomenko, A. P., & Zasyadko, A. A. (2008). *Dynamic synthesis in generalized vibration protection and vibration isolation problems for technical objects*. ISU.

Eliseev, S.V., Trofimov, A.N., & Kaimov, E.V. (2014). Forms of partial connectivity in mechanical system oscillations. International Journal of Applied and Fundamental Research, 5-1, 15-25.

Emelyanov, S. V., & Korovin, S. K. (1997). *New types of feedback: Control under uncertainty*. Nauka, Fizmatlit.

Frolov, K. V., & Furman, F. A. (1980). *Applied theory of vibration protection systems*. Mashinostroenie.

Frolov, K. V., & Furman, F. A. (1985). *Applied theory of vibration protection systems*. Mashinostroenie.

Galperin, I. I. (1964). *Automation as one-sided mechanics*. Energiya.

Ganiev, R. F., & Kononenko, V. O. (1976). *Vibrations of solid bodies*. Nauka.

Genkin, M. D., & Ryaboy, V. M. (1988). *Elastic-inertial vibration isolation systems: Ultimate capabilities, optimal structures*. Nauka.

Guskov, A. M., Panovko, G. Ya., & Shokhin, A. E. (2012). Calculation of a rod spatial system of vibration isolation for solid bodies under transport vibration. *Problems of Machine Engineering and Reliability of Machines*, *2*, 17–24.

Jie, F. S., Morse, I. E., & Hinkle, R. T. (1966). *Mechanical oscillations* (Obraztsov, I. F., Ed.). Mashinostroenie.

Karnovsky, I. A., & Lebed, E. (2016). *Theory of vibration protection*. Springer International Publishing. DOI: 10.1007/978-3-319-28020-2

Karnovsky, I. A., & Lebed, E. (2016). *Theory of vibration protection*. Springer International Publishing. DOI: 10.1007/978-3-319-28020-2

Kashuba, V. B., Eliseev, S. V., & Bolshakov, R. S. (2017). *Dynamic reactions in the connections of elements of mechanical oscillatory systems*. Nauka.

Khomenko, A. P., & Eliseev, S. V. (2016). Characteristic frequency equation: Structure, dynamic stiffness, and features of system element interaction. Modern Technologies. Systems Analysis. *Modeling*, *2*(50), 12–24.

Khomenko, A. P., & Eliseev, S. V. (2016). Lever connections in mechanical chains: Dynamic effects. Modern Technologies. Systems Analysis. *Modeling*, *4*(52), 8–16.

Khomenko, A. P., & Eliseev, S. V. (2016). Development of the energy method for determining the frequencies of free oscillations of mechanical systems. Modern Technologies. System Analysis. *Modeling*, *1*(49), 8–19.

Khomenko, A. P., & Eliseev, S. V. (2017). Structural formations in mechanical oscillatory systems: Diad as a form of interaction of elements. Modern Technologies. Systems Analysis. *Modeling*, *2*(54), 8–14.

Khomenko, A. P., Eliseev, S. V., & Yermoshenko, Y. V. (2012). System analysis and mathematical modeling in mechatronics of vibration protection systems. Irkutsk: IrGUPS. 288 p.

Kolovsky, M. Z. (1976). *Automatic control of vibration protection systems*. Nauka.

Kopylov, Y. R. (2011). *Dynamics of vibro-impact hardening processes*. IPC Scientific Book.

Kopylov, Y. R., & Eliseev, A. V. (2014). Features of the technical implementation of vibration strengthening technology. *Modern High Technologies*, *12*(2), 152–155.

Lenck, A. (1978). Electromechanical systems. Systems with concentrated parameters. Moscow: Mir. 288 p.

Lurie, A. I. (1961). *Analytical mechanics*. GIFML.

Naidenko, O. K., & Petrov, P. P. (1962). *Damping of ship engines and mechanisms*. Sudpromgiz.

Orlenko, A. I., Eliseev, A. V., & Eliseev, S. V. (2018). Additional connections: Possibilities for changing dynamic states of mechanical oscillatory systems. In Proceedings of the 1st International Scientific and Practical Conference "Machines, Units, and Processes: Design, Creation, and Modernization" (pp. 31–37). Saint Petersburg: SPbSF NII MS.

Panovko, G. Y. (2006). *Dynamics of vibrational technological processes (Text)*. NITS Regular and Chaotic Dynamics, Institute of Computer Technologies.

Panovko, Y. G. (1991). *Introduction to the theory of mechanical vibrations*. Nauka.

Rocard, Y. (1949). *General dynamics of vibrations*. Masson.

Sommerfeld, A. (2001). Mechanics. Nizhniy Novgorod: NITS RKhD.

Tarasik, V. P. (2004). Mathematical modeling of technical systems (2nd ed., revised and expanded). Minsk: Design PRO. 640 p.

Tse, F. S., Morse, I. E., & Hinkle, R. T. (1966). Mechanical vibrations (Edited by corresponding member of the USSR Academy of Sciences I.F. Obraztsov). Moscow: Mashinostroenie. 508 p.

Chapter 7
Development of Mathematical Models for Gapless Motion in Mechanical Systems

ABSTRACT

Methods of structural mathematical modeling are being developed in applications to solving problems of estimating, forming and correcting dynamic states of mechanical oscillatory systems that allow the implementation of unilateral interactions of elements.The purpose of the study is to determine the motion parameters of the system, at which there is no break in contact between the elements of the solid. The equations of static and dynamic equilibrium of the system are presented, as well as the contact conditions of the elements. The Dalembert principle is used to determine the dynamic component of the reaction. Particular attention is paid to assessing the range of change in the dynamic component of the contact reaction and constructing an estimate of the parameters at which the contact between the elements of the composite solid is not disturbed during the period of steady motion.

INTRODUCTION

A composite solid is considered, formed by two elements with masses $m_1$and m_2 in accordance with the scheme in Figure 1. Composite solid is placed on flexible element with rigidity $k_1$and length in unloaded statel_{01}. It is assumed that composite solid features have only one degree of freedom, being able to move only in the vertical direction. Elements of a composite solid with masses $m_1$and $m_2$form an irrepressible contact.

DOI: 10.4018/979-8-3373-0447-2.ch007

Figure 1. Composite solid on elastic oscillating base Z_1: P_i- viscous friction forces, Q_i-gravity forces, F_i - constant forces, N_{12}, N_{21} -full contact reactions, a_2,b_1- contact surfaces

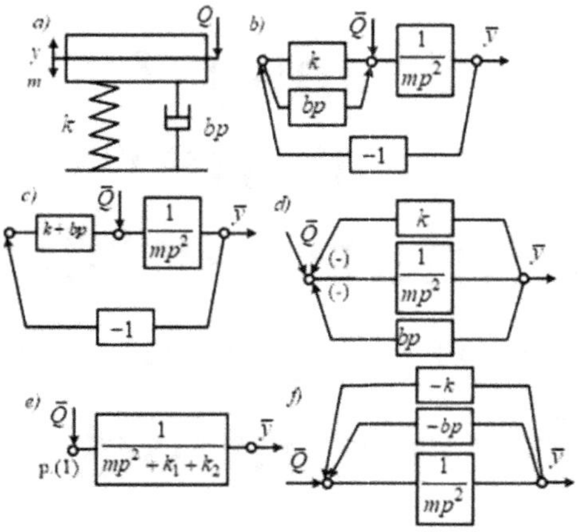

It is assumed that the supporting surface performs harmonic vibrations$Z_1 = A_1 \sin \omega_1 t$, causing movements of the composite solid. At low amplitudes and frequencies, the components of a solid move gaplessly.

X_iThe solid coordinate $m_i$is the surface coordinate of the mass $a_i$solid, $m_i$where $i = 1, 2$. The vertical dimensions of the elements are respectively $d_1$andd_2. In the process of movement, each composite element of a solid body is affected by a viscous friction force $P_i$from a stationary medium, weight $Q_i$and some constant force that $F_i$does not depend on the mass of the solid:

$$P_i = -p_i \dot{X}_i, Q_i = -m_i g F_i = f_{0i} where i = 1, 2 \quad (1)$$

The composite solid features have a non-retaining link of the form:

$$X_1 + d_1 \leq X_2 \quad (2)$$

The contact condition is:

$$X_1 + d_1 = X_2 \quad (3)$$

The condition for determining the critical state, which in some cases may mean the occurrence of a gap, is equal to zero of the complete contact reactionN_{21}.

To determine the conditions of a continuous mode at steady-state oscillations, the positive of the full contact reaction is used as a criterion for the absence of a gap. It is assumed that at the initial moment of time the system is in a position of static equilibrium and motionless.

Task consists in determination of parameters of gapless motion of composite elements of solid body in process of steady oscillation of mechanical system. Parameters are amplitude and frequency of oscillation of bearing surface, coefficients of viscous friction, mass of solid body, values of constant additional forces.

STATIC REACTION DEFINITION

To determine the driving conditions in a contact, you need to find the complete contact reaction. In turn, the complete contact reaction has constant and dynamic components. The static component is defined by static forces at the initial time. The dynamic component of a reaction is the difference between full and static.

For convenience, three superscripts are used in the distinction between static, full and dynamic components: *Σ*- static component, -*Π*complete and - *Δ*dynamic.

To determine the static reaction, write the static equilibrium equations:

$$\begin{cases} R_1^{\Sigma} + Q_1 + F_1 + N_{12}^{\Sigma} = 0, \\ Q_2 + F_2 + N_{21}^{\Sigma} = 0. \end{cases} \tag{4}$$

The static reaction of the elastic element, depending on the position of the static equilibrium, has the X_1^{Σ}form:

$$R_1^{\Sigma} = -k_1(X_1^{\Sigma} - X_{01}), where X_{01} = l_{01} \tag{5}$$

Let's enter the designations for the contact reaction:

$$N_{21}^{\Sigma} = N^{\Sigma}$$

$$N_{12}^{\Sigma} = -N^{\Sigma} \tag{6}$$

System of equations for static equilibrium positionX_1^{Σ}:

$$\begin{cases} -k_1(X_1^{\Sigma} - X_{01})+Q_1 + F_1 - N^{\Sigma} = 0 \\ Q_2 + F_2 + N^{\Sigma} = 0 \end{cases} \tag{7}$$

As a result, the coordinates of the static equilibrium position are:

$$\begin{cases} X_1^\Sigma = X_{01} + \dfrac{Q_1 + F_1 + Q_2 + F_2}{k_1} \\ X_2^\Sigma = X_1^\Sigma + d_1 \end{cases} \tag{8}$$

Thus, the static response of the bearing surface is:

$$N^\Sigma = N_{21}^\Sigma = -Q_2 - F_2 \tag{9}$$

The static component of the total reaction is the opposite of the sum of the weight force and the constant additional force.

TO DEFINE A DYNAMIC REACTION COMPONENT

The complete contact reaction is represented as the sum of the static and dynamic components

$$N_{21}^\Pi = N_{21}^\Sigma + N_{21}^\Delta \tag{10}$$

In turn, the total offsets $X_1^\Pi X_2^\Pi$are represented as the sum of the static and dynamic components:

$$X_1^\Pi = X_1^\Sigma + X_1^\Delta$$

$$X_2^\Pi = X_2^\Sigma + X_2^\Delta \tag{11}$$

The contact condition (3) for dynamic components takes the form:

$$X_1^\Delta = X_2^\Delta \tag{12}$$

To determine the dynamic component of N_{21}^Δ the complete reaction, N_{21}^Πwe use the Dalembert principle. In accordance with Figure 1, for each element of a composite solid, we write out the equilibrium equations of the applied reaction forces from the elastic elementR_1, the weight force, $Q_i$the constant force of the viscous friction F_i,force $P_i$and the inertia forceΦ_i:

$$\begin{cases} R_1 + Q_1 + F_1 + P_1 + \Phi_1 + N_{12}^\Pi = 0, \\ Q_2 + F_2 + P_2 + \Phi_2 + N_{21}^\Pi = 0. \end{cases} \tag{13}$$

For an element with mass, $m_1$the reaction from the elastic element, viscous friction and inertia forces have the following representation:

$$R_1^{\Pi} = -k_1(X_1^{\Pi} - (X_{01} + Z_1))$$

$$P_1 = -p_1\dot{X}_1^{\Pi}$$

$$\Phi_1 = -m_1\ddot{X}_1^{\Pi} \tag{14}$$

For an element with a mass, the viscous friction and inertia $m_2$forces are respectively:

$$P_2 = -p_2\dot{X}_2^{\Pi}$$

$$\Phi_2 = -m_2\ddot{X}_2^{\Pi} \tag{15}$$

The reaction from the elastic member can be represented as:

$$R_1^{\Pi} = R_1^{\Sigma} + R_1^{\Delta} + f_{z_1}, \tag{16}$$

where$R_1^{\Sigma} = -k_1(X_1^{\Sigma} - X_{01})$,$R_1^{\Delta} = -k_1X_1^{\Delta}$,$f_{z_1} = k_1Z_1$.

The system of equilibrium equations (13) with selected static and dynamic components of forces takes the form:

$$\begin{cases} R_1^{\Sigma} + R_1^{\Delta} + f_{z_1} + Q_1 + F_1 + P_1 + \Phi_1 + N_{12}^{\Sigma} + N_{12}^{\Delta} = 0, \\ Q_2 + F_2 + P_2 + \Phi_2 + N_{21}^{\Sigma} + N_{21}^{\Delta} = 0. \end{cases} \tag{17}$$

Given the static equilibrium equation (4), we obtain an equation for the dynamic components of displacement and contact reaction:

$$\begin{cases} R_1^{\Delta} + f_{z_1} + P_1 + \Phi_1 + N_{12}^{\Delta} = 0, \\ P_2 + \Phi_2 + N_{21}^{\Delta} = 0. \end{cases} \tag{18}$$

Viscous friction and inertia forces depend only on dynamic components:

$$P_1 = -p_1 \dot{X}_1^{\Delta}$$

$$\Phi_1 = -m_1 \ddot{X}_1^{\Delta}$$

$$P_2 = -p_2 \dot{X}_2^{\Delta}$$

$$\Phi_2 = -m_2 \ddot{X}_2^{\Delta} \quad (19)$$

After substituting the force expressions (19) from the dynamic displacement component in (18), we obtain a system of equations:

$$\begin{cases} -k_1 X_1^{\Delta} + f_{z_1} - p_1 \dot{X}_1^{\Delta} - m_1 \ddot{X}_1^{\Delta} + N_{12}^{\Delta} = 0, \\ -p_2 \dot{X}_2^{\Delta} - m_2 \ddot{X}_2^{\Delta} + N_{21}^{\Delta} = 0. \end{cases} \quad (20)$$

We assume that the initial conditions have zero values:

$$X_1^{\Delta}(0) = 0, X_2^{\Delta}(0) = 0$$

$$\dot{X}_1^{\Delta}(0) = 0, \dot{X}_2^{\Delta}(0) = 0 \quad (21)$$

The system of differential equations depending on the contact reaction N^{Δ}has the form:

$$\begin{cases} m_1 \ddot{X}_1^{\Delta} + p_1 \dot{X}_1^{\Delta} + k_1 X_1^{\Delta} = f_{z_1} - N^{\Delta}, \\ m_2 \ddot{X}_2^{\Delta} + p_2 \dot{X}_2^{\Delta} = N^{\Delta}, \\ \dot{X}_1^{\Delta}(0) = 0, \dot{X}_2^{\Delta}(0) = 0, \\ X_1^{\Delta}(0) = 0, X_2^{\Delta}(0) = 0. \end{cases} \quad (22)$$

The system (22) describes movement only at the contact interval.

We produce Laplace transform of the obtained system (22) of differential equations:

$$\begin{cases} m_1 s^2 \bar{X}_1 + p_1 s\bar{X}_1 + k_1 \bar{X}_1 = \bar{f}_{z_1} - \bar{N}^{\Delta}, \\ m_2 s^2 \bar{X}_2 + p_2 s\bar{X}_2 = \bar{N}^{\Delta}. \end{cases} \quad (23)$$

The functions $\bar{X}_1$, $\bar{X}_2$, $\bar{f}_{z_1}$, $\bar{N}^{\Delta}$ of the complex variable s are the images X_1^{Δ}, X_2^{Δ}, f_{z_1}, N^{Δ}, corresponding to the Laplace transform:

$$\bar{X}(s) = \int_0^{\infty} X(t)\, e^{-st}\, dt \quad (24)$$

We assume that all the necessary requirements for the existence of the corresponding integrals are fulfilled (Elyseev, Reznik, Khomenko, & Zasyadko, 2008).

The system (23) allows to express, $\bar{X}_1 \bar{X}_2$ via the reaction image $\bar{N}^{\Delta}$:

$$\begin{cases} \bar{X}_1 = \dfrac{\bar{f}_{z_1} - \bar{N}^{\Delta}}{m_1 s^2 + p_1 s + k_1}, \\ \bar{X}_2 = \dfrac{\bar{N}^{\Delta}}{m_2 s^2 + p_2 s}. \end{cases} \quad (25)$$

Let's perform Laplace transform of contact conditions for dynamic displacements (12) and use:

$$\bar{X}_1 = \bar{X}_2 \quad (26)$$

Conditions (26) allow you to determine the image of the dynamic contact reaction that ensures the connection of solid elements from the equality:

$$\frac{\bar{f}_{z_1} - \bar{N}^{\Delta}}{m_1 s^2 + p_1 s + k_1} = \frac{\bar{N}^{\Delta}}{m_2 s^2 + p_2 s} \quad (27)$$

Denote the denominators of the expression (27):

$$L_1(s) = m_1 s^2 + p_1 s + k_1$$

$$L_2(s) = m_2 s^2 + p_2 s. \quad (28)$$

We get an expression for the image of the dynamic component of the contact reaction:

$$\overline{N}^{\Delta} = \frac{L_1(s)L_2(s)}{L_1(s)+L_2(s)} \frac{\bar{f}_{z_1}}{L_1(s)} \tag{29}$$

Taking into account the representation $\bar{f}_{z_1} = k_1 \overline{Z}_1$(- $\overline{Z}_1$image), the transfer Z_1 function from kinematic perturbation to the dynamic component $\overline{Z}_1$of the contact reaction has the $\overline{N}^{\Delta}$form:

$$W_{\frac{\overline{N}^{\Delta}}{\overline{Z}_1}}(s) = \frac{k_1 L_2(s)}{L_1(s)+L_2(s)} \tag{30}$$

Thus, the dynamic contact reaction can also be characterized by a transfer function.

EVALUATION OF THE FLUCTUATION RANGE OF THE DYNAMIC COMPONENT OF THE CONTACT REACTION

The state of the mechanical system in the phase of steady motion is considered, assuming that the viscous friction forces are non-zero. Non-zero friction provides attenuation of natural oscillations. The amplitude-frequency characteristic of the transfer function $W_{\frac{\overline{N}^{\Delta}}{\overline{Z}_1}}(s)$has the form:

$$A(\omega_1) = \left| \frac{k_1 L_2(j\omega_1)}{L_1(j\omega_1)+L_2(j\omega_1)} \right| \tag{31}$$

where, $j^2 = -1$- $\omega_1$frequency of kinematic disturbance of the support surface. Expressions$L_1(s)$, $L_2(s)$after replacement, are $s = j\omega_1$respectively equal to:

$$\begin{cases} L_1(j\omega_1) = -m_1\omega_1{}^2 + jp_1\omega_1 + k_1, \\ L_2(j\omega_1) = -m_2\omega_1{}^2 + jp_2\omega_1. \end{cases} \tag{32}$$

Amplitude-frequency characteristic (31) is determined:

$$A(\omega_1) = \left| \frac{-m_2 k_1 \omega_1{}^2 + jp_2 k_1 \omega_1}{k_1 - (m_1+m_2)\omega_1{}^2 + j(p_1+p_2)\omega_1} \right| \tag{33}$$

Let's write $A(\omega_1)$it explicitly:

$$A(\omega_1) = \sqrt{\frac{(m_2 k_1 \omega_1{}^2)^2 + (p_2 k_1 \omega_1)^2}{(k_1 - (m_1 + m_2)\omega_1{}^2)^2 + ((p_1 + p_2)\omega_1)^2}} \tag{34}$$

We estimate the range of the dynamic component of the contact reaction using the inequality:

$$\max_{t>\delta} |N^{\Delta}(t)| < A_1 \cdot A(\omega_1) + \varepsilon \tag{35}$$

where $\varepsilon > 0$ is a certain number, is the $\delta > 0$ point in time from which the process of movement of the system can be considered established. The values and ε depend on δ the system parameters. Let the interval of the phase of steady motion be chosen in such a way that for the considered system parameters the estimate (35) is correct for a relatively small one. ε Based on the inequality (35), we formulate sufficient conditions for the absence of time points when the complete contact reaction takes N_{21}^{Π} only positive values:

$$A_1 \cdot A(\omega_1) + \varepsilon < N^{\Sigma} \tag{36}$$

Indeed, suppose (36), then given inequality (35) we obtain that:

$$|N^{\Delta}(t)| < N^{\Sigma}$$

$$t > \delta \tag{37}$$

Accordingly, the complete contact reaction will be positive:

$$N_{21}^{\Pi} = N^{\Sigma} + N^{\Delta} > 0$$

$$t > \delta \tag{38}$$

Thus, the frequency response can be used to construct an estimate of the range of change of the dynamic component of the contact reaction and, as a result, to estimate the parameters at which the contact between the elements of the composite solid is not disturbed during the period of steady motion.

EXAMPLE OF DETERMINING THE PARAMETERS OF A GAPLESS MOVEMENT SYSTEM

Mechanical system with parameters according to Table 1 is considered. The static contact reaction N_{21}^{Σ} at the equilibrium position $X_1^{\Sigma} = 0{,}096$m is $N_{21}^{\Sigma} = 2{,}98$H.

Table 1. Mechanical System Data

Parameter Name\System Name	Solid 1	Solid 2
Weight m_i, kg	0,1	0,1
Length of elastic element l_{0i}, m	0,1	-
Element stiffness, k_iN/m	1000	-
Coefficient of viscous friction p_i, kg/s	1	1
Constant force F_i, N	0	–2

The amplitude of the dynamic component of the total response is shown in accordance with Figure 2 using the frequency response ordinate. Graphs *1, 2, 3, 4* are products of the ordinate of the amplitude-frequency characteristic $A(\omega_1)$ and the amplitude of the kinematic disturbance A_1, taking the values 1, 2, 3, 4 mm.

Figure 2. Dynamic component vibration amplitude estimates: 1, 2, 3, 4 - dynamic component estimates for kinematic disturbance amplitudes A_1 from 1 to 4 mm, 5 - static reaction level

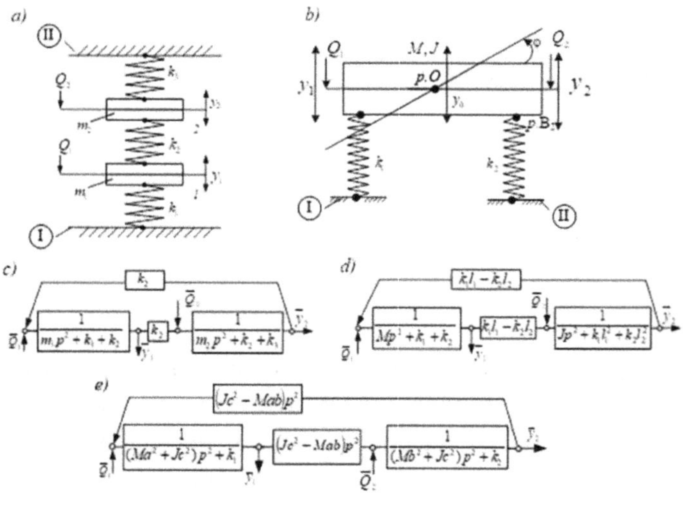

Frequency range of oscillation of bearing surface from 0 to 150 rad/s is considered. With a fixed amplitude of oscillation, the frequency range under consideration can be conditionally divided into two regions. In one area, the dynamic component amplitude estimation plot is below the static component level (Figure 2, line *5*). In this area, the overall contact response can be guaranteed to be positive at the steady state interval.

Figure 3. Amplitude-frequency contact area: I - contact area, II - possible gap area, G - level line of N^{Σ}_{21} for function $A_1 \cdot A(\omega_1)$, C - point of parameters of gapless oscillation, D - point of parameters of possible separation or gap occurrence

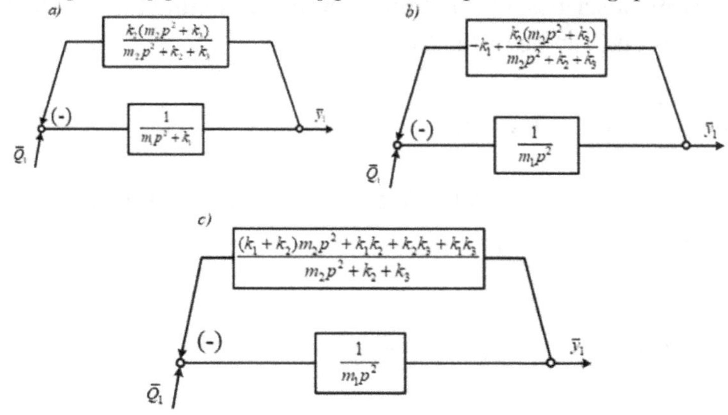

In another frequency region, the amplitude estimation curve (Figure 2, lines *1-4*) is higher than the total contact response level (Figure 2, line *5*). At this frequency, contact failure and clearance may occur.

Consideration of the amplitude range makes it possible to determine the area of amplitude and frequency pairs, which ensure that there is no gap between the elements of the composite solid body at the interval of steady motion. Figure 3 shows the range of parameters of I movement of the mechanical system without a gap.

The amplitude of oscillation of the support surfaceA_1, for which $A_1 \cdot A(\omega_1) = N^{\Sigma}$ we will call critical.

COMPOSITE SOLID TO SUPPORT SURFACE JOINT STIFFNESS LIMIT OPTION

Assigning certain finite values to the parameters of the system or striving for infinity allows us to consider special cases. The mathematical model of the system with two solids and an elastic element shown in Figure 1, with a stiffness factor,

$k_1 \rightarrow \infty$approaches the mathematical model of a solid placed on a vibrating surface. The peculiarity of the mathematical model of oscillation of a solid on a vibrating surface is to reduce the critical amplitude with increasing frequency. For a solid variant placed on a spring-loaded solid, the situation differs in the pattern of amplitude variation as the frequency increases.

In general, many motion parameters in contact are characterized by inequality:

$$\sqrt{\frac{(m_2 k_1 \omega_1{}^2)^2 + (p_2 k_1 \omega_1)^2}{(k_1 - (m_1 + m_2)\omega_1{}^2)^2 + ((p_1 + p_2)\omega_1)^2}} \times A_1 < m_2 g + f_c. \tag{39}$$

Inequality (39) describes the conditions for implementing the contact interaction of solid elements and can be of the form:

$$\omega_1 A_{12} \sqrt{m_2{}^2 \omega_1{}^2 + p_2{}^2} < m_2 g + f_c \tag{40}$$

where $A_{12}$is considered as amplitude depending on various characteristics of the system, including amplitudeA_1:

$$A_{12} = \frac{A_1 k_1}{\sqrt{(k_1 - (m_1 + m_2)\omega_1{}^2)^2 + ((p_1 + p_2)\omega_1)^2}} \tag{41}$$

If$k_1 \rightarrow \infty$, then the problem $A_{12} \rightarrow A_1$under consideration is reduced to the problem of determining the conditions for the separation of a solid body from the oscillation surface in the presence of viscous friction and an additional pressing force. It is assumed that the surface performs harmonic vibrations with amplitude $A_1$and frequencyω_1.

Investigation of contact of a solid body with a vibration surface is determined by the existence of a dependence of the implementation of contact for solids on the values of the masses of composite elements. Limit contact condition (40) at $k_1 \rightarrow \infty$has the form:

$$\omega_1 A_1 \sqrt{m_2{}^2 \omega_1{}^2 + p_2{}^2} < m_2 g + f_c \tag{42}$$

Squaring both sides of the inequality results in the expression:

$$\left((A_1 \omega_1)^2 \omega_1{}^2 - g^2\right) m_2{}^2 - 2 g f_c m_2 + (A_1 \omega_1)^2 p_2{}^2 - f_c{}^2 < 0 \tag{43}$$

In general, this inequality allows us to establish the relationship between the masses of solids, the acting forces and the parameters of surface vibration that provide contact. In a particular case, inequality (43) allows you to determine the mass variants of a composite solid $m_2$element that will provide gapless vibrations of the elements (in this case, simultaneous with the oscillation surface) taking into account the forces of pre-compression and viscous friction. The left side of the inequality (43) is a square polynomial. This means that, depending on various parameters, a plurality of contact masses are an interval or a half-interval. By varying the characteristics of the system, it is possible to achieve the situation that the contact condition will be fulfilled for a certain special mass range. For solids with masses from such a special interval, you can confirm the presence of contact or the absence of a gap. To determine such masses, consider the values:

$$\mu = (A_1\omega_1)^2\omega_1^{\,2} - g^2\lambda = 1 - \left(\frac{A_1\omega_1}{\gamma}\right)^2 \gamma = \left|\frac{f_c}{p_2}\right|$$

$$D_4 = (A_1\omega_1)^2 p_2^{\,2}\left(\frac{\omega_1^{\,2} f_c^{\,2}}{p_2^{\,2}} + g^2 - \left(A_1\omega_1^{\,2}\right)^2\right) \qquad (44)$$

These quantities characterize the position of the roots of the quadratic equation:

$$\left((A_1\omega_1)^2\omega_1^{\,2} - g^2\right) m_2^{\,2} - 2g f_c m_2 + (A_1\omega_1)^2 p_2^{\,2} - f_c^{\,2} = 0 \qquad (45)$$

Analysis of the dependence of characteristicsμ,λ, D_4fromω_1,A_1,$f_c p_2$allows you to determine the range of values of contact masses.

The valuesμ,λ, $D_4$are functions of system parameters and can be used to describe special frequency and amplitude regions. For each region, the range of separation masses has its own characteristics. Figure 4 shows three curves *1, 2, 3* in the plane of parameters, which is the Cartesian product of the parameters of the frequency and amplitude of the surface oscillation.

Figure 4. Regions of characteristic modes of interaction of a solid body with a vibration surface, which are defined by curves 1, 2, 3

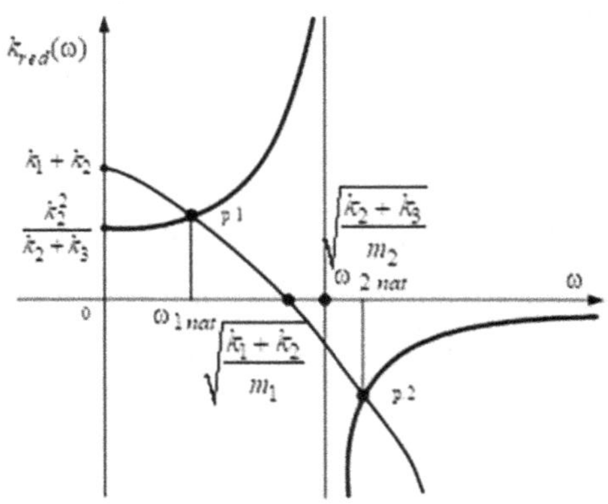

These curves denote a set of equation zeros with respect to amplitudes and frequencies for a fixed quantityγ:

$$D_4 = 0, \lambda = 0, \mu = 0. \tag{46}$$

Curve *1* (Figure 4) is a set of amplitudes and frequencies for which the equality is true:

$$D_4 = (A_1\omega_1)^2 p_2^{\,2}\left(\frac{\omega_1^{\,2} f_c^{\,2}}{p_2^{\,2}} + g^2 - \left(A_1\omega_1^{\,2}\right)^2\right) = 0 \tag{47}$$

At the points of this curve, the quadratic equation has a root match. This means that two options are conventionally possible. The first option - contact is implemented only for solids of fixed mass; the second option - contact is realized for all masses, except for a certain fixed mass. The transition of parameters through this curve will mean the appearance or disappearance of a certain interval of contact masses.

Curve *2* (Figure 4) denotes the roots of the equation$\lambda = 0$. At the points of this curve, one of the roots of the quadratic equation is zero. When crossing this curve, one of the roots changes sign. For solids, this means that there may be a range of masses for which contact is realized or there is no gap.

Curve *3* (Figure 4) denotes the roots of the equation. $\mu = 0$Formally, a given set means that there is one root, and the quadratic equation degenerates into a linear one. For the process of interaction of a solid with a vibration surface, the transition through the curve is equivalent to the implementation of the "inversion" of the contact. In the region of the parameters in which the contact conditions were met, a gap may occur, and vice versa, if for a solid body with a certain mass, the contact conditions were not met, then the "sticking" of such a solid body will occur.

For detailed analysis of contact features, the corresponding mass range is determined for each area and boundary indicated in accordance with Figure 4.

Table 2 shows the mass ranges for the areas (*c*), (*e*) and boundaries$\Gamma_0, \Gamma_1, \Gamma_2, \Gamma_3$, shown in Figure 4.

Table 2. Contact mass intervals

D_4	μ	λ	**Area**	$f_c > 0$	$f_c = 0$	$f_c < 0$
1	2	3	4	5	6	7
>0	>0	<0	(*c*)	(m'_1, m'_2)	No contact	No contact
>0	=0	=0	Γ_0	$(0, \infty)$	No contact	No contact

continued on following page

Table 2. Continued

D_4	μ	λ	**Area**	$f_c > 0$	$f_c = 0$	$f_c < 0$
=0	>0	<0	Γ_1	(m'_0, m'_0)	No contact	No contact
>0	=0	<0	Γ_2	(m''_1, ∞)	No contact	$(0, m''_2)$
>0	=0	=0	Γ_3	$(0, \infty)$	No contact	No contact
>0	<0	>0	(*e*)	$(0, \infty)$	(m''_0, ∞)	$(0, m'''_1) \cup (m'''_2, \infty)$

Table 2 contains 6 sets of 11 regions, boundaries and points Γ_0 described in accordance with Figure 4. Columns 5, 6, 7 describe a plurality of mass values providing contact for different directions of the additional constant force. $f_c > 0$The option where the constant force is directed towards gravity is shown in column 5. The option of no additional force $f_c = 0$is reflected in column 6. The option where the force is against gravity $f_c < 0$ and is $|f_c| < m_2 g$presented in column 7.

According to Table 2, in the row for the parameters of area (*c*), contact is realized provided that the constant force is directed towards gravity$f_c > 0$, at the mass interval(m'_1, m'_2):

$$m'_1 = \frac{gf_c - \sqrt{D_4}}{\mu}$$

$$m'_2 = \frac{gf_c + \sqrt{D_4}}{\mu} \tag{48}$$

To determine the mass interval, we fix the $\gamma = 0,216$m/s parameter, which characterizes the relationship between constant force and viscous friction force. Let the constant force be directed towards gravity: $f_c = 20H$. The amplitude of the oscillation is $A_1 = 0,004m$. The range of contact masses is realized only for the set (*c*) in accordance with Figure 4. Fix the oscillation frequency so that the point $(\omega_1, A_1)$is inside the area (*c*). Let it be the rad/s frequency$\omega_1 = 60$. To determine the interval of contact masses, we plot the characteristic:

$$\left((A_1\omega_1)^2\omega_1^{\,2} - g^2\right) m_2^{\,2} - 2gf_c m_2 + (A_1\omega_1)^2 p_2^{\,2} - f_c^{\,2} < 0 \tag{49}$$

Figure 5 (line 1) shows a graph of the contact characteristic.

Figure 5. Quadratic contact characteristic

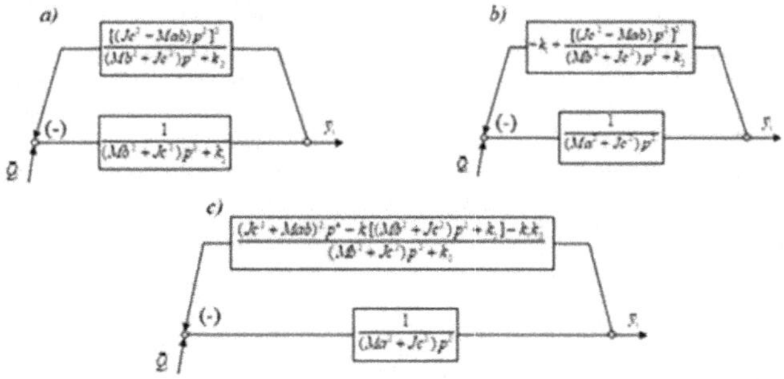

1 - characteristic graph, m'_1, m'_2- critical masses

The contact characteristic is seen as a quadratic polynomial in the mass of a solid. The range of arguments of negative values of the function defines a set of mass values that realize contact at given characteristics of forces and kinematic disturbance. Critical masses are taken as $m'_1 = 0{,}26\text{kg}$, $m'_2 = 3{,}26\text{kg}$. Accordingly, solids whose masses are in the range (m'_1, m'_2)will be in contact with the oscillation surface.

Changing the system parameters will affect this mass range. In particular, a change in frequency within regions (*d*), (*c*) will result in changes in the mass range. Figure 6 shows the mass range limits.

Figure 6. Change in mass range depending on oscillation frequency

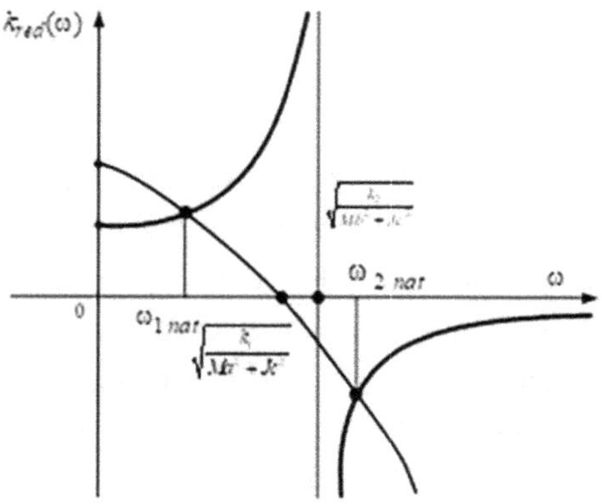

1 - change in critical massm'_1, 2 - change in critical massm'_2

This plot determines the "contact" mass range depending on the frequency of oscillation. In particular, as the frequency increases as the area (*e*) passes, all solids are in contact. When passing through the region (*d*), only solids not exceeding a certain mass are in contact. As the frequency increases, solids with a mass in the range are in contact. (m'_1, m'_2)When the frequency reaches 65 rad/s, only solids with a mass close to 0.89 kg will be in contact. With a higher frequency, there is no contact.

This is how the system can be characterized at sufficiently high stiffness values k_1. The reduction in stiffness changes the amplitude of oscillation of the solid elements in contact. The vibration amplitude of the composite solid begins to depend on the total mass of the elements and the viscous friction force arising during the movement. More detailed calculation of parameters and features of transformations is presented in works (Elyseev, Reznik, Khomenko, & Zasyadko, 2008; Elyseev & Elyseev, 2013; Elyseev & Elyseev, 2013).

LIMIT OF VISCOUS FRICTION FORCES AND CONSTANT ADDITIONAL FORCE

Conditions of gapless movement (39) with insignificant viscous friction forces and additional constant pressing force are obtained by replacing$p_2 << 1, p_1 << 1, f_c = 0$and are described by the inequality:

$$\frac{A_1 k_1 \omega_1^2 m_2}{|k_1 - (m_1 + m_2)\omega_1^2|} < g m_2 \tag{50}$$

When considering the limiting option of rigidity, the inequality $k_1 \to \infty$is obtained:

$$A_1 \omega_1^2 < g \tag{51}$$

Such a model describes the condition of contact of the material point with the oscillation surface. In this case, the contact conditions do not contain the mass of the composite element, which indicates that there is no influence of the mass on the contact.

Condition of absence of clearance (50) for final stiffness $k_1$takes the form:

$$A_1 \frac{\omega_0^2 \omega_1^2}{|\omega_0^2 - \omega_1^2|} m_2 < g m_2 \tag{52}$$

where $\omega_0 = \sqrt{\frac{k_1}{m_1 + m_2}}$is the natural frequency of the "whole" system. By "solid" system is meant a system derived from the original by replacing a two-member solid with a solid solid. The left side of the inequality is the amplitude of the dynamic coupling response, the right side is the static coupling response. Figure 7 shows variants of dynamic reactions of bonds with different amplitudes for a fixed frequency interval. The amplitude of the dynamic response increases as the frequency of the external disturbance approaches the natural frequency of $\omega_0$the system. Thus, the natural frequency of the "whole" system determines the frequency range of contact of the elements of the composite solid.

In the figure, 3.7 frequencies that maintain contact or lack of clearance are in the frequency range for which the corresponding dynamic response amplitude plot is below the static response level. As a result, each amplitude of external kinematic disturbance corresponds to either one or two critical frequencies.

Figure 7. Variants of ranges of dynamic reactions for different amplitudes of external disturbance: 1, 2, 3 - amplitudes of dynamic reactions for A_l1 to 3 mm, 4 - value of static reaction, 5 - natural frequency of the "whole" system

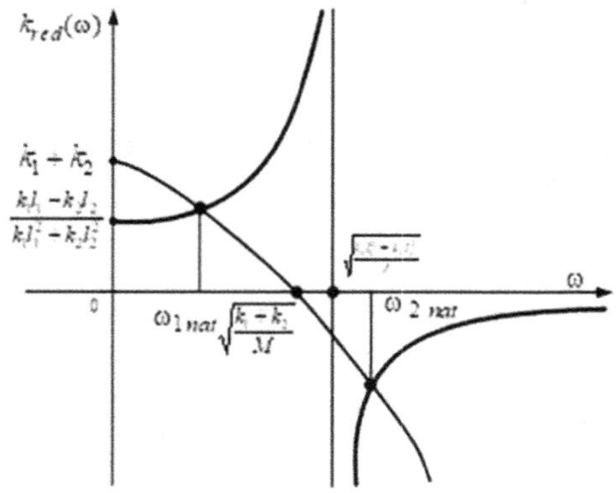

This correspondence allows you to construct a region of parameters on the Cartesian plane that (ω_1, A_1)ensure the absence of a gap. A characteristic graph of the boundary of the kinematic disturbance parameters is shown in Figure 8.

Figure 8. The range of parameters for maintaining contact

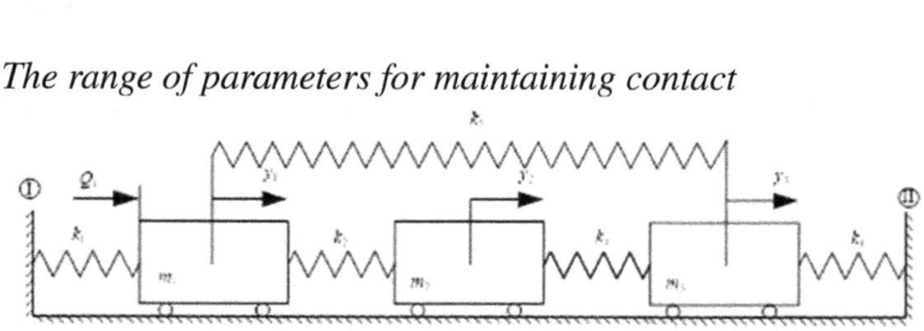

1 - the boundary of contact parameters for a system with a final stiffness coefficient, 2 - the boundary of the limit system at $k_l \to \infty$, 3 - the natural frequency of the "whole" system

The subgraph region 1 is a plurality of contact parameters. The curve 2 subgraph is the limit area of the curve 1 subgraph when striving $k_1 \to \infty$.

Thus, it is shown that the dependence of the contact conditions on the parameters of the system, in particular on the masses of the composite solid elements considered in the absence of friction forces and additional force, is realized through a parameter that is the natural frequency of the "whole" system.

The studies carried out allow us to draw some conclusions.

Within the proposed approach to investigation of contact interaction of composite elements in structure of mechanical oscillatory system under external kinematic action, condition of preservation of integrity of composite solid body is positive of complete contact reaction in phase of steady motion. As a critical state of the system preceding the breakdown of contact between the elements of a composite solid, the criterion of equality to zero of the complete contact reaction is proposed. To determine the modes that exclude contact disturbances and study the properties of the system when moving in contact modes, a representation of the full contact reaction in the form of the sum of the dynamic and static components is used.

In essence, a method has been developed for determining gapless motion modes for a system with a composite solid connected to the oscillation surface by an elastic element, taking into account additional compression forces and viscous friction forces. Relationship between areas of parameters of contact motion of composite solids and characteristics of kinematic disturbance is defined. It was found that the factors of maintaining contact are the mass of the solid, the characteristic of viscous friction, the value of additional pressing forces. It is shown that the addition of viscous friction forces to the system significantly changes the nature of the interaction modes of the elements. It has been established that the dependence of the contact conditions on the masses of the elements is realized through the natural frequency of the "whole" system. The possibility of adjusting the contact of bodies with either a fixed mass or masses from a certain range is revealed.

The novelty of the proposed approach lies in the fact that the task of assessing the dynamic interaction in the contacts of a composite solid is set and solved, as well as taking into account the features of interactions based on the use of the analytical apparatus of control theory. The possibilities of using concepts such as the transfer function are shown. The concept of the transfer function of contact pressure in relation to the input action, which has a kinematic character, is introduced and used, which is reflected in the works (Elyseev & Elyseev, 2014; Elyseev & Elyseev, 2014).

DYNAMIC CONTACT IN A SYSTEM WITH MULTIPLE DEGREES OF FREEDOM FROM COMPOSITE SOLIDS, TAKING INTO ACCOUNT NON-HOLDING BONDS SOME PROVISIONS OF THE GENERALIZED APPROACH

A mechanical system is considered, formed $s$by composite solids of fragments $2s$clamped by elastic elements between two oscillating support surfaces 1 and, $2s$as shown in Figure 9. Reference surface 1 will be called, surface $H_1^{OП}$-.$2sH_{2s}^{OП}$

A numbered solid $q \in [1..s]$body composed of fragments numbered $2q - 1$and $2q$, by means of resilient elements with stiffnesses $k_{2q-2,2q-1}k_{2q,2q+1}$and lengths in an unloaded state, $l_{2q-2,2q-1}l_{2q,2q+1}$is placed between support surfaces or other composite solids in accordance with Figure 9.

Figure 9. Composite solid with scontacts and 2s fragments: a) Construction: $\Gamma_{\beta,i}$, $\Gamma_{\alpha,i}$- contact surfaces of fragments: Z_1, Z_{2s}- coordinates of vibration surfaces; m_{2q}, m_{2q-1} -fragments of q-th composite solid; $k_{2q-2,2q-1}$ - stiffness of elastic elements, b) forces taken into account: $F_{b,i}$-viscous friction forces, $F_{g,i}$-gravity forces, $F_{0,i}$- constant forces; $R_{2q-1,2q-2}$, $R_{2q,2q+1}$ -reaction from elastic elements, Φ_i-inertia forces, $i = 1..n$; $N_{2q,2q-1}$, $N_{2q-1,2q}$- total contact reactions; support surfaces 1 and 2s are movable, $q = 1..s,, i = 1..n, n = 2s$

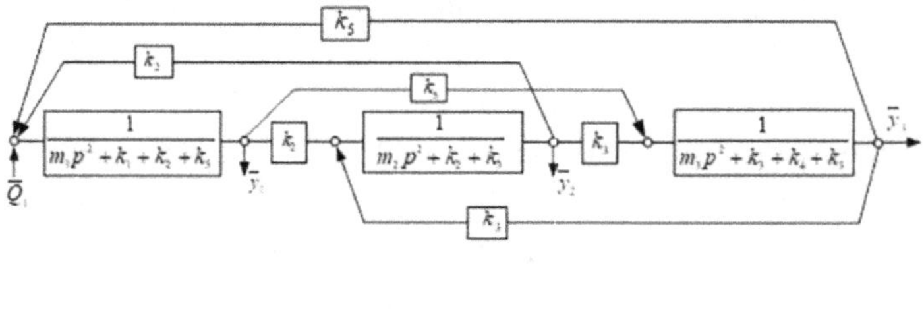

The longitudinal dimensions of the designated fragments ared_i, $i = 1..n$. The number of fragments is$n = 2s$. The fragments $2q - 1$and$2q$, forming $q$the composite solid, have non-holding bonds or non-holding contact. Lines of action of all considered forces are perpendicular to contact planes. Projections of contact planes $\Gamma_{\alpha,i}$on coordinate axis $OX$with origin at point are selected as coordinates of $X_i$solid body fragments$O$. Each fragment of the solid bodies in question, moving in the vertical direction, has only one degree of freedom.

It is assumed that up to a point in time, the $t = 0$system is in a static equilibrium position and stationary. Starting from the moment of time, $t = 0$the support surfaces $H_1^{OП}$and $H_{2s}^{OП}$make harmonic vibrations $Z_1 = A_1 \sin \omega_1 t Z_n = A_1 \sin \omega_1 t$relative to the points$H_1$, $H_n$respectively, causing the vibration of the composite solids. The system parameters are selected in such a way that at zero or sufficiently small amplitudes and frequencies of vibrations, fragments of the composite solid are in contact.

Each fragment of solids is affected by a viscous friction force, $F_{b,i}$gravity, $F_{g,i}$and an additional constant force $F_{0,i}$that is independent of the mass of the solid:

$$F_{b,i} = -b_i\dot{X}_i ... F_{g,i} = -m_i g F_{0,i} = f_{0i} wherei = 1..n. \quad (53)$$

Non-holding link expressions for fragment coordinates are:

$$X_{2q} + d_{2q} \leq X_{2q-1}, q = 1...s \quad (54)$$

In the state of contact of fragments, the coordinates satisfy the conditions:

$$X_{2q} + d_{2q} = X_{2q-1}, q = 1...s. \quad (55)$$

If equality is fulfilled $X_{2q} + d_{2q} = X_{2q-1}$for some under $q$the assumption that all other contact relations are of a holding nature, then it is considered that a simple contact is realized for a composite $q$solid. If condition (55) is met under the assumption that all connections are non-holding, then the contact is considered complex.

Harmonic oscillation of the support surface causes a transient. It is assumed that the movement of the system is stable and over time the system goes into the stage of steady oscillation.

An increase in the frequency and amplitude of oscillation of an external harmonic disturbance can cause a violation of contacts 1, 2,..., s between fragments of composite solids. The stage of the steady-state process is considered.

To determine the continuous mode of movement in the phase of steady-state oscillation, the criterion for maintaining contact is the simultaneous positivity of all full contact reactions $N_{2q-1,2q} > 0$, $q = 1...s$at the interval of steady-state movement. As a condition for determining the critical state, which in some cases is a necessary condition for the appearance of a gap, there is equality to zero of at least one of the complete contact reactions$N_{2q-1,2q} = 0$.

The task is to determine the conditions for continuous oscillation of composite solids during steady-state movement, depending on the parameters of the system. Parameters are amplitude and frequency of oscillation of support surfaces, coefficients of viscous friction, stiffness, mass of fragments of composite solids, constant positive forces. Of interest is the dependence of contact oscillation conditions on generalized system parameters.

To determine the conditions for the implementation of the contact during movement, inequality is built based on the static reaction and the assessment of the dynamic component. To determine the static components of displacements and reactions, a system of static equilibrium equations is constructed taking into account the coupling equations. Based on the Dalembert principle, a system of equilibrium equations is constructed for the complete components of displacements and complete contact reactions. Static components satisfying the static equilibrium equation are excluded from the resulting system of differential equations. Initial conditions are assumed to be zero. After the Laplace transform of a system of differential equations, a system of linear equations for images of dynamic components of displacements and contact reactions is obtained. Adding the coupling conditions in the images to the resulting system of equations forms a final, unambiguously solvable, linear system of equations for images of dynamic displacements and reactions. The resolution of the obtained linear system allows obtaining the desired images of the dynamic components of reactions depending on the images of the kinematic effect. The ratio of the dynamic contact reaction image to the kinematic disturbance image is interpreted as a transfer function (Elyseev, Reznik, Khomenko, & Zasyadko, 2008; Elyseev, Reznik, & Khomenko, 2011). The dynamic contact condition with a number of assumptions is formulated on the basis of the amplitude-frequency characteristic corresponding to the transfer function and the value of the static component of the contact reaction in the form of inequality from the system parameters. The resulting dynamic contact inequality is the amplitude constraint of the ratio of the static contact response to the frequency response values of the transfer function of the corresponding contact (with the input signal as the displacement of the reference surface $z(t)$ and the output signal as the dynamic component of the contact response $N(t)$).

Thus, the kinematic perturbation characteristics providing dynamic contact are a plurality of points in the amplitude-frequency domain of the parameters.

To Define a Static Reaction Component

The system is assumed to have parameters such that all fragments of the respective composite solids are in contact. The system is considered in the static equilibrium position in the absence of kinematic disturbance.

The static component of the reaction is determined by constant forces at the initial moment of time on the basis of the static equilibrium equation written out for each fragment of composite solids in accordance with Figure 9, *b*:

$$
\begin{cases}
R_{10}^{\Sigma} + F_1 + N_{12}^{\Sigma} = 0, \\
R_{23}^{\Sigma} + F_2 + N_{21}^{\Sigma} = 0, \\
R_{32}^{\Sigma} + F_3 + N_{34}^{\Sigma} = 0, \\
.. \\
R_{2q-1,2q-2}^{\Sigma} + F_{2q-1} + N_{2q-1,2q}^{\Sigma} = 0, \\
R_{2q,2q+1}^{\Sigma} + F_{2q} + N_{2q,2q-1}^{\Sigma} = 0, \\
R_{2q+1,2q}^{\Sigma} + F_{2q+1} + N_{2q+1,2q+2}^{\Sigma} = 0, \\
.. \\
R_{n-2,n-1}^{\Sigma} + F_{n-2} + N_{n-2,n-3}^{\Sigma} = 0, \\
R_{n-1,n-2}^{\Sigma} + F_{n-1} + N_{n-1,n}^{\Sigma} = 0, \\
R_{n,n+1}^{\Sigma} + F_n + N_{n,n-1}^{\Sigma} = 0.
\end{cases}
\tag{56}
$$

Depending on the displacements n, the reactions of the elastic elements are expressed as:

$$
\begin{cases}
R_{1,0}^{\Sigma} = -k_{01} \Delta X_1^{\Sigma}, \\
R_{2,3}^{\Sigma} = -k_{23}(\Delta X_2^{\Sigma} - \Delta X_3^{\Sigma}), \\
R_{3,2}^{\Sigma} = k_{23}(\Delta X_2^{\Sigma} - \Delta X_3^{\Sigma}), \\
.. \\
R_{2q-1,2q-2}^{\Sigma} = k_{2q-2,2q-1}(\Delta X_{2q-2}^{\Sigma} - \Delta X_{2q-1}^{\Sigma}), \\
R_{2q,2q+1}^{\Sigma} = -k_{2q,2q+1}(\Delta X_{2q}^{\Sigma} - \Delta X_{2q+1}^{\Sigma}), \\
R_{2q+1,2q}^{\Sigma} = k_{2q,2q+1}(\Delta X_{2q}^{\Sigma} - \Delta X_{2q+1}^{\Sigma}), \\
\ldots \\
R_{n-2,n-1}^{\Sigma} = -k_{n-2,n-1}(\Delta X_{n-2}^{\Sigma} - \Delta X_{n-1}^{\Sigma}), \\
R_{n-1,n-2}^{\Sigma} = k_{n-2,n-1}(\Delta X_{n-2}^{\Sigma} - \Delta X_{n-1}^{\Sigma}), \\
R_{n,n+1}^{\Sigma} = -k_{n,n+1} \Delta X_n^{\Sigma}.
\end{cases}
\tag{57}
$$

Offsets ΔX_i from points X_{0i} are expressions:

$$
\Delta X_i^{\Sigma} = X_i^{\Sigma} - X_{0i}, i = 1..n. \tag{58}
$$

Values X_{0i} are points of design positions, defined as:

$$\begin{cases} X_{0,1} = H_1 - l_{0,1} - d_1, \\ X_{0,2} = H_{2,3} + l_{2,3}, \\ X_{0,3} = H_{2,3} - d_3, \\ \dots \\ X_{0,2q} = H_{2q,2q+1} + l_{2q,2q+1}, \\ X_{0,2q+1} = H_{2q,2q+1} - d_{2q+1}, \\ \dots \\ X_{0,n-2} = H_{n-2,n-1} + l_{n-2,n-1}, \\ X_{0,n-1} = H_{n-2,n-1} - d_{n-1}, \\ X_{0,n} = H_n + l_{n,n+1}. \end{cases} \quad (59)$$

Values $H_{2q,2q+1}$ can be selected as $H_{2q,2q+1} = H_1 - q(H_1 - H_n)/s$.

After substituting the reactions (57) into the equilibrium equation (56), a system of equations is obtained regarding displacements:

$$\begin{cases} -k_{01}\Delta X_1^{\Sigma} + F_1 + N_{12}^{\Sigma} = 0, \\ -k_{23}(\Delta X_2^{\Sigma} - \Delta X_3^{\Sigma}) + F_2 + N_{21}^{\Sigma} = 0, \\ k_{23}(\Delta X_2^{\Sigma} - \Delta X_3^{\Sigma}) + F_3 + N_{34}^{\Sigma} = 0, \\ .. \\ k_{2q-2,2q-1}(\Delta X_{2q-2}^{\Sigma} - \Delta X_{2q-1}^{\Sigma}) + F_{2q-1} + N_{2q-1,2q}^{\Sigma} = 0, \\ -k_{2q,2q+1}(\Delta X_{2q}^{\Sigma} - \Delta X_{2q+1}^{\Sigma} + F_{2q} + N_{2q,2q-1}^{\Sigma} = 0, \\ k_{2q,2q+1}(\Delta X_{2q}^{\Sigma} - \Delta X_{2q+1}^{\Sigma}) + F_{2q+1} + N_{2q+1,2q+2}^{\Sigma} = 0, \\ .. \\ -k_{n-2,n-1}(\Delta X_{n-2}^{\Sigma} - \Delta X_{n-1}^{\Sigma}) + F_{n-2} + N_{n-2,n-3}^{\Sigma} = 0, \\ k_{n-2,n-1}(\Delta X_{n-2}^{\Sigma} - \Delta X_{n-1}^{\Sigma}) + F_{n-1} + N_{n-1,n}^{\Sigma} = 0, \\ -k_{n,n+1}\Delta X_n^{\Sigma} + F_n + N_{n,n-1}^{\Sigma} = 0. \end{cases} \quad (60)$$

To indicate static contact reactions equal in modulus and opposite in direction, $N_{2q,2q-1}^{\Sigma} N_{2q-1,2q}^{\Sigma}$ variables are introduced N_q^{Σ}:

$$N_{2q-1,2q}^{\Sigma} = N_q^{\Sigma}, N_{2q,2q-1}^{\Sigma} = -N_q^{\Sigma}, q = 1 \dots s. \quad (61)$$

Contact conditions (3) depending on displacements take the form:

$$\begin{cases} \Delta X_1 - \Delta X_2 = X_{02} + d_2 - X_{01} \\ \Delta X_3 - \Delta X_4 = X_{04} + d_4 - X_{03} \\ \Delta X_5 - \Delta X_6 = X_{06} + d_6 - X_{05} \\ \ldots \\ \Delta X_{2q-1} - \Delta X_{2q} = X_{0,2q} + d_{2q} - X_{0,2q-1} \\ \ldots \\ \Delta X_{n-5} - \Delta X_{n-4} = X_{0,n-4} + d_{n-4} - X_{0,n-5} \\ \Delta X_{n-3} - \Delta X_{n-2} = X_{0,n-2} + d_{n-2} - X_{0,n-3} \\ \Delta X_{n-1} - \Delta X_n = X_{0,n} + d_n - X_{0,n-1} \end{cases} \tag{62}$$

Displacements and contact reactions can be determined from a system of equations:

$$\begin{bmatrix}
-k_{01} & 0 & 0 & \ldots & & & .. & 0 & 1 & 0 & \ldots & \ldots & 0 \\
0 & -k_{23} & k_{23} & 0 & & & \ldots & 0 & -1 & 0 & & & 0 \\
0 & k_{23} & -k_{23} & 0 & \ldots & \ldots & & \ldots & 0 & 1 & & & 0 \\
\ldots & 0 & 0 & \ldots & 0 & 0 & & \ldots & 0 & -1 & & & 0 \\
 & \ldots & \ldots & 0 & -k_{2q,2q+1} & k_{2q,2q+1} & 0 & \ldots & \ldots & 0 & & 1 & 0 \\
 & & & 0 & k_{2q,2q+1} & -k_{2q,2q+1} & 0 & \ldots & & .. & & -1 & 0 \\
\ldots & \ldots & & \ldots & 0 & 0 & \ldots & 0 & \ldots & & & 0 & 1 \\
0 & 0 & \ldots & \ldots & .. & .. & 0 & -k_{n,n+1} & 0 & \ldots & \ldots & 0 & -1 \\
1 & -1 & 0 & 0 & & & .. & 0 & 0 & 0 & 0 & 0 & 0 \\
0 & 0 & 1 & -1 & 0 & \ldots & & .. & 0 & \ldots & & & \ldots \\
\ldots & \ldots & & & & & .. & .. & 0 & & & & \\
\ldots & \ldots & \ldots & .. & 1 & -1 & 0 & 0 & 0 & & & & \ldots \\
0 & 0 & 0 & 0 & 0 & 0 & 1 & -1 & 0 & .. & & \ldots & 0
\end{bmatrix} \times$$

$$\times \begin{bmatrix} \Delta X_1^{\Sigma} \\ \Delta X_2^{\Sigma} \\ \Delta X_3^{\Sigma} \\ \ldots \\ \Delta X_i^{\Sigma} \\ \ldots \\ \Delta X_{n-1}^{\Sigma} \\ \Delta X_n^{\Sigma} \\ N_1^{\Sigma} \\ \ldots \\ N_q^{\Sigma} \\ \ldots \\ N_s^{\Sigma} \end{bmatrix}$$

$$= \begin{bmatrix} -F_1 \\ -F_2 \\ -F_3 \\ \dots \\ -F_i \\ \dots \\ -F_{n-1} \\ -F_n \\ \Delta S_{12} \\ \Delta S_{34} \\ \dots \\ \Delta S_{2s-3,2s-2} \\ \Delta S_{2s-1,2s} \end{bmatrix}, \begin{cases} \Delta S_{12} = X_{02} + d_2 - X_{01} \\ \Delta S_{34} = X_{04} + d_4 - X_{03} \\ \Delta S_{56} = X_{06} + d_6 - X_{05} \\ \dots \\ \Delta S_{2q-1,2q} = X_{0,2q} + d_{2q} - X_{0,2q-1} \\ \dots \\ \Delta S_{n-5,n-4} = X_{0,n-4} + d_{n-4} - X_{0,n-5} \\ \Delta S_{n-3,n-2} = X_{0,n-2} + d_{n-2} - X_{0,n-3} \\ \Delta S_{n-1,n} = X_{0,n} + d_n - X_{0,n-1} \end{cases}. \tag{63}$$

The matrix in question has the structure:

$$\begin{bmatrix} K & U \\ V & 0 \end{bmatrix} \begin{bmatrix} \Delta X^{\Sigma} \\ N^{\Sigma} \end{bmatrix} = \begin{bmatrix} -\vec{F} \\ \Delta \vec{S} \end{bmatrix} \tag{64}$$

The elements K, U, V, $\vec{F}$, $\Delta\vec{S}$ that are matrices and vectors are:

$$[K]_{n\times n} = \begin{bmatrix}
-k_{01} & 0 & 0 & \dots & & & & & \dots & 0 \\
0 & -k_{23} & k_{23} & 0 & & & & & & \dots \\
0 & k_{23} & -k_{23} & 0 & \dots & & & & & \\
\dots & 0 & 0 & \dots & 0 & 0 & \dots & & & \\
 & & & 0 & -k_{2q,2q+1} & k_{2q,2q+1} & 0 & & & \\
 & & & 0 & k_{2q,2q+1} & -k_{2q,2q+1} & 0 & \dots & \dots & \\
 & & & & 0 & 0 & \dots & 0 & 0 & \dots \\
 & & & & & & 0 & -k_{n-2,n-1} & k_{n-2,n-1} & 0 \\
\dots & & & & & & 0 & k_{n-2,n-1} & -k_{n-2,n-1} & 0 \\
0 & \dots & \dots & \dots & \dots & \dots & \dots & 0 & 0 & -k_{n,n+1}
\end{bmatrix}$$

$$[U]_{2s\times s} = \begin{bmatrix} 1 & 0 & & \dots & 0 \\ -1 & 0 & \dots & & \\ 0 & 1 & 0 & & \\ 0 & -1 & 0 & & \\ \dots & 0 & \dots & 0 & \\ & 0 & \dots & 0 & \dots \\ & & 0 & 1 & 0 \\ & & 0 & -1 & 0 \\ & & \dots & 0 & 1 \\ 0 & \dots & & 0 & -1 \end{bmatrix} [V]_{s\times 2s}$$

$$= \begin{bmatrix} 1 & -1 & 0 & 0 & \dots & & & & & 0 \\ 0 & 0 & 1 & -1 & 0 & \dots & & & & \\ \dots & & 0 & 0 & \dots & \dots & 0 & 0 & \dots & \\ & & & & & 0 & 1 & -1 & 0 & 0 \\ 0 & \dots & & & & & 0 & 0 & 1 & -1 \end{bmatrix}$$

$$[\Delta X^{\Sigma}]_n = \begin{bmatrix} \Delta X_1^{\Sigma} \\ \dots \\ \Delta X_i^{\Sigma} \\ \dots \\ \Delta X_n^{\Sigma} \end{bmatrix}$$

$$[F]_n = \begin{bmatrix} F_1 \\ \dots \\ F_i \\ \dots \\ F_n \end{bmatrix}$$

$$[N^{\Sigma}]_s = \begin{bmatrix} N_1^{\Sigma} \\ \dots \\ N_q^{\Sigma} \\ \dots \\ N_s^{\Sigma} \end{bmatrix}$$

$$[\Delta S]_s = \begin{bmatrix} \Delta S_{12} \\ \dots \\ \Delta S_{2q-1,2q} \\ \dots \\ \Delta S_{2s-1,2s} \end{bmatrix} \tag{65}$$

The matrix $[K]_{n\times n}$ is degenerate. To determine the vector of static displacements and reactions, a system of equations is considered:

$$\begin{bmatrix} K^* & V^* \\ U^* & 0 \end{bmatrix} \begin{bmatrix} K & U \\ V & 0 \end{bmatrix} \begin{bmatrix} \Delta X^{\Sigma} \\ N^{\Sigma} \end{bmatrix} = \begin{bmatrix} K^* & V^* \\ U^* & 0 \end{bmatrix} \begin{bmatrix} -\vec{F} \\ \Delta \vec{S} \end{bmatrix} \tag{66}$$

where are $K^* V^* U^*$transpose matrices.

The system (66) is represented as:

$$\begin{bmatrix} K^* K + V^* V & K^* U \\ U^* K & U^* U \end{bmatrix} \begin{bmatrix} \Delta X^{\Sigma} \\ N^{\Sigma} \end{bmatrix} = \begin{bmatrix} -K^* F + V^* \Delta S \\ -U^* F \end{bmatrix} \tag{67}$$

This system is formed from conditions of equilibrium for displacements, conditions of contact in displacements and conditions of combination of contact reactions.

The desired static displacements and reactions are:

$$\Delta \vec{X}^{\Sigma} = [K_v - \tfrac{1}{2} K^* U U^* K]^{-1} \left(V^* \Delta \vec{S} - K^* \vec{F} + \tfrac{1}{2} K^* U U^* \vec{F} \right)$$

$$\vec{N}^{\Sigma} = -\tfrac{1}{2} U^* \vec{F} - \tfrac{1}{2} U^* K [K_v - \tfrac{1}{2} K^* U U^* K]^{-1} \left(V^* \Delta \vec{S} - K^* \vec{F} + \tfrac{1}{2} K^* U U^* \vec{F} \right) \tag{68}$$

Where:

$$K_v = K^* K + V^* V$$

The contact reaction may be determined in the following sequence. Coordinates $2s$ of fragments are represented as sum of value of structural position and displacement. A system of equilibrium n equations for each fragment is compiled depending on n the reactions of elastic elements, s contact reactions and applied forces. s Contact conditions are represented as expressions depending on n offsets. To take into account n contact reactions, s the main ones are selected, the values of which should be positive. Based on the system of equilibrium equations and contact conditions, a final system $3s$ of equations with 2s unknown displacements and contact reactions s is compiled.

The resolution of the resulting system of equations allows you to set the condition for the implementation of the contact or the absence of a gap in the form of the positivity of each component $N^{\Sigma}$of the static reaction.

DETERMINATION OF THE DYNAMIC COMPONENT OF THE CONTACT REACTION

The harmonic oscillation of support surfaces results in oscillations of composite solids, the motion of which causes oscillations of contact reactions between fragments of composite solids.

The dynamic reaction component is defined as the difference between the full reaction and the static component. To find the magnitude of the dynamic component, the total contact reactions are represented as sums of static and dynamic components:

$$N_q^{\Pi} = N_q^{\Sigma} + N_q^{\Delta}$$

$$q = 1...s \quad (69)$$

Similarly, full offsets $X_i^{\Pi}$are represented as the sum of the static and dynamic components:

$$X_i^{\Pi} = X_i^{\Sigma} + X_i^{\Delta}$$

$$i = 1...n \quad (70)$$

The dynamic component of a solid fragment offset is the difference between the full component and the static component.

Considering the representation (70), the contact realization condition (55) for the dynamic displacement components takes the form:

$$X_{2q-1}^{\Delta} - X_{2q}^{\Delta} = 0$$

$$q = 1...s \quad (71)$$

The Dalembert principle N_q^{Π}can be used to define complete reactions. In accordance with Figure 9, for each fragment of a composite solid, an equilibrium equation is written for the applied full reactions from elastic elements$R_{2q,2q+1}^{\Pi}$, viscous friction forces, $F_{b,i}$weight$F_{g,i}$, constant forces$F_{0,i}$, inertia forces of $\Phi_i$full contact reactions $N_{2q,2q-1}^{\Pi}$:

$$\begin{cases} R^{\Pi}_{10} + F_1 + F_{b,1} + \Phi_1 + N^{\Pi}_{12} = 0, \\ R^{\Pi}_{23} + F_2 + F_{b,2} + \Phi_2 + N^{\Pi}_{21} = 0, \\ R^{\Pi}_{32} + F_3 + F_{b,2} + \Phi_3 + N^{\Pi}_{34} = 0, \\ \dots \\ R^{\Pi}_{2q-1,2q-2} + F_{2q-1} + F_{b,2q-1} + \Phi_{2q-1} + N^{\Pi}_{2q-1,2q} = 0, \\ R^{\Pi}_{2q,2q+1} + F_{2q} + F_{b,2q} + \Phi_{2q} + N^{\Pi}_{2q,2q-1} = 0, \\ R^{\Pi}_{2q+1,2q} + F_{2q+1} + F_{b,2q+1} + \Phi_{2q+1} + N^{\Pi}_{2q+1,2q+2} = 0, \\ \dots \\ R^{\Pi}_{n-2,n-1} + F_{n-2} + F_{b,n-2} + \Phi_{n-2} + N^{\Pi}_{n-2,n-3} = 0, \\ R^{\Pi}_{n-1,n-2} + F_{n-1} + F_{b,n-1} + \Phi_{n-1} + N^{\Pi}_{n-1,n} = 0, \\ R^{\Pi}_{n,n+1} + F_n + F_{b,n} + \Phi_n + N^{\Pi}_{n,n-1} = 0. \end{cases} \tag{72}$$

Viscous friction forces and inertial forces are represented as:

$$F_{b,i} = -b_i \dot{X}^{\Pi}_i$$

$$\Phi_i = -m_i \ddot{X}^{\Pi}_i \tag{73}$$

In the form of the sum of the static and dynamic components, the complete reaction from the elastic elements has the following representation:

$$\begin{cases} R^{\Pi}_{1,0} = R^{\Sigma}_{1,0} + R^{\Delta}_{1,0} + f_1, \\ R^{\Pi}_{2,3} = R^{\Sigma}_{2,3} + R^{\Delta}_{2,3}, \\ R^{\Pi}_{3,2} = R^{\Sigma}_{3,2} + R^{\Delta}_{3,2}, \\ \dots \\ R^{\Pi}_{2q-1,2q-2} = R^{\Sigma}_{2q-1,2q-2} + R^{\Delta}_{2q-1,2q-2}, \\ R^{\Pi}_{2q,2q+1} = R^{\Sigma}_{2q,2q+1} + R^{\Delta}_{2q,2q+1}, \\ R^{\Pi}_{2q+1,2q} = R^{\Sigma}_{2q+1,2q} + R^{\Delta}_{2q+1,2q}, \\ R^{\Pi}_{n-2,n-1} = R^{\Sigma}_{n-2,n-1} + R^{\Delta}_{n-2,n-1}, \\ R^{\Pi}_{n-1,n-2} = R^{\Sigma}_{n-1,n-2} + R^{\Delta}_{n-1,n-2}, \\ R^{\Pi}_{n,n+1} = R^{\Sigma}_{n,n+1} + R^{\Delta}_{n,n+1} + f_n, \end{cases} \tag{74}$$

where $f_1 = k_{01} Z_1$, $f_n = k_{n,n+1} Z_n$

Dynamic components $R^{\Delta}_{2q,2q+1}$ on the side of elastic elements have the form:

$$
\begin{cases}
R^{\Delta}_{1,0} = -k_{01} X^{\Delta}_1, \\
R^{\Delta}_{2,3} = -k_{23}(X^{\Delta}_2 - X^{\Delta}_3), \\
R^{\Delta}_{3,2} = k_{23}(X^{\Delta}_2 - X^{\Delta}_3), \\
\ldots \\
R^{\Delta}_{2q-1,2q-2} = k_{2q-2,2q-1}(X^{\Delta}_{2q-2} - X^{\Delta}_{2q-1}), \\
R^{\Delta}_{2q,2q+1} = -k_{2q,2q+1}(X^{\Delta}_{2q} - X^{\Delta}_{2q+1}), \\
R^{\Delta}_{2q+1,2q} = k_{2q,2q+1}(X^{\Delta}_{2q} - X^{\Delta}_{2q+1}), \\
\ldots \\
R^{\Delta}_{n-2,n-1} = -k_{n-2,n-1}(X^{\Delta}_{n-2} - X^{\Delta}_{n-1}), \\
R^{\Delta}_{n-1,n-2} = k_{n-2,n-1}(X^{\Delta}_{n-2} - X^{\Delta}_{n-1}), \\
R^{\Delta}_{n,n+1} = -k_{n,n+1} X^{\Delta}_n.
\end{cases}
\tag{75}
$$

The system of equilibrium equations (72) with selected static and dynamic components of forces and contact reactions (69) takes the form:

$$
\begin{cases}
R^{\Sigma}_{1,0} + R^{\Delta}_{1,0} + f_1 + F_1 + F_{b,1} + \Phi_1 + N^{\Sigma}_{12} + N^{\Delta}_{12} = 0, \\
R^{\Sigma}_{2,3} + R^{\Delta}_{2,3} + F_2 + F_{b,2} + \Phi_2 + N^{\Sigma}_{21} + N^{\Delta}_{21} = 0, \\
R^{\Sigma}_{3,2} + R^{\Delta}_{3,2} + F_3 + F_{b,2} + \Phi_3 + N^{\Sigma}_{34} + N^{\Delta}_{34} = 0, \\
\ldots \\
R^{\Sigma}_{2q-1,2q-2} + R^{\Delta}_{2q-1,2q-2} + F_{2q-1} + F_{b,2q-1} + \Phi_{2q-1} + N^{\Sigma}_{2q-1,2q} + N^{\Delta}_{2q-1,2q} = 0, \\
R^{\Sigma}_{2q,2q+1} + R^{\Delta}_{2q,2q+1} + F_{2q} + F_{b,2q} + \Phi_{2q} + N^{\Sigma}_{2q,2q-1} + N^{\Delta}_{2q,2q-1} = 0, \\
R^{\Sigma}_{2q+1,2q} + R^{\Delta}_{2q+1,2q} + F_{2q+1} + F_{b,2q+1} + \Phi_{2q+1} + N^{\Sigma}_{2q+1,2q+2} + N^{\Delta}_{2q+1,2q+2} = 0, \\
\ldots \\
R^{\Sigma}_{n-2,n-1} + R^{\Delta}_{n-2,n-1} + F_{n-2} + F_{b,n-2} + \Phi_{n-2} + N^{\Sigma}_{n-2,n-3} + N^{\Delta}_{n-2,n-3} = 0, \\
R^{\Sigma}_{n-1,n-2} + R^{\Delta}_{n-1,n-2} + F_{n-1} + F_{b,n-1} + \Phi_{n-1} + N^{\Sigma}_{n-1,n} + N^{\Delta}_{n-1,n} = 0, \\
R^{\Sigma}_{n,n+1} + R^{\Delta}_{n,n+1} + f_n + F_n + F_{b,n} + \Phi_n + N^{\Sigma}_{n,n-1} + N^{\Delta}_{n,n-1} = 0.
\end{cases}
\tag{76}
$$

Taking into account the static equilibrium equation (56), the static components of all quantities are excluded from the decomposition (76).

The resulting system of equations contains only the dynamic components of the desired quantities:

$$\begin{cases} R^{\Delta}_{1,0} + f_1 + F_{b,1} + \Phi_1 + N^{\Delta}_{12} = 0, \\ R^{\Delta}_{2,3} + F_{b,2} + \Phi_2 + N^{\Delta}_{21} = 0, \\ R^{\Delta}_{3,2} + F_{b,2} + \Phi_3 + N^{\Delta}_{34} = 0, \\ .. \\ R^{\Delta}_{2q-1,2q-2} + F_{b,2q-1} + \Phi_{2q-1} + N^{\Delta}_{2q-1,2q} = 0, \\ R^{\Delta}_{2q,2q+1} + F_{b,2q} + \Phi_{2q} + N^{\Delta}_{2q,2q-1} = 0, \\ R^{\Delta}_{2q+1,2q} + F_{b,2q+1} + \Phi_{2q+1} + N^{\Delta}_{2q+1,2q+2} = 0, \\ .. \\ R^{\Delta}_{n-2,n-1} + F_{b,n-2} + \Phi_{n-2} + N^{\Delta}_{n-2,n-3} = 0, \\ R^{\Delta}_{n-1,n-2} + F_{b,n-1} + \Phi_{n-1} + N^{\Delta}_{n-1,n} = 0, \\ R^{\Delta}_{n,n+1} + f_n + F_{b,n} + \Phi_n + N^{\Delta}_{n,n-1} = 0. \end{cases} \tag{77}$$

Viscous friction and inertia forces depend only on dynamic components:

$$F_{b,i} = -b_i \dot{X}^{\Pi}_i = -b_i \dot{X}^{\Delta}_i$$

$$\Phi_i = -m_i \ddot{X}^{\Pi}_i = -m_i \ddot{X}^{\Delta}_i$$

$$i = 1...n \tag{78}$$

After substituting the equations of equilibrium (76) with the expressions of the corresponding forces (77), (78), depending on the dynamic components of displacements $X^{\Delta}_i$and reactions, a system of differential equations with zero initial conditions $N^{\Delta}_q$is obtained:

$$\begin{cases} -k_{01} X^{\Delta}_1 + f_1 - b_1 \dot{X}^{\Delta}_1 - m_1 \ddot{X}^{\Delta}_1 + N^{\Delta}_1 = 0, \\ -k_{23}(X^{\Delta}_2 - X^{\Delta}_3) - b_2 \dot{X}^{\Delta}_2 - m_2 \ddot{X}^{\Delta}_2 - N^{\Delta}_1 = 0, \\ k_{23}(X^{\Delta}_2 - X^{\Delta}_3) - b_3 \dot{X}^{\Delta}_3 - m_3 \ddot{X}^{\Delta}_3 + N^{\Delta}_2 = 0, \\ .. \\ k_{2q-2,2q-1}(X^{\Delta}_{2q-2} - X^{\Delta}_{2q-1}) - b_{2q-1} \dot{X}^{\Delta}_{2q-1} - m_{2q-1} \ddot{X}^{\Delta}_{2q-1} + N^{\Delta}_q = 0, \\ -k_{2q,2q+1}(X^{\Delta}_{2q} - X^{\Delta}_{2q+1}) - b_{2q} \dot{X}^{\Delta}_{2q} - m_{2q} \ddot{X}^{\Delta}_{2q} - N^{\Delta}_q = 0, \\ k_{2q,2q+1}(X^{\Delta}_{2q} - X^{\Delta}_{2q+1}) - b_{2q+1} \dot{X}^{\Delta}_{2q+1} - m_{2q+1} \ddot{X}^{\Delta}_{2q+1} + N^{\Delta}_{q+1} = 0, \\ .. \\ -k_{n-2,n-1}(X^{\Delta}_{n-2} - X^{\Delta}_{n-1}) - b_{n-2} \dot{X}^{\Delta}_{n-2} - m_{n-2} \ddot{X}^{\Delta}_{n-2} - N^{\Delta}_{s-1} = 0, \\ k_{n-2,n-1}(X^{\Delta}_{n-2} - X^{\Delta}_{n-1}) - b_{n-1} \dot{X}^{\Delta}_{n-1} - m_{n-1} \ddot{X}^{\Delta}_{n-1} + N^{\Delta}_s = 0, \\ -k_{n,n+1} X^{\Delta}_n + f_n - b_n \dot{X}^{\Delta}_n - m_n \ddot{X}^{\Delta}_n - N^{\Delta}_s = 0, \end{cases} , \begin{cases} X^{\Delta}_1(0) = 0, \dot{X}^{\Delta}_1(0) = 0, \\ ... \\ X^{\Delta}_i(0) = 0, \dot{X}^{\Delta}_i(0) = 0, \\ ... \\ X^{\Delta}_n(0) = 0, \dot{X}^{\Delta}_n(0) = 0. \end{cases} \tag{79}$$

To indicate dynamic contact reactions equal in modulus and opposite in direction, $N^{\Delta}_{2q,2q-1}N^{\Delta}_{2q-1,2q}$variables are introduced$N^{\Delta}_{q}$:

$$N^{\Delta}_{2q-1,2q} = N^{\Delta}_{q}$$

$$N^{\Delta}_{2q,2q-1} = -N^{\Delta}_{q}$$

$$q = 1...s \tag{80}$$

The system (79) represents motion depending on contact reactions of N^{Δ}_{q}bonds imposed as conditions on dynamic components (71). It is assumed that the solution of the system (79) is considered only at the contact interval of fragments of a composite solid.

To determine the characteristics of motion in contact, the Laplace transform of the resulting system (79) of differential equations is performed:

$$\begin{cases} -(m_1p^2 + b_1p + k_{01})\overline{X}^{\Delta}_1 + \overline{N}^{\Delta}_1 = -\overline{f}_1, \\ -(m_2p^2 + b_2p + k_{23})\overline{X}^{\Delta}_2 + k_{23}\overline{X}^{\Delta}_3 - \overline{N}^{\Delta}_1 = 0, \\ -(m_3p^2 + b_3p + k_{23})\overline{X}^{\Delta}_3 + k_{23}\overline{X}^{\Delta}_2 + \overline{N}^{\Delta}_2 = 0, \\ .. \\ -(m_{2q-1}p^2 + b_{2q-1}p + k_{2q-2,2q-1})\overline{X}^{\Delta}_{2q-1} + k_{2q-2,2q-1}\overline{X}^{\Delta}_{2q-2} + \overline{N}^{\Delta}_q = 0, \\ -(m_{2q}p^2 + b_{2q}p + k_{2q,2q+1})\overline{X}^{\Delta}_{2q} + k_{2q,2q+1}\overline{X}^{\Delta}_{2q+1} - \overline{N}^{\Delta}_q = 0, \\ -(m_{2q+1}p^2 + b_{2q+1}p + k_{2q,2q+1})\overline{X}^{\Delta}_{2q+1} + k_{2q,2q+1}\overline{X}^{\Delta}_{2q} + \overline{N}^{\Delta}_{q+1} = 0, \\ .. \\ -(m_{n-2}p^2 + b_{n-2}p + k_{n-2,n-1})\overline{X}^{\Delta}_{n-2} + k_{n-2,n-1}\overline{X}^{\Delta}_{n-1} - \overline{N}^{\Delta}_{s-1} = 0 \\ -(m_{n-1}p^2 + b_{n-1}p + k_{n-2,n-1})\overline{X}^{\Delta}_{n-1} + k_{n-2,n-1}\overline{X}^{\Delta}_{n-2} + \overline{N}^{\Delta}_{s} = 0, \\ -(m_np^2 + b_np + k_{n,n+1})\overline{X}^{\Delta}_n - \overline{N}^{\Delta}_s = -\overline{f}_n. \end{cases} \tag{81}$$

The functions$\overline{X}^{\Delta}_i$,$\overline{N}^{\Delta}_q$,$\overline{f}_1 = k_{01}\overline{Z}$,$\overline{f}_n = k_{n,n+1}\overline{Z}$of the complex variable$p$are images of the functions,,,,, of the Laplace $X^{\Delta}_iN^{\Delta}_qf_1f_nZ = Z_1$transform:

$$\overline{X}(p) = \int_0^{\infty} X(t)\,e^{-pt}\,dt \tag{82}$$

It is assumed that all the necessary requirements for the conversion are met.

In the images, the contact conditions (71) take the form:

$$\overline{X}^{\Delta}_{2q-1} = \overline{X}^{\Delta}_{2q} \tag{83}$$

After combining the contact conditions (83) and the system of equations in the images (81), a system of *3s* equations with *3s* unknowns is obtained:

$$\begin{bmatrix} -K_1(p) & 0 & & & & 0 & 1 & 0 & & & \\ 0 & \dots & 0 & 0 & & & -1 & 0 & 0 & & \\ & 0 & -K_{2q}(p) & k_{2q,2q+1} & 0 & & 0 & \dots & 1 & \dots & \\ & 0 & k_{2q,2q+1} & -K_{2q+1}(p) & 0 & & \dots & \dots & -1 & \dots & 0 \\ & & 0 & 0 & \dots & 0 & \dots & & 0 & 0 & 1 \\ & & & & 0 & -K_n(p) & 0 & & & 0 & -1 \\ 1 & -1 & & & \dots & 0 & 0 & 0 & 0 & 0 & 0 \\ 0 & & \dots & \dots & & & 0 & & & & \dots \\ \dots & & 1 & -1 & 0 & & 0 & & & & \\ \dots & & \dots & \dots & 0 & 0 & 0 & & & & \dots \\ 0 & & 0 & 0 & 1 & -1 & 0 & \dots & & \dots & 0 \end{bmatrix} \begin{bmatrix} \overline{X}^{\Delta}_1 \\ \dots \\ \overline{X}^{\Delta}_{2q} \\ \overline{X}^{\Delta}_{2q+1} \\ \dots \\ \overline{X}^{\Delta}_n \\ \overline{N}^{\Delta}_1 \\ \dots \\ \overline{N}^{\Sigma}_q \\ \dots \\ \overline{N}^{\Sigma}_s \end{bmatrix} = \begin{bmatrix} -\vec{f}_1 \\ 0 \\ \dots \\ \dots \\ 0 \\ -\vec{f}_n \\ 0 \\ \dots \\ \\ \dots \\ 0 \end{bmatrix} \tag{84}$$

The matrix coefficients in the system (84) depend on *p*the complex variable according to the expression:

$$\begin{cases} K_1(p) = m_1 p^2 + b_1 p + k_{01}, \\ \dots \\ K_{2q}(p) = m_{2q} p^2 + b_{2q} p + k_{2q,2q+1}, \\ K_{2q+1}(p) = m_{2q+1} p^2 + b_{2q+1} p + k_{2q,2q+1}, \\ \dots \\ K_n(p) = m_n p^2 + b_n p + k_{n,n+1}. \end{cases} \tag{85}$$

The matrix of equation (84) has the structure:

$$\begin{bmatrix} K(p) & U \\ V & 0 \end{bmatrix} \begin{bmatrix} \overline{X}^{\Delta} \\ \overline{N}^{\Delta} \end{bmatrix} = \begin{bmatrix} -\vec{k}_f \overline{Z} \\ 0 \end{bmatrix} \tag{86}$$

The matrices $[K(p)]_{n\times n}$, $[U]_{2s\times s}$, $[V]_{s\times 2s}$ and vectors $[\overline{X}^{\Delta}]_n$, $[\overline{N}^{\Delta}]_s$, $\left[\vec{k}_f\right]_n$ are:

$$[K(p)]_{n\times n} = \begin{bmatrix} -K_1(p) & 0 & \dots & & & 0 \\ 0 & \dots & 0 & 0 & \dots & \\ \dots & 0 & -K_{2q}(p) & k_{2q,2q+1} & 0 & \\ & 0 & k_{2q,2q+1} & -K_{2q+1}(p) & 0 & \dots \\ & \dots & 0 & 0 & \dots & 0 \\ 0 & & & \dots & 0 & -K_n(p) \end{bmatrix} \tag{87}$$

$$[U]_{2s\times s} = \begin{bmatrix} 1 & 0 & & \dots & 0 \\ -1 & 0 & \dots & & \\ 0 & 1 & 0 & & \\ 0 & -1 & 0 & & \\ \dots & 0 & \dots & 0 & \\ & 0 & \dots & 0 & \dots \\ & & 0 & 1 & 0 \\ & & 0 & -1 & 0 \\ & & \dots & 0 & 1 \\ 0 & \dots & & 0 & -1 \end{bmatrix} [V]_{s\times 2s}$$

$$= \begin{bmatrix} 1 & -1 & 0 & 0 & \dots & & & & & & 0 \\ 0 & 0 & 1 & -1 & 0 & \dots & & & & & \\ \dots & & 0 & 0 & \dots & \dots & 0 & 0 & \dots & & \\ & & & & & 0 & 1 & -1 & 0 & 0 \\ 0 & \dots & & & & & 0 & 0 & 1 & -1 \end{bmatrix}$$

$$[\overline{X}^{\Delta}]_n = \begin{bmatrix} \overline{X}_1^{\Delta} \\ \dots \\ \overline{X}_i^{\Delta} \\ \dots \\ \widehat{X}_n^{\Delta} \end{bmatrix} [\overline{N}^{\Delta}]_s = \begin{bmatrix} \overline{N}_1^{\Delta} \\ \dots \\ \overline{N}_q^{\Delta} \\ \dots \\ \overline{N}_s^{\Delta} \end{bmatrix} \left[\vec{k}_f\right]_n = \begin{bmatrix} k_{01} \\ 0 \\ \dots \\ 0 \\ k_{n,n+1} \end{bmatrix}$$

The matrix $[K(p)]_{n\times n}$ is degenerate. To determine the images of displacements and reactions, a system of equations is considered:

$$\begin{bmatrix} K(p)^* & V^* \\ U^* & 0 \end{bmatrix} \begin{bmatrix} K(p) & U \\ V & 0 \end{bmatrix} \begin{bmatrix} \overline{X}^{\Delta} \\ \overline{N}^{\Delta} \end{bmatrix} = \begin{bmatrix} K(p)^* & V^* \\ U^* & 0 \end{bmatrix} \begin{bmatrix} -\vec{k}_f\overline{Z} \\ 0 \end{bmatrix} \tag{88}$$

where, V^* are U^* transpose matrices, is $K(p)^*$ a conjugate transpose matrix. Equation (88) can be represented as:

$$\begin{bmatrix} K(p)^*K(p)+V^*V & K(p)^*U \\ U^*K(p) & U^*U \end{bmatrix} \begin{bmatrix} \overline{X}^{\Delta} \\ \overline{N}^{\Delta} \end{bmatrix} = \begin{bmatrix} -K(p)^*\vec{k}_f\overline{Z} \\ -U^*\vec{k}_f\overline{Z} \end{bmatrix} \tag{89}$$

The desired values of the displacement and reaction image are as follows:

$$\vec{\overline{X}}^{\Delta} = [K_v(p) - \tfrac{1}{2}K(p)^*UU^*K(p)]^{-1}\left(-K(p)^* + \tfrac{1}{2}K(p)^*UU^*\right)\vec{k}_f\overline{Z},$$

$$\vec{N}^{\Delta}$$

$$= \left(-\frac{1}{2}U^{*} - \frac{1}{2}U^{*}K(p)[K_{v}(p)-\frac{1}{2}K(p)^{*}UU^{*}K(p)]^{-1}\left(-K(p)^{*} + \frac{1}{2}K(p)^{*}UU^{*}\right)\right)\vec{k}_{f}\bar{Z} \quad (90)$$

Where:

$$K_{v}(p) = K(p)^{*}K(p) + V^{*}V$$

The resulting reaction image solution can be represented depending on the kinematic perturbation image:

$$\vec{N}^{\Delta} = \vec{W}(p)\bar{Z} \quad (91)$$

where the contact reaction image is a vector $\vec{W}(p)$ multiplied by $\bar{Z}$:

$$\vec{W}(p) = \begin{bmatrix} W_{1}(p) \\ \dots \\ W_{q}(p) \\ \dots \\ W_{s}(p) \end{bmatrix}$$

$$\vec{W}(p) = \left(-\frac{1}{2}U^{*} - \frac{1}{2}U^{*}K[K_{v} - \frac{1}{2}K^{*}UU^{*}K]^{-1}\left(-K^{*} + \frac{1}{2}K^{*}UU^{*}\right)\right)\vec{k}_{f} \quad (92)$$

The obtained representation of the image of contact reactions $\bar{N}^{\Delta}$ depending on the external disturbance $\bar{Z}$ allows introducing transfer functions $W_{q}(p)$ from the kinematic disturbance $\bar{Z}$ to the dynamic component of the contact reaction $\bar{N}^{\Delta}_{q}$.

Thus, the dynamic interaction of fragments in contact can be characterized by transfer functions $W_{q}(p)$.

The amplitude-frequency characteristics $A_{(q)}(\omega_{1})$ of the transfer functions $W_{q}(p)$ are determined by the relationship:

$$A_{(q)}(\omega_{1}) = \left|W_{q}(j\omega_{1})\right|$$

$$q = 1 \dots s \quad (93)$$

where $j^{2} = -1$.

The function $W_q(j\omega_1)$is a complex number of the form:

$$W_q(j\omega_1)=\frac{C_q(\omega_1)+D_q(\omega_1)j}{E_q(\omega_1)+F_q(\omega_1)j} \tag{94}$$

Where $C_q(\omega_1)D_q(\omega_1)E_q(\omega_1)F_q(\omega_1)$ are real frequency functions.

The general structure of the corresponding frequency response as a modulus of a complex number is:

$$A_{(q)}(\omega_1)=\sqrt{\frac{C_q(\omega_1)^2+D_q(\omega_1)^2}{E_q(\omega_1)^2+F_q(\omega_1)^2}} \tag{95}$$

Thus, the properties of the dynamic components of the contact reactions can be represented by the amplitude-frequency characteristics of the transfer functions, considered as the response characteristics of the mechanical system to the input signal in the form of kinematic disturbance of the support surfaces. The proposed approach makes it possible to investigate contact interactions in the frequency domain.

DETERMINATION OF DRIVING CONDITIONS WITHOUT CLEARANCE

Amplitude-frequency characteristic of transfer function makes it possible to determine characteristics of system motion in stage of steady motion.

Total contact reactions are sums of dynamic and static components:

$$N_q^{\Pi} = N_q^{\Sigma}+N_q^{\Delta} > 0$$

$$q = 1...s \tag{96}$$

As sufficient conditions for the implementation of the contact, inequality is used, the validity of which can be accepted under some assumptions during the period of steady motion:

$$A_1 \cdot A_{(q)}(\omega_1)+\varepsilon_q < N_q^{\Sigma}$$

$$q = 1 \ldots s \tag{97}$$

where, ε_q, $q = 1 \ldots s$ - values characterizing the attenuation of the transient process.

The inequalities (97) reflect the fact that the frequency response can be used in estimating the amplitude of the dynamic component of the corresponding response. With some coarsening, you can put ε_q zero.

Taking into account static contact reactions, sufficient contact conditions take the form:

$$A_1 \cdot A_{(q)}(\omega_1) < N_q^{\Sigma}$$

$$q = 1 \ldots s \tag{98}$$

Given system parameters and defined static contact reactions allow, based on inequalities (98), to characterize the amplitude-frequency region of a complex contact by inequality:

$$A_1 < \min_q \left\{ \frac{N_q^{\Sigma}}{A_{(q)}(\omega_1)} \right\} \tag{99}$$

The right-hand side of the presented inequality is interpreted as a function that matches each oscillation frequency of the support surfaces with a minimum of s values that are a constraint on the contact amplitudes. Considered separately as functions of frequencies, values:

$$A_1^{KP,q} = \frac{N_q^{\Sigma}}{A_{(q)}(\omega_1)}$$

$$q = 1 \ldots s \tag{100}$$

are the critical - maximum allowable contact retention amplitudes of fragments of composite solids.

Thus, the critical amplitude of the complex contact interaction can be given through the critical amplitudes of the individual contacts $A_1^{KP,q}$:

$$A_1^{KP} = \min_q \left\{ A_1^{KP,q} \right\} \tag{101}$$

Building a critical amplitude function allows you to determine the amplitude-frequency region of a complex contact.

EXAMPLE OF AMPLITUDE-FREQUENCY CONTACT AREA

The static contact condition sets a limit on the static parameters of the system and allows you to determine the magnitude of the static reaction. The value of the static reaction together with the amplitude-frequency characteristic sets the contact condition for all parameters of the mathematical model in form of inequality. The fixed parameters of the applied forces and the metric characteristics of the system elements make it possible to determine the amplitude-frequency contact region - a set of frequencies and amplitudes at which the corresponding kinematic disturbance provides contact oscillation of fragments of all composite solids. The conditions for holding a complex contact for s joints are determined by the intersection of the amplitude-frequency $s$regions of individual contacts.

As an example, a mechanical system of 10 fragments is considered, the main characteristics of which are presented in Table 3. Characteristics of 6 elastic elements pressing fragments 5 of solid bodies are given in Table 4. The bearing surfaces 1 and 10 oscillate in law $A_1 \sin \omega_1 t$with respect to coordinates $H_1$and $H_{10} = 0$m.

Table 3. Solid Fragment Data

Parameter name	Fragment$i$,$i = 1...10$
Weightm_i, kg	0,1
Fragment dimensions, d_im	0,01
Coefficient of viscous frictionb_i, kg/s	1
Constant force$F_{0,i}$, N	0

Table 4. Characteristics of elastic elements

Parameters	Elastic element$q = 1...6$
Element stiffness, $k_{2q-2,2q-1}$N/m	1000
Length of elastic element$l_{2q-2,2q-1}$, m	0,01

The position of the support surfaces, along with many system parameters, determines the static positions of fragments of composite solids. Assuming that the system parameters are fixed, and the position of the support surface $H_1$is considered in a certain range, you can construct a function that maps each value of the surface coordinate to the values of the static positions $H_1$of the system fragments. Figure 10 shows graphs of 1-5 functions of positions of fragments of a composite solid and coordinates of support surfaces 6, 7.

When you increase the datum surface coordinate, $H_1$the distance between the compressed solids changes. The weakening of the elastic elements leads to a decrease in the contact response. When the reference surface reaches H_1a critical value, the static contact reaction turns to zero.

Figure 10. Static equilibrium positions depending on the coordinate of the supporting surface H_1: 6 - the coordinate of the supporting surface H_{10}, 7 - the coordinate of the supporting surface H_1, 1-5 - the position of the static equilibrium of solid fragments 1-5, respectively

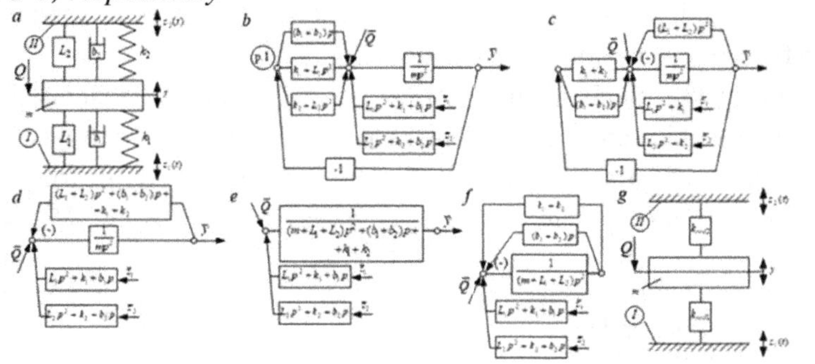

Figure 11 shows graph *1* of the dependence of static contact reactions on the coordinate of the supporting surfaceH_1.

Figure 11. Reaction in contacts depending on the coordinate of the supporting surface H_1: 1 - contact reactions, 2 - critical coordinate

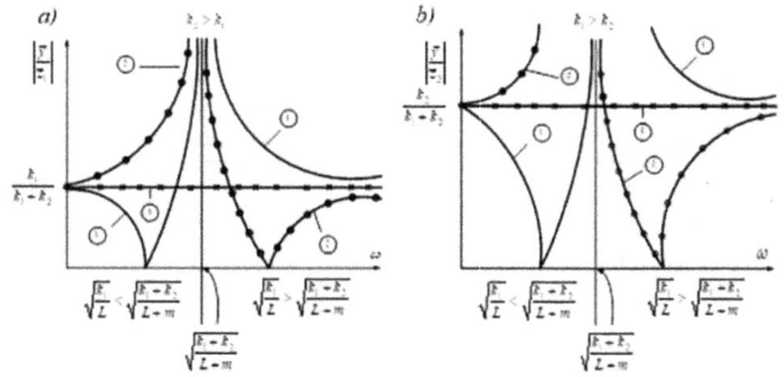

When point *2* is reached, the contact reaction approaches zero. Thus, based on the graphical representation, it can be argued that to ensure static contact, the coordinate of the H_1support surface must be less than the critical coordinate of 0.7 m (in accordance with the example data for numerical modeling presented in Table 3). Here, the coordinate of the reference surface is assumed to be $H_1 = 0,3$m.

Figure 12 shows the amplitude-frequency characteristics $A_{(q)}(\omega_1)$for contacts 1-5 of the mechanical system. Plots 1 are the frequency responses for pins 1 and 5. Graphs 2 represent the frequency response of contacts 2 and 4. Graph 3 - amplitude-frequency characteristic of contact 3. Based on Figure 12, it can be assumed that the frequency response of contact 3 is less than that of contacts 1, 2, 4, 5.

Figure 12. Amplitude-frequency characteristic

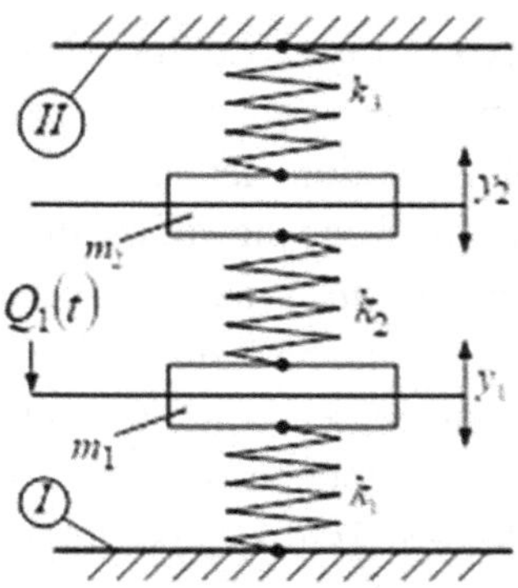

1 - amplitude-frequency characteristics $A_{(1)}(\omega_1)$, $A_{(5)}(\omega_1)$for contacts 1, 5; 2 - amplitude-frequency characteristics $A_{(2)}(\omega_1)$, $A_{(4)}(\omega_1)$ for contacts 2, 4; 3 - amplitude-frequency characteristics $A_{(3)}(\omega_1)$ for contact 3

Figure 13 shows the critical amplitude plots for each individual contact. The subgraph of each critical amplitude function defines a plurality of amplitudes and frequencies providing contact. To determine the contact oscillation conditions for a complex contact, it is necessary to determine the intersection of all amplitude-frequency regions for each simple contact. In accordance with Figure 13, it can be assumed that the contact condition is determined only by contacts 1, 2, 4, 5. The critical amplitude plot 3 is above plots 1, 2, 4, 5 and does not affect the frequency domain of the complex mechanical system contact.

Figure 13. Critical amplitudes of simple contacts

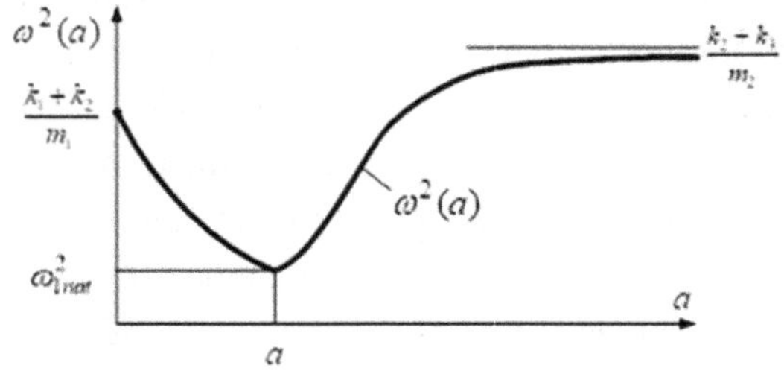

graphs 1 - critical amplitudes $A_1^{KP.1}, A_1^{KP.5}$ for contacts 1, 5; graphs 2 - critical amplitudes $A_1^{KP.2}$, $A_1^{KP.4}$ for contacts 2, 4; graphs 3 - critical amplitude $A_{(3)}(\omega_1)$ for contact 3

The resulting frequency domain is shown in Figure 14. Graph *1* shows the minimum values of critical amplitudes.

According to Figure 14, it can be assumed that when the frequency tends to zero, contact is realized for any arbitrarily large amplitude of oscillation of the support surfaces. As the frequency of oscillation increases, the critical amplitude decreases to 0.015 m. This means that at any frequency of oscillation of support surfaces with oscillation amplitude less than 0.015 m, no rupture occurs. From the relative position of the critical amplitude graphs, it can be assumed that at the frequency range from 0 to 225 rad/s, with a continuous increase in the amplitude of the oscillation from zero, the simple contact conditions for contacts 1 and 5 are initially violated. In the frequency range from 225 rad/s to 385 rad/s, the simple contact condition for contacts 2 and 4 is first violated. If the oscillation frequency exceeds 385 rad/s, then with a gradual increase in the oscillation amplitude from zero, the contact conditions are violated in contacts 2 and 4.

Figure 14. Critical amplitude of the compound contact

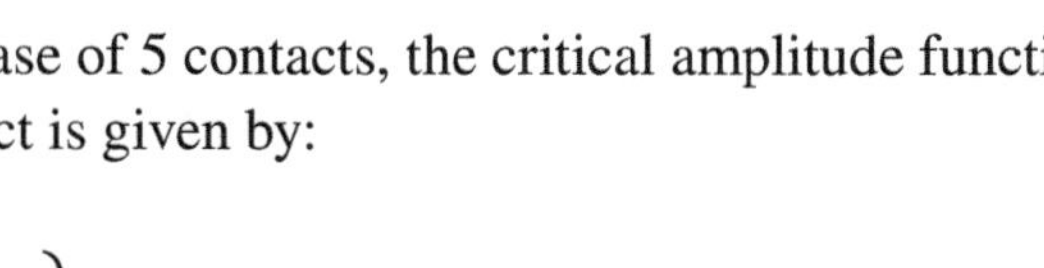

1 - graph of the function of the critical amplitude of the compound contact

In the general case of 5 contacts, the critical amplitude function corresponding to a complex contact is given by:

$$A_1^{KP} = \min_{q=1...5}\left\{\frac{N_q^{\Sigma}}{A_{(q)}(\omega_1)}\right\} \tag{102}$$

The type of function shows that the preservation of contact is determined by static reactions and values inverse to the amplitude-frequency characteristics. Accordingly, it can be assumed that the resonant frequencies form the minima of the critical frequency functions, and the zeros of the amplitude-frequency characteristics lead to the maxima of the corresponding critical amplitude functions.

The presented example demonstrates a general technique for constructing the amplitude-frequency domain of a complex contact based on the amplitude-frequency characteristics and the magnitude of static reactions for a model mechanical system with 5 contacts. It is shown that the amplitude-frequency region of the complex contact is formed based on the intersection of the amplitude-frequency regions of individual simple contacts.

Thus, the use of the amplitude-frequency characteristic and the magnitude of the static reaction allows us to formulate sufficient conditions for dynamic contact in the frequency domain and obtain conditions for the parameters of the system that ensure contact oscillation of fragments of a solid body in the stage of steady motion.

DYNAMIC CONTACT DEPENDING ON SYSTEM PARAMETERS

The critical amplitude is determined by the parameters of the mechanical system and the dynamic effect characterized by the frequency of oscillation of the support surfacesω_1. Mechanical system parameters are masses of solid body fragments, stiffness of elastic elements, viscous friction forces and additional constant forces.

The parameters entered in the system affect the contact conditions in a certain way. If the critical amplitude increases when the selected parameter changes, it means that the parameter change helps to hold the contact. If the critical amplitude decreases, it means that the parameter contributes to the contact failure.

Viscosity, $b_i$stiffness of elements, mass of $k_{2q-2,2q-1}$fragments specified $m_i$in accordance with Tables 3 and 4 are considered as key values of parameters. Generalized parameters are entered into the system - dimensionless coefficients of viscosity, stiffnessλ_1, mass $\lambda_2$in accordance $\lambda_3$with the expressions:

$$b'_i = \lambda_1 b_i$$

$$k'_{2q-2,2q-1} = \lambda_2 k_{2q-2,2q-1}$$

$$m'_i = \lambda_3 m_i \tag{103}$$

Depending on the generalized viscosity parameters, stiffness $\lambda_1$factors, $\lambda_2$mass, the system $\lambda_3$parameters are equal to the values$b'_{i},,.\ k'_{2q-2,2q-1} m'_i$Changing one of the parametersλ_i,$i = 1,2,3$, at a fixed frequency of oscillation of $\omega_1$the support surface leads to a change in the critical amplitude. The critical amplitude versus parameters can be used $\lambda_i$to visualize the effect of viscosity, stiffness and mass on the contact interaction. The construction of several functions of critical amplitudes from generalized parameters for different frequencies allows us to evaluate different dependence options.

Figure 15 shows three options for dependence of the critical amplitude on the generalized viscosity parameter $\lambda_1$for three different frequenciesω_1, taking values of 50 rad/s, 100 rad/s, 150 rad/s.

Figure 15. Critical amplitude depending on viscosity coefficient

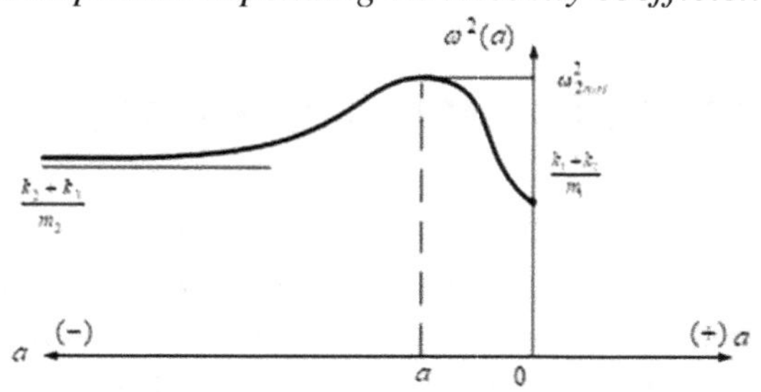

graphs 1, 2, 3 correspond to frequencies $\omega_i = 50i$rad/s,$i = 1,2,3$

The presented graph shows various variants of the effect of generalized viscosity on the contact interaction condition depending on the frequency of oscillation of the support surfaces.

Figure 16 shows the dependence of the critical amplitude on the generalized parameter $\lambda_2$for three different frequencies $\omega_1$of 50 rad/s, 100 rad/s, 150 rad/s. The contact is implemented only for the parameter$\lambda_2 > 0,28$. Depending on the frequency of the oscillation, an increase in the generalized parameter can either contribute to the contact or lead to a violation of the contact conditions.

Figure 16. Critical amplitude depending on the stiffness factor: graphs 1, 2, 3 correspond to frequencies $\omega_i = 50i rad/s, i = 1, 2, 3$

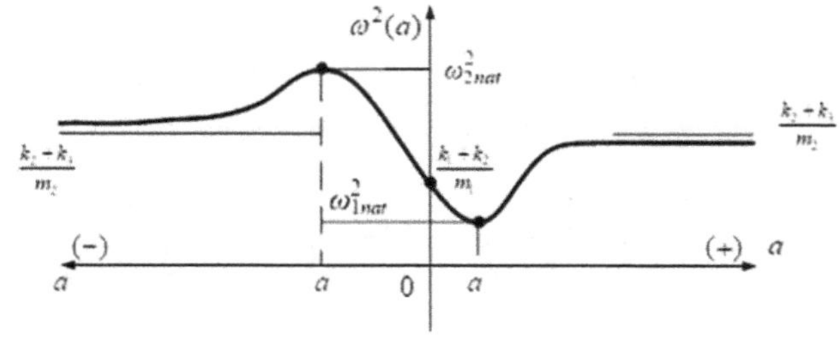

Dependence on the generalized mass parameter is more complex in accordance with Figure 17.

Figure 17. Critical amplitude depending on mass factor:

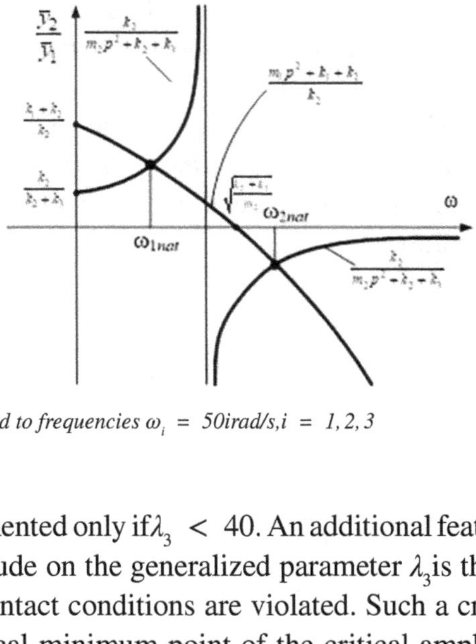

graphs 1, 2, 3 correspond to frequencies $\omega_i = 50i rad/s, i = 1, 2, 3$

Contact is implemented only if $\lambda_3 < 40$. An additional feature of the dependence of the critical amplitude on the generalized parameter λ_3 is the existence of a mass interval for which contact conditions are violated. Such a critical mass interval is determined at the local minimum point of the critical amplitude function. When the oscillation frequency changes, the critical mass interval changes in accordance with Figure 17.

Figure 18 shows *1* - diagram of critical amplitudes and *2* - level of oscillation amplitude of support surfaces. The intersection of plots *1* and *2* at points *3* and *4* determines the (m_{01}, m_{02}) range of masses with violation of contact conditions. In this case, it can be assumed that with a vibration amplitude of 8 mm and a vibration frequency of 100 rad/s, the contact conditions will be violated for the mass range of fragments from 0.08 kg to 0.11 kg.

Figure 18. Contact mass range

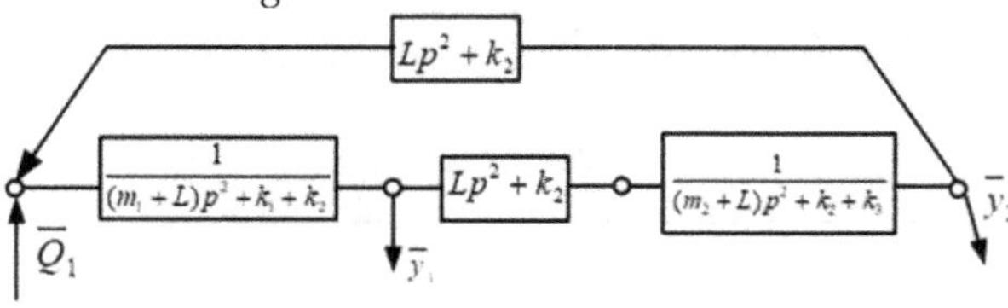

1 - critical amplitude, 2 - vibration amplitude level, 3, 4 - intersection points, m_{01} = 8, 3,m_{02} = 11, 6- boundaries of the parametric range of contact masses

The presented examples allow us to build preliminary hypotheses about the nature of the influence of system parameters on the conditions of contact interaction. Such hypotheses include the existence of a range of separation masses determined by the frequency of external kinematic influence from the support surfaces, the existence of frequency ranges at which viscosity contributes to separation, and the frequency range for which viscosity contributes to contact.

The main properties of a mechanical system are the realization of unconditional contact at frequencies close to zero, the existence of finite critical amplitudes at frequencies tending to infinity, and the existence of intermediate frequencies of violation of contact conditions.

REFERENCES

Elyseev, A. V., & Elyseev, S. V. (2014). *Determining contact reactions in composite solid bodies under dynamic loads considering non-holding connections. Modern Technologies. System Analysis. Modeling, 1, 45–54.* Irkutsk State University of Railway Transport.

Elyseev, S. V., & Elyseev, A. V. (2013). Methodological foundations for determining contact static and dynamic reactions considering non-holding connections. Irkutsk: Irkutsk State University of Railway Transport. 138 pages. Illustrated. Dep. in VINITI 16.12.13 No. 369.

Elyseev, S. V., & Elyseev, A. V. (2013). *Generalized approaches in the determination of contact reactions in solid bodies under static loads considering non-holding connections. Modern Technologies. System Analysis. Modeling, 4, 51–60.* Irkutsk State University of Railway Transport.

Elyseev, S. V., & Elyseev, A. V. (2014). Conditions for contact vibrations of composite solid bodies under dynamic loads considering non-holding connections. Irkutsk: Irkutsk State University of Railway Transport. 42 pages. Illustrated. Dep. in VINITI 27.01.2014 No. 31.

Elyseev, S. V., Reznik, Yu. N., & Khomenko, A. P. (2011). *Mechatronic approaches in the dynamics of mechanical systems*. Nauka.

Elyseev, S. V., Reznik, Yu. N., Khomenko, A. P., & Zasyadko, A. A. (2008). *Dynamic synthesis in generalized problems of vibration protection and vibration isolation of technical objects*. Irkutsk State University.

ADDITIONAL READING

de Silva, C. W. (2000). *Vibration: Fundamentals and practice*. CRC Press.

Eliseev, A. V. (2014). Model problems of dynamics in mechanical vibrating systems with unretained connections. In Mathematics, Its Applications, and Mathematical Education (MPMO 14): V International Conference (pp. 111-115).

Eliseev, A.V. (2014, February 27). Dynamic contact in composite rigid bodies, considering unretained connections. Deposition manuscript No. 61-V.

Eliseev, A. V. (2014). Conditions for dynamic contact in composite solids with non-holding connections depending on parameters. Deposited manuscript, 56-B, 25 February 2014.

Eliseev, A. V. (2014). Dynamics of vibrational interaction: Non-holding connections, trajectory features, and measuring devices. In Kulaginsky readings: Techniques and technologies of production processes. XIV International Scientific and Practical Conference (pp. 287-293).

Eliseev, A. V. (2014). Dynamics of vibrational bouncing: Non-holding connections, parameters, and trajectory features. *Reshetnev Readings*, *1*, 284–286.

Eliseev, A. V. (2015). Vibration regimes of interaction between bulk media and a vibrating surface considering unretained connections: Mathematical models, forms of interaction, realization factors. Modern Technologies. System Analysis. *Modeling*, *1*(45), 29–37.

Eliseev, A. V. (2015). Features of mathematical modeling in tasks of vibrational interactions of elements in technological systems considering non-holding connections. In V. K. Astashev, V. L. Krupenin, G. Ya. Panovko, & K. B. Salamandra (Eds.), Dynamics of vibroimpact (highly nonlinear) systems: XVIII International Symposium dedicated to the 100th anniversary of Professor A. E. Kobrinsky (pp. 107-114).

Eliseev, A. V. (2015). Dynamics of vibrational interactions of elements in technological systems considering non-holding connections (Abstract of dissertation for the degree of Candidate of Technical Sciences). Southwest State University, Kursk.

Eliseev, A. V. (2015). Dynamics of vibrational interactions of elements in technological systems considering non-holding connections (Dissertation for the degree of Candidate of Technical Sciences). Southwest State University, Kursk.

Eliseev, A. V. (2018). Non-holding connections in tasks of dynamics of technological and transport systems. In Problems of Mechanics of Modern Machines: Materials of the VII International Scientific Conference (pp. 189-192).

Eliseev, A. V. (2024). Frequency ratios in the formation of continuous bouncing modes of solid bodies on vibrating surfaces considering non-holding connections. In Current Problems in Applied Mathematics, Informatics, and Mechanics: Proceedings of the International Scientific Conference (pp. 426-430). Voronezh.

Eliseev, A. V., Artyunin, A. I., Eliseev, S. V., & Sitov, I. S. (2014). Scientific and methodological justification of the technology for vibration hardening processes in dynamic interactions with bulk material and vibrating surfaces considering non-holding connections. *Reshetnev Readings*, *1*, 394–396.

Eliseev, A. V., Eliseev, S. V., & Kashuba, V. B. (2015). Methodological approaches to modeling contact interactions in vibrational technological processes and machines considering non-holding connections. In V. A. Iljukhina, V. I. Zhukovskij, N. P. Ketova, A. M. Gazaliev, & G. S. Mal (Eds.), Science, Technology, and Life - 2014: Proceedings of the International Scientific Conference (pp. 146-155).

Eliseev, A. V., Kashuba, V. B., & Kaimov, E. V. (2014). Measuring devices for fixing conditions of vibrational interaction under unretained connections. Systems. Methods. *Technologies*, *4*(24), 13–20.

Eliseev, A. V., Khomenko, A. P., Eliseev, S. V., & Pnev, A. G. (2016). Theoretical foundations of vibration technologies: Dynamic interactions in non-holding connections. Deposited manuscript, 16-B, 12 January 2016.

Eliseev, A. V., Kinas, N. Z., & Pneev, A. G. (2015). Mathematical modeling of interactions of elements in vibrational technological systems considering non-holding connections. In Fundamental and applied scientific research: Proceedings of the International Scientific and Practical Conference (pp. 21-24).

Eliseev, A. V., & Kopylov, Yu. R. (2015). Some issues in the dynamics of vibration interaction between bulk media and a surface, considering unretained connections. *Modern High-Tech Technologies*, *3*, 28–31.

Eliseev, A. V., & Kopylov, Y. R. (2015). Modeling of contact interactions in vibrational technological processes and machines considering non-holding connections. Modern Science-intensive Technologies, 1-1, 19-22.

Eliseev, A. V., Kopylov, Y. R., & Eliseev, S. V. (2015). Features of the dynamics of vibrational interactions of elements in technological systems considering non-holding connections. In Aerospace technologies (ACT-2015): Proceedings of the XVI International Scientific and Technical Conference and School for Young Scientists, Postgraduates, and Students (pp. 133-135).

Eliseev, A. V., Kopylov, Y. R., & Eliseev, S. V. (2015). Features of the dynamics of vibrational interactions of elements in technological systems considering non-holding connections. In Aerospace technologies (ACT-2015): Proceedings of the 1st round of the XVI All-Russian Scientific and Technical Conference and School for Young Scientists, Postgraduates, and Students (pp. 27-29).

Eliseev, A. V., Kopylov, Y. R., & Mironov, A. S. (2023). Frequency ratios in the formation of regimes with multiple bouncing in elastic interaction of material particles with a vibrating surface considering non-holding connections. *Journal of Advanced Research in Technical Science*, *38*, 15–23.

Eliseev, A. V., & Mironov, A. S. (2024). Features of elastic interactions of elements of vibrational technological machines considering non-holding connections: Calculation schemes, mathematical models, applications. *Current Problems in Machine Theory*, *17*, 74–79.

Eliseev, A. V., Pneev, A. G., Sigachev, N. P., & Nguyen, D. H. (2015). Features of the distribution of contact pressures in composite solids under static loading with non-holding connections. In Kulaginsky readings: Technology and processes in production, XV International Scientific and Practical Conference: Proceedings in 3 volumes (pp. 283-288).

Eliseev, A. V., Sitov, I. S., & Eliseev, S. V. (2015). Mathematical models of the dynamics of vibrational interactions of elements in technological systems considering non-holding connections. Deposited manuscript, 14-B, 21 January 2015.

Eliseev, A. V., Sitov, I. S., & Eliseev, S. V. (2016). Method for determining conditions for maintaining contact with non-holding connections in dynamic interactions. In Mechanics and Tribology of Transport Systems (MechTriboTrans-2016): Proceedings of the International Scientific Conference (Vol. 2, pp. 120-124).

Eliseev, A. V., Sitov, I. S., & Vjong, K. C. (2016). Dynamics of vibrational interactions of elements in technological systems considering non-holding connections: Mathematical modeling, control approaches, and measuring instruments. *Mechanics of the XXI Century*, *15*, 285–290.

Eliseev, S.V., Eliseev, A.V. (2014, January 27). Conditions for contact vibrations of composite rigid bodies under dynamic loads, considering unretained connections. Deposition manuscript No. 31-V.

Eliseev, S.V., Eliseev, A.V. (2014, January 27). Conditions for contact vibrations of composite rigid bodies under dynamic loads considering unretained connections. Deposition manuscript No. 31-V.

Eliseev, S. V., & Eliseev, A. V. (2015). Mathematical models of the dynamics of vibrational interactions of elements in technological systems considering non-holding connections. Deposited manuscript, 14, 21 January 2015.

Eliseev, S. V., & Eliseev, A. V. (2015). Mathematical models of the dynamics of vibrational interactions of elements in technological systems considering non-holding connections. Deposited manuscript, 14-B, 21 January 2015.

Eliseev, S. V., Eliseev, A. V., Orlenko, A. I., Ermoshenko, Y. V., & Bolshakov, R. S. (2019). Investigation and development of methodological foundations of the interaction of solid bodies under vibrational interactions with non-holding connections in applications to problems of dynamic movement of technical objects and technological processes based on the effects of vibrational interactions (theoretical and experimental approaches). Research report on scientific work.

Eliseev, S.V., Eliseev, A.V., Orlenko, A.I., Ermoshenko, Yu.V., Bolshakov, R.S., Nikolaev, A.V., Mironov, A.S., Trofimov, A.N. (2018, September 17). Methodological foundations for solving dynamic interaction problems of technological system elements with unretained connections. Deposition manuscript No. 105-V.

Eliseev, S. V., Eliseev, A. V., & Sigachev, N. P. (2015). Mathematical models of dynamic interactions in contacts with unretained connections under vibrational loads. *Bulletin of the All-Russian Research and Design Institute of Electric Locomotive Engineering*, *2*(70), 156–167.

Elyseev, A. V. (2015). Dynamics of vibrational interactions of elements of technological systems considering non-holding connections. In V.S. Balbarov (Ed.), Problems of Mechanics of Modern Machines: Materials of the VI International Conference (pp. 93–98). Irkutsk.

Elyseev, A. V., Artyunin, A. I., & Sitov, I. S. (2017). Method for determining contact disruption conditions under vibrational loads considering non-holding connections. Modern Technologies. System Analysis. *Modeling*, *1*(53), 26–34.

Elyseev, A. V., Selvinsky, V. V., & Elyseev, S. V. (2015). *Dynamics of vibrational interactions of elements of technological systems considering non-holding connections.*

Elyseev, A.V., Selvinsky, V.V., & Elyseev, S.V. (2015). Dynamics of vibrational interactions of elements of technological systems considering non-holding connections. Ecology. Analytical Reviews of World Literature, 332.

Elyseev, S. V., & Elyseev, A. V. (2014). Determining contact reactions in composite solid bodies under dynamic loads with non-holding connections. Modern Technologies. System Analysis. *Modeling*, *1*(41), 45–54.

Elyseev, S. V., Kashuba, V. B., Pnev, A. G., Elyseev, A. V., & Sitov, I. S. (2014). Non-holding connections in dynamic interactions between granular media and a vibrating surface: Scientific and methodological basis for the technology of vibrational hardening processes. Systems. Methods. *Technologies*, *3*(23), 17–31.

Khomenko, A. P., Eliseev, S. V., & Eliseev, A. V. (2015). Unretained connections in the dynamics of interaction problems. In Safety and Survivability of Technical Systems. *Cailiao Daobao*, *3*, 224–229.

Khomenko, A. P., Eliseev, S. V., & Eliseev, A. V. (2015). Dynamics of vibrational interactions in technological system elements, considering unretained connections. Modern Technologies. System Analysis. *Modeling*, *3*(47), 8–17.

Orlenko, A. I., Bulavko, O. A., & Eliseev, A. V. (2017). Mathematical modeling in dynamic problems of interaction between transport vehicle elements, considering unretained connections. In Problems of Safety in Transport: Materials of the VIII International Scientific and Practical Conference, Dedicated to the Year of Science (pp. 205-207).

Orlenko, A. I., & Eliseev, A. V. (2017). Features of the dynamics of the collector-brush node of the traction electric motor of an electric locomotive, considering unretained connections. In *Kulaginsky Readings* (pp. 148–154). Technology and Technology of Production Processes. Collection of Articles.

Orlenko, A. I., & Eliseev, A. V. (2017). Dynamics of the collector-brush node of the traction motor of a locomotive: Considering unretained connections. In Problems of Safety in Transport: Materials of the VIII International Scientific and Practical Conference, Dedicated to the Year of Science (pp. 207-208).

Orlenko, A. I., & Eliseev, A. V. (2017). Mathematical model of collector-brush interactions considering non-holding connections. *Fundamental Issues of Mechanics*, *2*, 42–44.

Orlenko, A. I., & Eliseev, A. V. (2017). Generalized gap function as the basis for estimating dynamic systems with non-holding connections. *Science and Education in Transport*, *2*, 168–172.

Orlenko, A. I., Eliseev, A. V., & Eliseev, S. V. (2018). Non-holding connections: Forms and manifestations of contact interaction. In Machines, Assemblies, and Processes: Design, Creation, and Modernization: Proceedings of the I International Scientific-Practical Conference (pp. 38-42).

Schurger, B., Frankovsky, P., Janigova, S., Bocko, J., & Kolodziej, A. (2023). Mechanical metamaterials: Properties and classification. *Acta Mechatronica*, *8*(3), 29–35. DOI: 10.22306/am.v8i3.98

Sitov, I. S., Elyseev, A. V., & Kaimov, E. V. (2015). Generalized approaches to the determination of contact reactions in solid bodies under static loads considering non-holding connections: A multidimensional approach. Systems. Methods. *Technologies*, *1*(25), 19–29.

Yang, T., Jiang, X., Huang, Y., Tian, Q., Zhang, L., Dai, Z., & Zhu, H. (2022). Mechanical sensors based on two-dimensional materials: Sensing mechanisms, structural designs and wearable applications. *iScience*, *25*(1), 103728. DOI: 10.1016/j.isci.2021.103728 PMID: 35072014

Chapter 8
Vibration Field Formation for Transport and Techno Objects Using Variation Approach

ABSTRACT

The concept of a generalized lever as an approach to the formation of the vibration field of technical objects of transport and technological purpose is discussed. The issues of safety and dynamic quality of vibration processes, as well as the tasks of dynamic interaction of technical facilities are considered. Special attention is paid to changing the dynamic states of a mechanical oscillatory system by changing the coupling coefficients of kinematic disturbances. It is proposed to use variational principles to find these coefficients and ensure that the frequencies of zeroing the oscillation amplitudes coincide with the natural frequencies of the system.

CHANGE IN THE DYNAMIC STATES OF A MECHANICAL OSCILLATORY SYSTEM WITHIN THE FRAMEWORK OF THE CONCEPT OF A GENERALIZED SYSTEM LEVER.

Currently, significant attention is paid to the issues of ensuring the safety of technical facilities under conditions of intense vibration loads. In terms of a systematic approach, safety issues are very complex, complex, interdisciplinary and other features. At the same time, from a local point of view, the safety issues of technical facilities under conditions of intense loads can be correlated with the tasks of machine dynamics, vibration interactions, vibration protection and vibration isolation. Of particular note are the issues of ensuring the dynamic quality of vibration

DOI: 10.4018/979-8-3373-0447-2.ch008

processes implemented in technical facilities operating under vibration loads. Along with the security problem, the tasks of dynamic interactions of technical objects can be considered. Along with the safety tasks of technical objects, dual tasks can be considered, which are associated with intense vibration loads, but display the "useful" side of vibration loads and are rather associated with the effects occurring in technical objects under conditions of vibration loads. From that point of view, a technical object under conditions of vibration loads is considered not from the side of safety threats, but from the side of the usefulness of certain useful dynamic effects that are of great importance for the processes of transportation, hardening, separation of materials, separation of bulk materials into different fractions, orientation of parts in space and much more. Thus, it can be noted that the safety issues of technical objects under vibration loads pay attention to the tasks of assessing, forming and correcting the dynamic states of technical objects from a certain general point of view, where safety is a summary characteristic.

Along with the generalized safety characteristic, local assessment of dynamic states is considered as an assessment of dynamic interactions of individual elements of technical objects. Thus, the safety of technical objects under vibration loads implies the development of methods that allow conducting system studies to assess, form and correct the dynamic states of technical objects. At the same time, the dynamic states of technical objects are associated with the idea of the vibration states of technical objects in the so-called technical states. Thus, the dynamic state itself has sufficient specificity when considering dynamics problems. However, for a number of assumptions, the dynamic state can be fairly accurately characterized and assumed.

In particular, the range of problems of machine dynamics is known when the design scheme is a mechanical oscillatory system with a finite number of degrees of freedom. A certain part or link of the technical system can also be considered, which in turn also has a design scheme in the form of a mechanical oscillatory system. For a number of technical objects under conditions of intense vibration loads, it can be assumed that the vibrations in question are small and occur near the position of static equilibrium or a steady-state form of motion. For such systems, options are possible when external kinematic disturbances are interconnected. An example of connected perturbations are kinematic perturbations transmitted to an object from a swinging platform. For a wide class of technical objects, it can be assumed that external disturbances are in-phase harmonic disturbances.

Thus, in the tasks of evaluating, forming and correcting dynamic states of technical objects, it can be assumed that the dynamic state is understood as the established form of small forced vibrations of the generalized coordinates of the mechanical vibrational system.

To solve the problems of assessment, formation and correction of dynamic states, a structural approach has spread, within the framework of which a structural diagram of a dynamically equivalent automatic control system is associated with a mechanical oscillatory system. Connected kinematic disturbances are considered as input signal, and amplitudes of oscillations of generalized coordinates serve as output signals. For evaluation of lever links between system links and support surfaces forming kinematic disturbances on the system. A transfer function can be used to estimate the linkage, which is the ratio of the amplitudes of the output signal amplitude to the amplitude of the input signal. In a physical sense, such a transfer function conveys ideas about leverage. The values of the transfer function can take positive, negative, zero and infinite values. The set of received values of amplitude-frequency characteristics of linkage transfer functions can be interpreted by levers of the 1st or 2nd kind, as well as forced levers having arms of zero or infinite length. Depending on the frequency of the external disturbance, the dynamic state can be characterized by lever coefficients. At the same time, the total set of dynamic states can be characterized by an amplitude-frequency characteristic, the generalization of which can be characterized by the so-called dynamic invariant, which displays generalized properties in the form of the number of positive branches, negative branches, as well as the number of frequencies that zero the amplitude-frequency characteristic, and the number of frequencies for which the amplitude-frequency characteristic turns into infinity. Thus, from a generalized point of view, the set of dynamic states can be represented by the number of frequency intervals in which the frequency amplitude is positive, the number of frequency intervals in which the frequency amplitude characteristic is negative, and the number of frequencies in which the frequency amplitude characteristic takes a zero value or infinity. Thus, a collection of fixed external perturbations can be represented by an invariant representing significant dynamic features. Changing the connectivity of kinematic perturbations allows you to determine different sets of generalized dynamic states. The choice of kinematic perturbation connectivity factors allows significant images to affect dynamic states. In the works (Eliseev, Kuznetsov, & Mironov, 2023; Eliseev & Mironov, 2022), maps of dynamic invariants were built, giving ideas about the formation of a variety of dynamic states. In (Eliseev, Sitov, & Kuznetsov, 2024; Eliseev & Kuznetsov, 2023), it was shown that the amplitude-frequency can be displayed using the so-called generalized lever. Work (Eliseev, Sitov, & Kuznetsov, 2024) shows that the dependence of interval characteristics on connectivity coefficients is discontinuous. At the same time, issues related to the change in generalized dynamic states have not yet received proper development.

The chapter is devoted to the development of a methodology for changing generalized dynamic states by changing the connectivity coefficients of kinematic perturbations, is aimed at changing the dynamic diversity of the set of dynamic states associated with the combination of the frequency of zeroing the vibration amplitudes of an object whose dynamic state is estimated and the natural frequency of the system. It should be noted that external perturbations of the zeroing frequency can coincide with their own frequencies, which leads to a change in the dynamic invariant. This chapter discusses the use of variational principles that allow finding connectivity coefficients that ensure the convergence of zeroing frequencies with natural frequencies.

Dynamic States And Lever Relations In Mechanical Oscillatory Systems

A mechanical oscillatory system is considered, formed by two solid bodies in the form of rods, installed with the help of elastic elements on support surfaces, making connected in-phase harmonic disturbances of kinematic nature. It is assumed that connected kinematic disturbances cause small steady-state vibrations of elements near the static equilibrium position (Figure 1).

Figure 1. Mechanical oscillation system. Design scheme

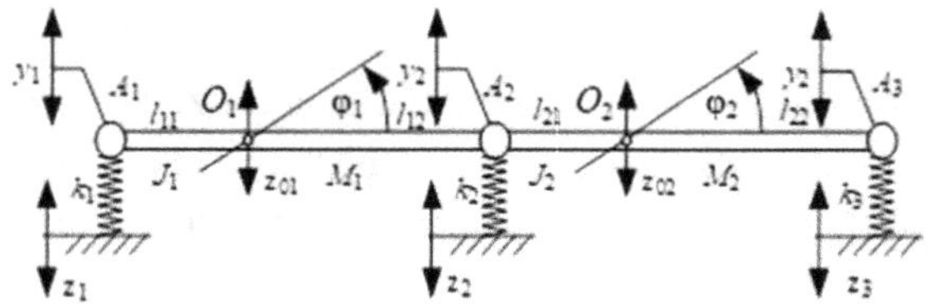

Connected kinematic perturbations form a set of dynamic states, which can be characterized on the basis of ideas about lever links, which appear in the display of dynamic interactions using levers of the 1st and 2nd kind, as well as levers of degenerate forms. Dynamic states are displayed based on three linkages. The works (Eliseev, Sitov, & Kuznetsov, 2024; Eliseev & Kuznetsov, 2023) show that a set of lever relations can be characterized using a generalized lever.

The task is to develop a method for determining critical dynamic states within the variation supply.

Mathematical Model And Dynamic States Of a Mechanical Oscillatory System

Considered is a mechanical oscillatory system formed by two solids with masses M_1, M_2, corresponding moments of inertia J_1, J_1; dimensions of arms for solid bodies are values l_{11}, l_{12} and l_{21}, l_{22} (Figure1).

The coordinates of the mechanical oscillatory system are the values y_1, y_2, y_3, which display the displacements of points A_1, A_2, A_3 relative to the position of static equilibrium. It is assumed that the external in-phase harmonic z_1, z_2, z_3 are related by the relations:

$$z_2=\gamma_2 z_1, z_3=\gamma_2 z_1 \quad (1)$$

where γ 2, γ 3 - coupling coefficients of external kinematic disturbances. To construct a mathematical model within the framework of the formalism of Lagrange equations of the 2nd kind, equations of potential and kinetic energy in the coordinates y_1, y_2, y_3 and $z_{01}, z_{02}, \varphi_1, \varphi_2$ can be written. Coordinates z_{0i} -denote displacements by the center of gravity O_1 and O_2 of solids, and φ_1, φ_2 - small angles of rotation of solids relative to static equilibrium positions. The coordinates $\{y_i\}, i = 1..3$ and $\{z_{0i}, \varphi_i\}, i = 1..2$ are related by the relations:

$$z_{01}=a_1y_1+b_1y_2 \quad (2)$$

$$\varphi_1=c_1(y_2-y_1)$$

(3)

$$z_{02}=a_2y_2+b_2y_3$$

(4)

$$\varphi_2=c_2(y_3-y_2) \quad (5)$$

In the coordinates $\{y_i\}, \{z_{0i}\}, \{\varphi_i\}$, the potential and kinetic energy of the system can be represented by the expressions:

$$\Pi = \sum_{i=1}^{n}\frac{1}{2}k_i(y_i - z_i)^2 \quad (6)$$

$$T = \sum_{i=1}^{n} \frac{1}{2}\left(M_i \dot{z}_{0i}^{\;2} + \frac{1}{2} J_i \dot{\varphi}_i^{\;2}\right) \quad (7)$$

The potential and kinetic energy can be represented in matrix form:

$$\Pi = \frac{1}{2}\langle K(y - z),(y - z)\rangle \quad (8)$$

$$T = \frac{1}{2}\langle M\dot{z}_0, \dot{z}_0\rangle + \frac{1}{2}\langle J\dot{\varphi}, \dot{\varphi}\rangle \quad (9)$$

where $K = \begin{pmatrix} k_1 & & 0 \\ & k_2 & \\ 0 & & k_3 \end{pmatrix}$ is the matrix of stiffness coefficients, $M = \begin{pmatrix} M_1 & 0 \\ 0 & M_2 \end{pmatrix}$ is the matrix of masses, $J = \begin{pmatrix} J_1 & 0 \\ 0 & J_2 \end{pmatrix}$ is the matrix of moments of inertia, $z = \begin{pmatrix} z_1 \\ z_2 \\ z_3 \end{pmatrix}$ is the vector of external kinematic perturbations, $y = \begin{pmatrix} y_1 \\ y_2 \\ y_3 \end{pmatrix}$ is the vector of coordinates of the system, $z_0 = \begin{pmatrix} z_{01} \\ z_{02} \\ z_{03} \end{pmatrix}$ is the vector of displacements of the centers of mass relative to the positions of static equilibrium, $\varphi = \begin{pmatrix} \varphi_1 \\ \varphi_2 \\ \varphi_3 \end{pmatrix}$ is the vector of small angles of rotation, $\dot{\varphi}$, $\dot{z}_0$ are derivatives of the corresponding vectors.

The z and φ coordinates can be expressed in terms of y coordinates by expressions

$$z = Uy \quad (10)$$

$$\varphi = Vy \quad (11)$$

where
$U = \begin{pmatrix} a_1 & b_1 & 0 \\ 0 & a_2 & b_2 \end{pmatrix}$, $V = \begin{pmatrix} -c_1 & c_1 & 0 \\ 0 & -c_2 & c_2 \end{pmatrix}$ are transition matrices.

In the $\{y_i\}$ coordinate system, potential and kinetic energy can be expressed as:

$$\Pi = \frac{1}{2}\langle K(y - z),(y - z)\rangle \quad (12)$$

$$T = \frac{1}{2}\langle (U^T M U + V^T J V)y, y\rangle \quad (13)$$

In turn, the system of differential Lagrange equations of the 2nd kind takes the form:

$$(U^T MU + V^T JV)\ddot{y} + Ky \;=\; K^T z \tag{14}$$

After integral Laplace transformations, taking into account the initial zero conditions, the system takes the form of algebraic equations:

$$((U^T MU + V^T JV)p^2 + K)\bar{y} \;=\; K^T \bar{z} \tag{15}$$

where U^T, V^T and K^T are transposed matrices U, V, K, $p = j\omega$, j is an imaginary unit.

Based on the known methods (Eliseev & Eliseev, 2020; Eliseev, 2023), the system of algebraic equations (15) can be represented by a structural diagram of a dynamically equivalent automatic control system (Figure 2).

Figure 2. Structural diagram of mechanical oscillatory system (Figure1)

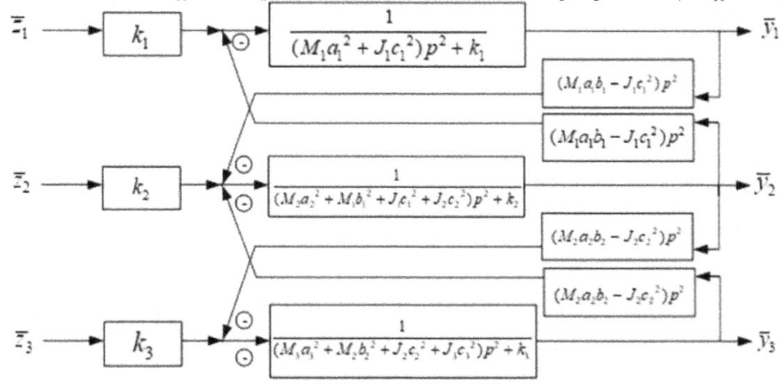

On the basis of the structural diagram (Figure 2), transfer functions can be built that display leverage:

$$W_1(p) = \frac{\bar{y}_1}{\bar{z}_1}\Big|_{\bar{z}_1 \neq 0,\, \bar{z}_{i\neq1} = 0} \tag{16}$$

$$W_2(p) = \frac{\bar{y}_2}{\bar{z}_1}\Big|_{\bar{z}_2 \neq 0,\, \bar{z}_{i\neq2} = 0} \tag{17}$$

$$W_3(p) = \frac{\bar{y}_3}{\bar{z}_1}\Big|_{\bar{z}_3 \neq 0,\, \bar{z}_{i\neq3} = 0} \tag{18}$$

The presented transfer functions (16) - (18) contain as parameters the coupling coefficients (1) of kinematic disturbances and represent families of transfer functions.

In general, for fixed connectivity coefficients γ_2, γ_3 transfer functions (16) - (18) can be represented by amplitude-frequency characteristics representing dynamic features of linkage by sign-defined forms of dynamic interactions, displayed by positive and negative branches of amplitude-frequency characteristics, which are interpreted as levers of the 1st or 2nd kind, and critical dynamic states represented by amplitude-frequency characteristics in the form of frequencies, in which oscillation amplitudes of the corresponding coordinates are zeroed systems, and resonant frequencies, in which amplitude-frequency characteristics suffer a break of the 2nd kind. Within lever interpretation, critical states can be mapped by special levers.

A change in connectivity coefficients can lead to a redistribution of the zeroing frequency positions relative to the natural frequencies and form a set of different dynamic states of the system.

Evaluation Of Dynamic States By Means Of Generalized Lever

The frequency response of the transfer function can be mapped to the circular arc by converting:

$$X_i = \frac{W_i(j\omega)}{1 + W_i^2(j\omega)} \tag{19}$$

$$Y_i = \frac{W_i^2(j\omega)}{1 + W_i^2(j\omega)} \tag{20}$$

Conversion (X_i, Y_i) converts amplitude-frequency characteristics (Figure 3) into generalized levers (Figure 4).

Figure 3. Amplitude-frequency characteristics: a -y_1/z_1; b-y_2/z_1; c-y_3/z_1. Resonances σ_1 = 6.85 rad/s, σ_2 = 8.75 rad/s, σ_3 = 23.88 rad/s

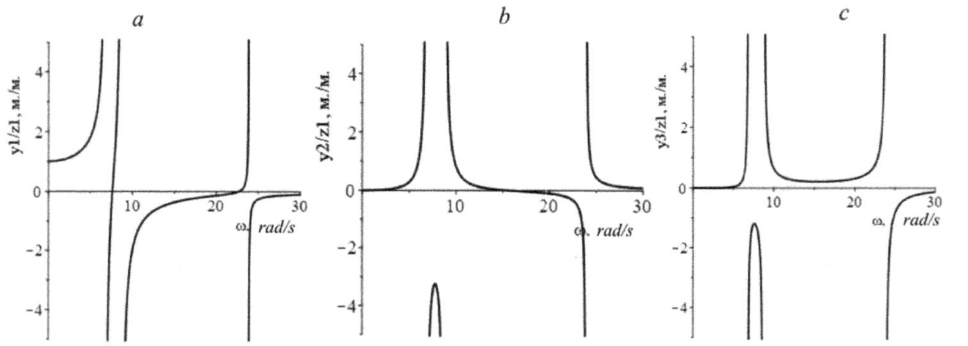

Figure 4. Forms of generalized levers for connectivity coefficients $\gamma_2 = 0$, $\gamma_3 = 0$ amplitude-frequency characteristics: a - y_1/z_1; b - y_2/z_1; c - y_3/z_1

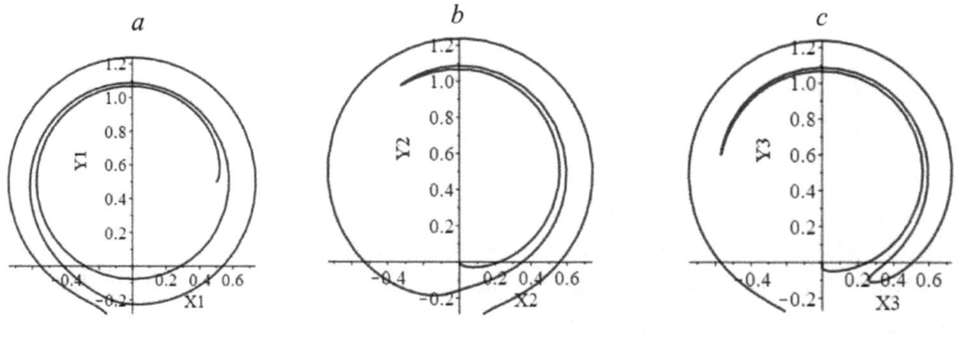

Convergence of zeroing frequencies with natural frequencies. There are options for approximation of σ_0 zeroing frequency with eigenfrequencies of σ_i. In this case, the neighborhood of the natural frequency point goes into a circle (Figure 5).

Figure 5. Projection of the amplitude-frequency characteristic (a) on a circle (b) with radius $R = 1/2$ in the vicinity of $\sigma_1 = 6.85$ rad/s in the interval (6.6.. 7) rad/s

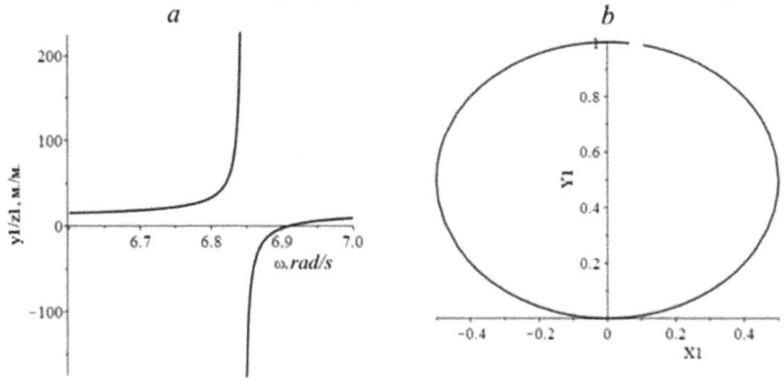

In the vicinity of the natural frequency of the σ_1, when the zeroing frequency of the σ_0 approaches the frequency of the σ_1, it can be assumed that the projection of the amplitude-frequency characteristic onto the circle of the unit diameter is realized. It can be shown that the approximate relation:

$$\int_{\sigma_i-\varepsilon}^{\sigma_i+\varepsilon} \sqrt{\left(\frac{\partial X_i}{\partial \omega}\right)^2 + \left(\frac{\partial Y_i}{\partial \omega}\right)^2}\, d\omega \approx \pi \qquad (21)$$

Therefore, it can be assumed that the connectivity coefficient γ_3, which ensures the proximity of the zeroing frequency and the natural frequency, with the known γ_2, you can find at least the function:

$$\varphi_i(\gamma_3) = \left(\int_{\sigma_i-\varepsilon}^{\sigma_i+\varepsilon} \sqrt{\left(\frac{\partial X_1}{\partial \omega}\right)^2 + \left(\frac{\partial Y_1}{\partial \omega}\right)^2} d\omega - \pi \right)^2 \tag{22}$$

To determine the connectivity factor of γ_3 providing the proximity of the zeroing frequency to the natural frequency $\sigma_1 = 6.85$ rad/s, the function $\varphi_1(\gamma_3)$ can be constructed for the fixed connectivity factor $\gamma_2 = 0$ (Figure 6). In particular, the function $\varphi_1(\gamma_3)$ reaches its minimum at the point $\gamma_3 = -2.6423$ (Figure 6, *a*). The corresponding amplitude-frequency characteristic has only two of three breaks of the 2nd kind (Figure 6, *b*). Break of the 2nd type at frequency $\sigma_1 = 6.85$ rad/s is eliminated as a result of selection of critical parameters $\gamma_2 = 0$, $\gamma_3 = -2.6423$, which provide matching of zeroing frequency with natural frequency. The generalized lever corresponding to the amplitude-frequency characteristic (Figure 6, *b*) has a special shape and length (Figure 6, *c*) in comparison with the generalized lever for shifted parameters $\gamma_2 = 0$, $\gamma_3 = -2.3423$ (Figure 6, *d*).

Figure 6. Determination of mode of zeroing frequency alignment with natural frequency σ_1 = 6.85 rad/s for critical connectivity factor γ_3 = -2.6423, γ_2 = 0. a) graph of function φ_1 (γ_3), γ_2 = 0; b) amplitude-frequency characteristic for the variant of zeroing frequency coincidence with natural frequency σ_1 = 6.85 rad/s; c) generalized lever for critical connectivity coefficients γ_2 = 0, γ_3 = -2.6423; d) generalized lever for connectivity coefficients γ_2 = 0, γ_3 = -2.3423.

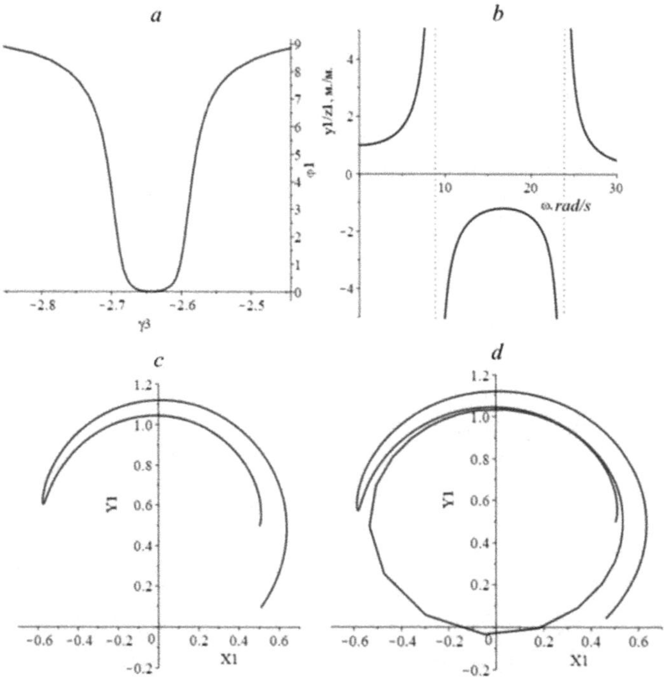

Along with the frequency of σ_1, the resonance effect can be extinguished at the natural frequency of σ_2 by combining the zeroing frequency with the natural frequency of σ_2 (Figure 7). To determine the critical connectivity factor of γ_3, which ensures that the zeroing frequency coincides with the natural frequency of σ_2, the function $\varphi_2(\gamma_3)$ can be constructed, the minimum of which allows determining the critical value of the connectivity factor of $\gamma_3^{(2)}$ (Figure 7, *a*). The corresponding frequency response for the connectivity factor $\gamma_3^{(2)}$ = 1.4920 (Figure 7, *b*).

Figure 7. Determination of mode of zeroing frequency alignment with natural frequency $\sigma_1 = 8.75$ rad/s for critical connectivity factor $\gamma_3 = 1.4920$, $\gamma_2 = 0$. a) graph of φ function $_2(\gamma_3)$, $\gamma_2 = 0$; b) amplitude-frequency characteristic for the variant of zeroing frequency coincidence with natural frequency $\sigma_1 = 8.75$ rad/s; c) generalized lever for critical connectivity coefficients $\gamma_2 = 0$, $\gamma_3 = 1.4920$; d) generalized lever for connectivity coefficients $\gamma_2 = 0$, $\gamma_3 = 0.992$.

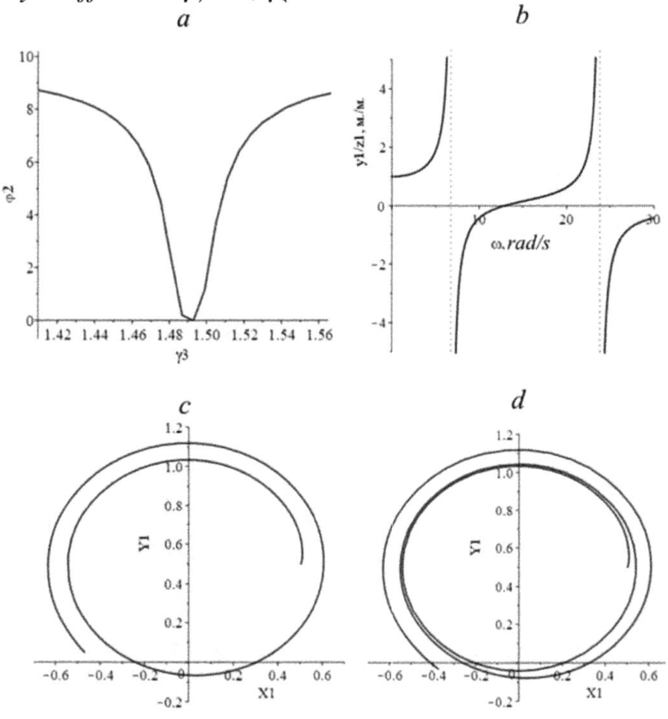

Break of the 2nd type at frequency $\sigma_1 = 8.75$ rad/s is eliminated as a result of selection of critical parameters $\gamma_2 = 0$, $\gamma_3 = 1.4920$, which provide matching of zeroing frequency with natural frequency. The generalized lever corresponding to the amplitude-frequency characteristic (Figure 7, *b*) has a special shape and length (Figure 7, *c*) in comparison with the generalized lever for shifted parameters $\gamma_2 = 0$, $\gamma_3 = 1.4920$ (Figure7, *d*).

Figure 8. Determination of mode of zeroing frequency alignment with natural frequency σ_1 = 23.88 rad/s for critical connectivity factor γ_3 = 1.003, γ_2 = 0. a) graph of function $\varphi_3(\gamma_3)$, γ_2 = 0; b) amplitude-frequency characteristic for the variant of zeroing frequency coincidence with natural frequency σ_1 = 23.88 rad/s; c) generalized lever for critical connectivity coefficients γ_2 = 0, γ_3 = -1.003; d) generalized lever for connectivity coefficients γ_2 = 0, $\gamma_3^{(3)}$ = 0.11.

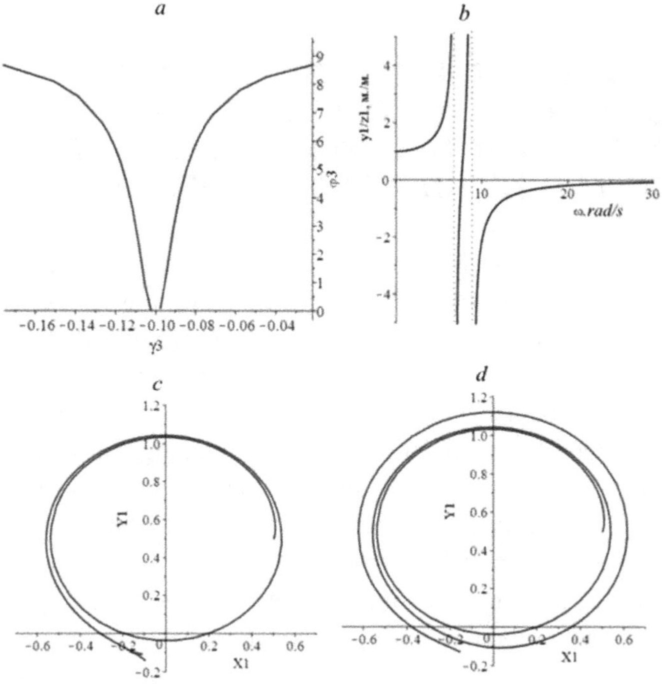

Break of the 2nd type at frequency σ_1 = 23.88 rad/s is eliminated as a result of selection of critical parameters γ_2 = 0, γ_3 = -1.003, which provide matching of zeroing frequency with natural frequency.

To determine the critical connectivity factor of γ_3, which ensures that the zeroing frequency coincides with the natural frequency of σ_3, the function $\varphi_3(\gamma_3)$ can be constructed, the minimum of which allows determining the critical value of the connectivity factor of $\gamma_3^{(2)}$ (Figure 8, *a*).

The generalized lever corresponding to the amplitude-frequency characteristic (Figure 8, b) has a special shape and length (Figure 8, *c*) in comparison with the generalized lever for shifted parameters γ_2 = 0, γ_3 = 1.003 (Figure 8, d). Thus, for the connectivity factor $\gamma_3^{(3)}$, the frequency response has a "dead" resonance at the natural frequency of σ_3.

Thus, a methodology has been developed for critical connectivity coefficients that make it possible to change the set of dynamic states of a mechanical oscillatory system based on the concept of the length of a generalized lever. The concept of change of generalized dynamic states on the basis of change of length of generalized lever is developed. At the same time, it has been shown that tasks have a number of signs of instability.

FUNDAMENTALS OF THE ENERGY APPROACH TO THE FORMATION AND CORRECTION OF VIBRATION FIELDS OF TECHNICAL FACILITIES FOR TRANSPORT AND TECHNOLOGICAL PURPOSES

Technologies harnessing the extensive potential of vibration interactions have been developed in recent years. Vibration technologies are widespread and interdisciplinary, closely linked to machine dynamics. In machine dynamics, vibration interactions are integral to vibration isolation and protection tasks, aimed at assessing, forming, and adjusting specific dynamic states of technical objects under intense vibration loads. Within the framework of evaluating, forming, and correcting dynamic states, emphasis is placed on the effects related to vibration interactions, which are applied in processes like rock transportation, separating bulk materials, surface strengthening, or spatial orientation of parts.

To generalize the tasks of assessing, forming, and correcting dynamic states of technical objects under vibration loads, the concept of a vibration field of dynamic features, expressed as amplitudes of displacement vibrations, reactions, and their relationships, was developed. A wide range of dynamic effects can be represented through expressions involving the vibrational amplitudes of object points, dynamic reactions, and their relationships.

By selecting specific vibration interaction parameters, dynamic effects can either jeopardize the functioning of a technical object or facilitate its intended production function. An example of a harmful effect is the excess of permissible vibration amplitudes at object coordinates. In contrast, a beneficial effect in vibration protection tasks might involve significantly reducing or nullifying the amplitude of vibration at the coordinate of the protected object. In vibration hardening tasks, dynamic effects are understood as the alignment of vibration shapes at the working element's points, creating a uniform vibration field that ensures the required dynamic quality of interactions between the working medium and the surface of the hardened part. The required movement form of the working element can be defined by the condition of equal oscillation amplitudes at the points of the working element. In vibration interaction problems, dynamic effects can be mapped by evaluating leverage and inter-partial connections, while interactions with non-holding bonds can be reflected by considering the amplitudes of dynamic reactions relative to static characteristics.

Thus, the dynamic state of a technical object under vibration loads is defined by a system of conditions, representing dynamic effects caused by external disturbances of either power or kinematic nature. To form and adjust dynamic states that meet these conditions, a methodological foundation is being developed, enabling the construction of mathematical models of vibration fields, taking into account the interaction features of the elements.

The methodological foundation is expected to be clear and equipped with an advanced mathematical framework, capable of performing analytical calculations, structural interpretations, and employing analogies to expand the developed approaches to a broad class of models. This foundation can be based on a structural approach that compares mechanical oscillatory systems used as design diagrams for technical objects with structural diagrams of dynamically equivalent automatic control systems. In this structural approach, dynamic state assessment is conducted using transfer functions, where external disturbances serve as input signals and output oscillations of object coordinates indicate the estimated dynamic state. The transfer function is chosen based on the physical characteristics used to assess the system's dynamic properties. For example, the transfer function might represent linkage or inter-partial relationships, with the ratio of oscillation amplitudes for the system's coordinates, or dynamic compliance, expressed as the ratio of oscillation amplitude of displacement to the amplitude of force disturbances. The transfer function can also be used to evaluate generalized linkage, representing the ratio of force disturbances, moments, and dynamic reactions as harmonic vibration forms.

However, the methodological foundation for forming and correcting vibration fields with specified characteristics has yet to be sufficiently developed in detail. This chapter is dedicated to the development of a methodology for structural mathematical modeling of vibration fields in technical objects, taking into account a system of conditions that reflect physical effects.

Determining External Kinematic Disturbances In a Mechanical Oscillatory System

Mechanical oscillatory system formed by a set of solid bodies under conditions of vibration loading of kinematic nature is considered. It is assumed that for the vibration field of a technical object, a system of conditions should be performed that displays restrictions in the form of linear equations for the values of vibration amplitudes. The mechanical oscillatory system is formed by solid bodies in the form of rods hingedly connected to each other at extreme points, serving as attachment points for elastic elements installed on support surfaces that perform in-phase harmonic vibrations (Figure9).

Figure 9. A mechanical oscillatory system formed by solids.

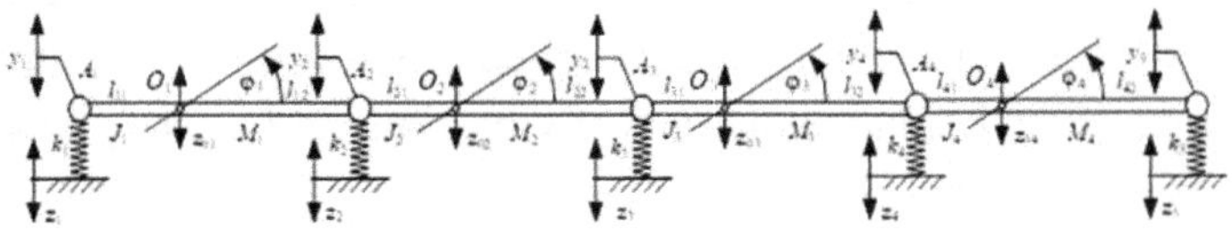

It is assumed that under the influence of external kinematic disturbances, the system makes small steady-state vibrations relative to the position of static equilibrium. System of conditions consists in the fact that at certain points of technical object amplitudes of oscillations of displacements relative to position of static equilibrium should take fixed values regardless of frequency of external disturbances.

The task is to determine external disturbances of kinematic nature, which provide a vibration field that satisfies a system of given conditions.

Development of Differential Equations of Motion for a Mechanical System with Five Degrees of Freedom

We consider a mechanical oscillatory system (Figure 9) with 5 degrees of freedom, formed with the help of 4 links in the form of solid rods with masses M_i and moments of inertia J_i, for which the values of arms l_{1i} and l_{2i}, and the lengths are $l_{1i} + l_{2i}$, where $i = 1,.., 5$.

To compile a system of differential equations, coordinates $\{y_1, y_2, y_3, y_4, y_5\}$ of displacements of nodes A_i, $i = 1..5$ relative to the positions of static equilibrium are considered. On an equal basis with the system of coordinates $\{y_i\}$ the system of the coordinates $\{z_{01}, z_{02}, z_{03}, z_{04}, z_{05}\}$ displaying shifts of the centers of gravity of O_i of cores concerning provisions of static balance, and the coordinate $\{\varphi_1, \varphi_2, \varphi_3, \varphi_4, \varphi_5\}$ representing small to an angle of rotation around the centers of gravity is considered. The coordinates $\{z_{0i}, \varphi_i\}$ can be expressed in terms of the coordinates $\{y_i\}$:

$$\begin{cases} z_{0i} = a_i y_i + b_i y_{i+1} \\ \varphi_i = c_i(y_{i+1} - y_i) \end{cases} \tag{23}$$

where

$$a_i = \frac{l_{i2}}{l_{i1} + l_{i2}} \tag{24}$$

$$b_i = \frac{l_{i1}}{l_{i1} + l_{i2}} \tag{25}$$

$$c_i = \frac{1}{l_{i1} + l_{i2}} \tag{26}$$

To compile a mathematical model of system movements in the form of differential Lagrange equations of the 2nd kind, we present expressions of potential and kinematic energy:

$$\Pi = \sum_{i=1}^{n} \frac{1}{2} k_i (y_i - z_i)^2 \tag{27}$$

$$T = \sum_{i=1}^{n} \frac{1}{2} \left(M_i \dot{z}_{0i}{}^2 + \frac{1}{2} J_i \dot{\varphi}_i{}^2 \right) \tag{28}$$

To determine the matrix structure of differential Lagrange equations of the 2nd kind in the $\{y_i\}$ coordinate system, we represent the kinetic and potential energies using the scalar product of vectors$\langle , \rangle$:

$$\Pi = \frac{1}{2} \langle K(\vec{y} - \vec{z}), (\vec{y} - \vec{z}) \rangle \tag{29}$$

$$T = \frac{1}{2} \langle M \vec{z}_0, \vec{z}_0 \rangle + \frac{1}{2} \langle J \vec{\varphi}, \vec{\varphi} \rangle \tag{30}$$

where

$$J = \begin{pmatrix} J_1 & & & 0 \\ & J_2 & & \\ & & J_3 & \\ 0 & & & J_4 \end{pmatrix} \tag{31}$$

$$M = \begin{pmatrix} M_1 & & & 0 \\ & M_2 & & \\ & & M_3 & \\ 0 & & & M_4 \end{pmatrix} \tag{32}$$

$$\vec{\varphi} = \begin{pmatrix} \varphi_1 \\ \varphi_2 \\ \varphi_3 \\ \varphi_4 \end{pmatrix} \tag{33}$$

$$\vec{z}_0 = \begin{pmatrix} z_{01} \\ z_{02} \\ z_{03} \\ z_{04} \end{pmatrix} \tag{34}$$

$$\vec{y} = \begin{pmatrix} y_1 \\ y_2 \\ y_3 \\ y_4 \\ y_5 \end{pmatrix} \tag{35}$$

$$\vec{z} = \begin{pmatrix} z_1 \\ z_2 \\ z_3 \\ z_4 \\ z_5 \end{pmatrix} \tag{36}$$

Coordinate vectors $\vec{z}_0$and $\vec{\varphi}$are expressed in terms of a vector$\vec{y}$:

$$\vec{z}_0 = U\vec{y} \tag{37}$$

$$\vec{\varphi} = V\vec{y} \tag{38}$$

where

$$U = \begin{pmatrix} a_1 & b_1 & & & 0 \\ & a_2 & b_2 & & \\ & & a_3 & b_3 & \\ 0 & & & a_4 & b_4 \end{pmatrix} \tag{39}$$

$$V = \begin{pmatrix} -c_1 & c_1 & & & 0 \\ & -c_2 & c_2 & & \\ & & -c_3 & c_3 & \\ 0 & & & -c_4 & c_4 \end{pmatrix} \tag{40}$$

In $\{y_i\}$ coordinates, kinetic energy (30) takes the form:

$$T = \frac{1}{2}\langle (U^T MU + V^T JV)\vec{y}, \vec{y} \rangle \tag{41}$$

The Lagrange equation of the 2nd kind in the matrix representation has the form:

$$(U^T MU + V^T JV)\ddot{\vec{y}} + K\vec{y} = K\vec{z} \quad (42)$$

After Laplace integral transformations, the system of differential equations (42), taking into account zero initial conditions, is reduced to a system of algebraic equations of the form:

$$\left(M_0 p^2 + K\right)\bar{y} = K\bar{z} \quad (43)$$

where is the matrix $M_0 = U^T MU + V^T JV$of the system.

Based on the algebraic system (43), using known methods (Eliseev & Eliseev, 2020; Eliseev, 2023), a structural diagram of a dynamically equivalent automatic control system can be built (Figure 10).

Figure 10. Mechanical oscillation system. Block diagram

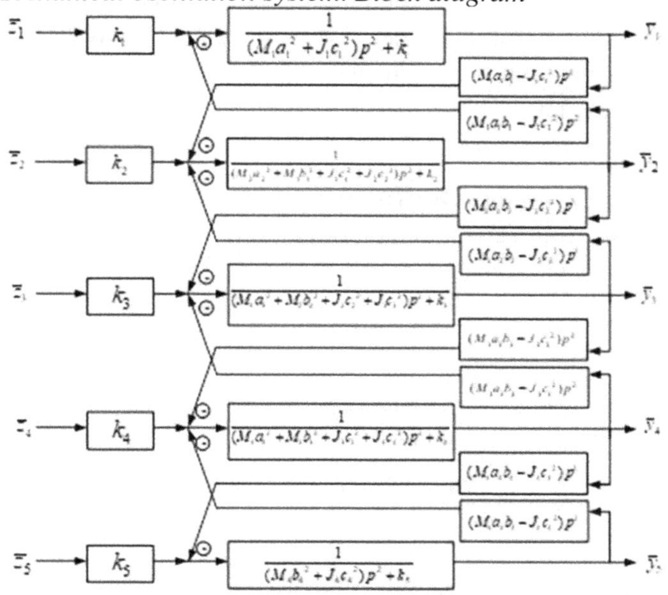

The dynamic states of the system can be characterized by a set of transfer functions representing linkage:

$$W_{11}(p) = \frac{\bar{y}_1}{\bar{z}_1}\bigg|\bar{z}_1 \neq 0, \bar{z}_{i\neq 1} = 0 \quad (44)$$

$$W_{22}(p) = \frac{\bar{y}_2}{\bar{z}_2}\bigg|\bar{z}_2 \neq 0, \bar{z}_{i\neq 2} = 0 \quad (45)$$

$$W_{33}(p) = \frac{\bar{y}_3}{\bar{z}_3}\bigg|_{\bar{z}_3 \neq 0, \bar{z}_{i \neq 3} = 0} \tag{46}$$

$$W_{44}(p) = \frac{\bar{y}_4}{\bar{z}_4}\bigg|_{\bar{z}_4 \neq 0, \bar{z}_{i \neq 4} = 0} \tag{47}$$

$$W_{55}(p) = \frac{\bar{y}_5}{\bar{z}_5}\bigg|_{\bar{z}_5 \neq 0, \bar{z}_{i \neq 5} = 0} \tag{48}$$

Transfer functions (44) - (48) in physical sense represent leverage, characterizing ratio of amplitudes of oscillations of generalized coordinates to amplitudes of kinematic disturbances of support surfaces. To estimate dynamic states, the amplitude-frequency characteristic of the transfer function W_{11} (41) can be used, which has five break points of the second kind, representing resonance states, and four zero values, representing dynamic damping modes of the amplitude of oscillation of the generalized coordinate y_1 (Figure11).

In the framework of lever interpretation, positive branches of the amplitude-frequency characteristic display levers of the 1st kind, negative - of the 2nd, resonance or dynamic blanking modes display the position of the turning point at the extreme points of the lever.

Figure 11. Frequency response $W_{11}(p)$, $p = j\omega$

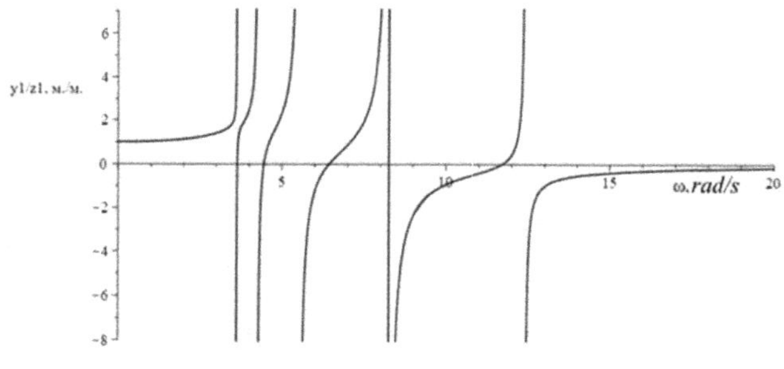

Selection of oscillation amplitudes of external kinematic disturbances unambiguously determines oscillation amplitudes of generalized coordinates of the system. However, the necessary conditions can be imposed only on a part of the generalized coordinates, while arbitrariness or uncertainty remains for the rest of the coordinates. In such a situation, it becomes necessary to develop an approach that allows rational criteria for selecting such external disturbances so that the corresponding dynamic state or vibration field is unambiguously formed. To construct external kinematic perturbations providing given amplitude values at fixed points of a technical object, an energy approach can be used, which consists in constructing a vibration field based on the ideas of the theory of variational spline interpolation (Rozhenko, 2005;

Vasilenko, 1983; Vasilenko, V. A., & Eliseev, A. V., 1998; Bezhaev & Vasilenko, 2001).

Elements of the Theory Of Variational Spline Approximation Applied To The Formation Of Vibrational Fields Of Mechanical Vibrational Systems

It is assumed that at points $A_{i1}, A_{i2}, .. A_{ik}$ of vibration field $\{y_i\}$, $i = 1..n$ of mechanical oscillatory system are set conditions, which consist in the fact that amplitudes $\{y_{ik}\}$, $k = 1..m$ take fixed values $\{y_{01}, .. y_{0m}\}$:

$$y_{ik} = y_{0k}, k=1..m \tag{49}$$

The necessary requirements (49) can be presented in the form of an analytical expression of the form:

$$Ay = b \tag{50}$$

where A is the operator of conditions for the shape of the vibration field at given points, b is the vector of values of the characteristics of the vibration field.

At the remaining points $\{y_i, i = 1..\text{n}\} \backslash \{y_{ik}, k = 1..m\}$, the requirements for vibration field values are not specified, which means the possibility of many variants of vibration fields. Additional requirements may be proposed to construct a single vibration field. Such requirements can provide a variational or energy approach, which has become widespread in approximation problems, consists in the fact that from the whole set of vibration fields that satisfy the necessary conditions, a vibration field is selected that delivers an extreme value to certain energy ratios (Rozhenko, 2005; Vasilenko, 1983; Vasilenko & Eliseev, 1998; Bezhaev & Vasilenko, 2001). Within the framework of the idea that the values of $y_1, y_2, .., y_n$ mean vibration amplitudes, for each fixed frequency of ω external disturbances, the variational task of constructing a vibration field that delivers a minimum to a certain energy functional can be formally set. The energy functional can be a characteristic which displays the shape of the vibration field, potential or kinetic energy. In general, the energy functionality can display the specifics of the requirements in the form of a vibration field aimed at realizing a given physical effect.

Within the framework of the task under consideration, the energy functionality was selected, which contributes to the formation of a uniform vibration field:

$$T_e(\bar{y}) = \sum_{i=1}^{n-1} (\bar{y}_{i+1} - \bar{y}_i)^2 \tag{51}$$

Energy functional (51), reaching zero value, determines uniform vibration field, amplitudes of oscillations of displacements of neighbouring points of which are equal. Minimization of energy functional (51) taking into account limitations (50) determines vibration field close in uniform within the special metric.

The formal formulation of the problem of constructing a vibration field for a fixed frequency of external disturbances delivering a minimum to the energy functional (51), provided that the amplitude of the vibration field at the control points meets the necessary conditions (49, 50), can be presented in the form of:

$$\bar{\sigma}(\omega) = \arg \min_{A\bar{y}=b, \bar{y}=(M_0 p^2+K)^{-1}K\bar{z}} T_e(\bar{y}) \tag{52}$$

where y is the vibration field defined by external kinematic disturbances z.

For a fixed frequency, the ω vibration field $\bar{\sigma}(\omega)$ is formally solved as the problem of conditional minimization of the quadratic function:

$$L(\bar{y}, \Lambda) = T_e(\bar{y}) + \sum_{k=1}^{m} \lambda_k (\bar{y}_{i_k} - y_{0k}) \rightarrow \min_{\lambda_k, \bar{y}=(M_0 p^2+K)^{-1}K\bar{z}} \tag{53}$$

However, the vibration field $\bar{\sigma}(\omega)$ may be approximated by a smoothing vibration field $\bar{\sigma}_\alpha(\omega)$, that is, a field approximately equal to the necessary values (49), (50):

$$\bar{\sigma}_\alpha(\omega) = \arg \min_{\bar{y}=(M_0 p^2+K)^{-1}K\bar{z}} \left(\alpha T_e(\bar{y}) + \sum_{k=1}^{m} (\bar{y}_{i_k} - y_{0k})^2 \right) \tag{54}$$

where the α smoothing coefficient is determined from the conditions of sufficient proximity of the vibration field values to the required values (49), (50).

In matrix representations, a vibration field satisfying the necessary conditions (49, 50) can be found from the conditions for minimizing the Lagrange function of the conditional extremum problem:

$$L(\bar{y}, \Lambda) = \langle T_{e0}\bar{y}, \bar{y} \rangle + \langle A\bar{y} - b, \Lambda \rangle \rightarrow \min_{\Lambda, \bar{y}=(M_0 p^2+K)^{-1}K\bar{z}} \tag{55}$$

where y - vibration field determined by external kinematic disturbance z is the solution to the problem

$$(M_0 p^2 + K)\bar{y} = K\bar{z} \tag{56}$$

where $M_0 = U^T MU + V^T JV$ is the system matrix, T_{0e} *is* the matrix defining the quadratic form (51) of the variables y. The interpolation vibration field y can be determined based on solving the conditional minimum problem using the Lagrange function:

$$L(\bar{z},\Lambda) = \langle T_{e0}\left(M_0p^2+K\right)^{-1}K\bar{z}, \left(M_0p^2+K\right)^{-1}K\bar{z}\rangle + \langle A\left(M_0p^2+K\right)^{-1}K\bar{z}-b,\Lambda\rangle \rightarrow \min_{\Lambda,\bar{z}}$$

where Λ = (λ 1, λ 2,. .· λ k) T is a vector of Lagrange factors $\bar{z}$ a vector of oscillation amplitudes of external kinematic perturbations. External $\bar{z}_\sigma$ perturbations and Lagrange multipliers can Λ be found as a solution to the system:

$$\begin{pmatrix} K^T\left(\left(M_0p^2+K\right)^{-1}\right)^T T_{e0}\left(M_0p^2+K\right)^{-1}K & \left(A\left(M_0p^2+K\right)^{-1}K\right)^T \\ A\left(M_0p^2+K\right)^{-1}K & 0 \end{pmatrix}\begin{pmatrix}\bar{z}\\ \Lambda\end{pmatrix} = \begin{pmatrix}0\\ b\end{pmatrix} \quad (58)$$

The corresponding vibration field will look like:

$$\bar{y}_\sigma = \left(M_0p^2+K\right)^{-1}K\bar{z} \quad (59)$$

However, a vibration field satisfying the interpolation conditions or constraints (49), (50) can be approximated by a vibration field smoothing predetermined values of vibration amplitudes at reference points. The vibration field smoothing the given values of vibration amplitudes at fixed points is determined by a system of equations obtained by generating Lagrange function minimum conditions:

$$L_\alpha(\bar{z}) = \alpha\langle T_{e0}\bar{y},\bar{y}\rangle + \|A\bar{y}-b\|^2 \rightarrow \min_{\bar{y}} \quad (60)$$

where $\bar{y}$ is a vibration field satisfying the equation:

$$\left(M_0p^2+K\right)\bar{y} = K\bar{z} \quad (61)$$

After replacement:

$$\bar{y} = \left(M_0p^2+K\right)^{-1}K\bar{z} \quad (62)$$

the Lagrange function (60) is reduced to:

$$L_\alpha(\bar{z}) = \alpha\langle T_{e0}\left(M_0p^2+K\right)^{-1}K\bar{z}, \left(M_0p^2+K\right)^{-1}K\bar{z}\rangle + \|A\left(M_0p^2+K\right)^{-1}K\bar{z}-b\|^2 \rightarrow \min_{\bar{z}} \quad (63)$$

where $\|u\|^2 = \langle u, u \rangle$ is the length of the vector in R^n.

External kinematic disturbances z_α, forming a vibration field, smoothing the specified conditions, are determined by the system of expressions:

$$K^T\left(\left(M_0p^2 + K\right)^{-1}\right)^T\left(\alpha T_{e0} + A^TA\right)\left(M_0p^2 + K\right)^{-1}K\bar{z}$$

$$= K^T\left(\left(M_0p^2 + K\right)^{-1}\right)^TA^Tb \tag{64}$$

After simplifications, the desired amplitudes of kinematic disturbances are determined by the expression:

$$\bar{z}_\alpha = K^{-1}\left(M_0p^2 + K\right)\left(\alpha T_0 + A^TA\right)^{-1}A^Tb \tag{65}$$

External disturbances of the $\bar{z}_\alpha$, which are the solution to the problem (64), taking into account the choice of the α coefficient, provide for the construction of a smoothing vibration field y_α:

$$\bar{y}_\alpha = \left(M_0p^2 + K\right)^{-1}K\bar{z}_\alpha \tag{66}$$

It can be shown that the smoothing vibration field has the form clearly independent of the frequencies of external disturbances:

$$\bar{y}_\alpha = \left(\alpha T_{e0} + A^TA\right)^{-1}A^Tb \tag{67}$$

The external perturbation of the $\bar{z}_\alpha$, which provides for the construction of a smoothing vibration field y_α, can be used to construct an interpolating vibration field as $\alpha \to 0$. The choice of α is determined from the conditions of the smallness of the so-called residual:

$$\varphi(\alpha) = \|A\bar{y}_\alpha - b\|^2 \tag{68}$$

where, $\bar{y}_\alpha = \left(M_0p^2 + K\right)^{-1}K\bar{z}_\alpha\bar{z}_\alpha$ - kinematic disturbance (65).

Construction Of Vibration Field For Mechanical Vibration System. Computational Experiment

Let the parameters of the mechanical oscillatory system (Figure 9) be given by specific stiffnesses k_i, masses M_i, moments of inertia J_i and geometric characteristics l_{1i}, l_{2i} (Table 1).

Table 1. Parameters of mechanical oscillatory system

N°	M, kg.	J, kg m2	k, N/m.	l_1, m.	l_2, m.
1	100	10	1000	1	1
2	200	20	2000	1	1
3	300	30	3000	1	1
4	400	40	4000	1	1

The requirements for vibration field amplitudes are set for the values at points A_i (Table 2).

Table 2. Vibration amplitude conditions

	A_1	A_2	A_3	A_4	A_5	
y, m.	0.002	-	-	-	0.005	Required values
y, m.	0.002	0.0027	0.0035	0.0043	0.005	Smoothed values

As the energy functional, the expression (51) is used, which can be represented using the T_1 operator:

$$T:\begin{pmatrix}\overline{y}_1\\\overline{y}_2\\\overline{y}_3\\\overline{y}_4\\\overline{y}_5\end{pmatrix}\rightarrow\begin{pmatrix}\overline{y}_2-\overline{y}_1\\\overline{y}_3-\overline{y}_2\\\overline{y}_4-\overline{y}_3\\\overline{y}_5-\overline{y}_4\end{pmatrix}=\begin{pmatrix}1 & -1 & & & 0\\ & 1 & -1 & & \\ & & 1 & -1 & \\ 0 & & & 1 & -1\end{pmatrix}\begin{pmatrix}\overline{y}_1\\\overline{y}_2\\\overline{y}_3\\\overline{y}_4\\\overline{y}_5\end{pmatrix} \qquad (69)$$

A expression <*Ty* , *Ty* > can be represented as:

$$\langle T\overline{y},T\overline{y}\rangle = \langle T^T T\overline{y},\overline{y}\rangle = \begin{pmatrix}1 & -1 & & & 0\\ -1 & 2 & -1 & & \\ & -1 & 2 & -1 & \\ & & -1 & 2 & -1\\ 0 & & & -1 & 1\end{pmatrix}\begin{pmatrix}\overline{y}_1\\\overline{y}_2\\\overline{y}_3\\\overline{y}_4\\\overline{y}_5\end{pmatrix} = \langle T_{e0}\overline{y},\overline{y}\rangle \qquad (70)$$

where T_{oe} is the matrix defining the quadratic form. Operator T_e is equal to zero on constant vibration field, amplitudes of oscillations of points of which are equal to each other for external kinematic disturbance at arbitrary frequency.

To set conditions (49), (50) on the vibration field, a data operator A_{15} can be used, which compares the amplitude values at points A_1, A_5 to the vibration field *y*:

$$A_{15}: \begin{pmatrix} \bar{y}_1 \\ \bar{y}_2 \\ \bar{y}_3 \\ \bar{y}_4 \\ \bar{y}_5 \end{pmatrix} \rightarrow \begin{pmatrix} 1 & 0 & 0 & 0 & 0 \\ 0 & 0 & 0 & 0 & 1 \end{pmatrix} \begin{pmatrix} \bar{y}_1 \\ \bar{y}_2 \\ \bar{y}_3 \\ \bar{y}_4 \\ \bar{y}_5 \end{pmatrix} = \begin{pmatrix} \bar{y}_1 \\ \bar{y}_5 \end{pmatrix} \tag{71}$$

The vibration field interpolating the predetermined values at points A_1, A_5 (Table 2) can be approximated by a smoothing vibration field y_α for a smoothing factor of $\alpha = 0.01$. The difference between the vibration field values and the required values for the selected α smoothing factor is determined by the residual function:

$$\varphi(\alpha) = \|A\bar{y}_\alpha - b\|^2 \tag{72}$$

the value of which is approximately 0.00001 m, where b = (0.002, 0.005) ^{T}. The determined vibration field has a linear shape (Table 2) and smoothed data. For a fixed coefficient $\alpha = 0.01$, the set of vibration amplitudes of the points of the smoothing vibration field, which ensures proximity to the necessary requirements φ (α) = 0.00001 m (Table 2), has the form of constant functions of the frequency of external disturbance of the ω (Figure 12).

Figure 12. Values of vibration amplitudes of smoothing vibration field for frequencies of external disturbances within the range of ω (0.100) rad/s.

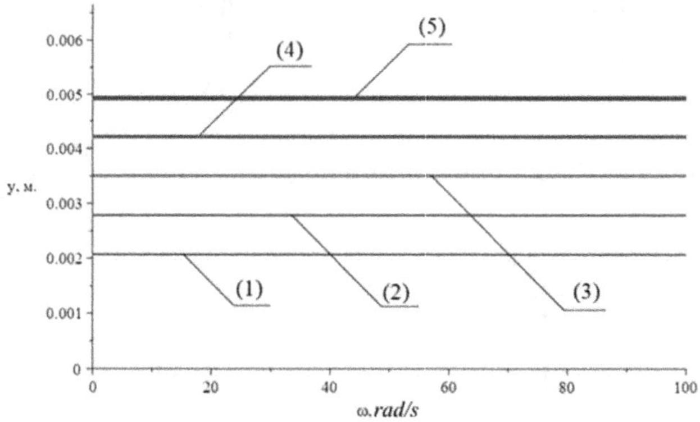

Values of oscillation amplitudes of generalized coordinates in this case are kept constant at change of frequency of external kinematic disturbances. In turn, the external kinematic disturbances required to provide the required amplitudes at fixed points of the vibration field are significantly dependent on the frequency of external disturbances (Figure 13).

Figure 13. Amplitudes of external kinematic disturbances required to provide required vibration amplitudes depending on disturbance frequency

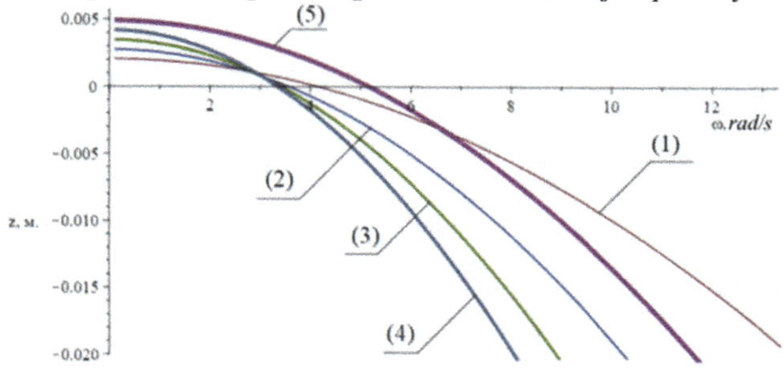

The vibration field for a fixed frequency or some frequency interval in which it remains unchanged can be represented by a graph of a function depending on the number of the generalized coordinate or the coordinate of the point of the system in which the vibration field is evaluated (Figure 14).

Figure 14. The shape of the vibration field for frequencies in the interval ω (0.. 100): (1), (5) - the required values of the vibration field, (2), (3), (4) - values based on the energy approach

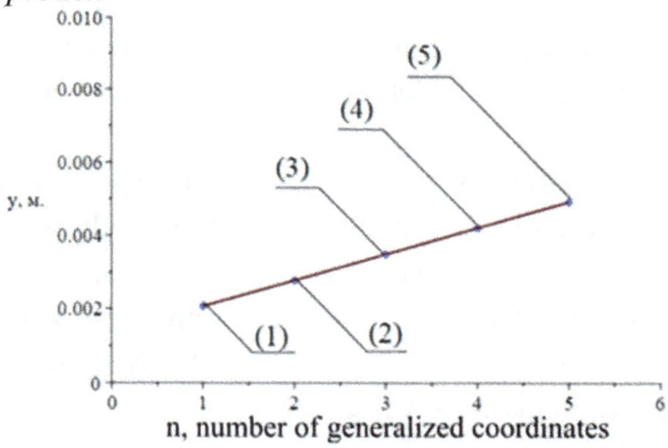

It should be noted that the «interpolation» vibration field is constructed using a "smoothing" vibration field corresponding to the α smoothing factor. In this case, the deviation of the "smoothing" vibration field should be insignificant. In particular, for α = 0.1, the deviation of the smoothing vibration field is φ (α) = 0.0001 m. (Figure 15).

Figure 15. Deviation of the vibration field φ (ω) from the required amplitude values at fixed points of the technical object for a constant smoothing factor

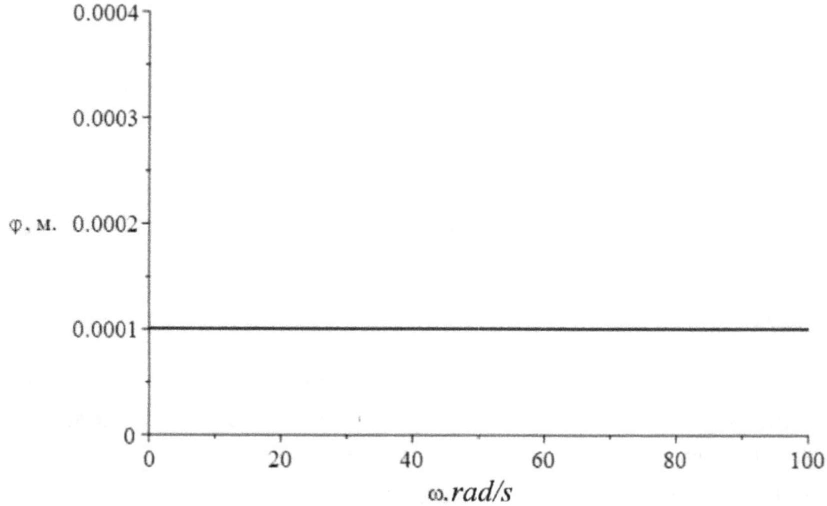

$\alpha = 0.1$

However, it can be assumed that in the general case of energy functionality, the value of the smoothing factor α, which provides a fixed φ level (α) = φ *, depends on the frequency.

Thus, variants of formation of dynamic states of mechanical vibration systems in the form of vibration field satisfying the system of conditions by interpolation conditions are determined on the basis of energy functional reflecting features of vibration field and operator displaying set of requirements to vibration field.

At the same time, other options for generating a vibration field based on the choice of the energy operator are possible.

Peculiarities Of Vibration Field Formation Based On Energy Operator Selection

Along with the shape of the vibration field, which displays the possibility of equalizing the amplitudes of oscillations of the coordinates of a technical object, approaches are possible aimed at minimizing the amplitudes of generalized coordinates by selecting the energy operator T_E in the form of an identical display:

$$T_E : \bar{y} \rightarrow E\bar{y} \tag{73}$$

where E is the identity operator.

The formation of a vibration field based on the energy operator $_{TE}$ is associated with minimization or zeroing of generalized coordinates of the vibration field, considered as amplitudes of in-phase harmonic vibrations of a fixed frequency. As a result of solving the smoothing problem for the energy functional corresponding to a single T_E operator (73), a vibration field is formed, the vibration amplitudes of which at points A_2, A_3, A_4 are zero (Figure 16).

Figure 16. Vibration field shape for energy functional (73) of vibration amplitude minimization considering the required requirements (49), (50) for amplitude at fixed points

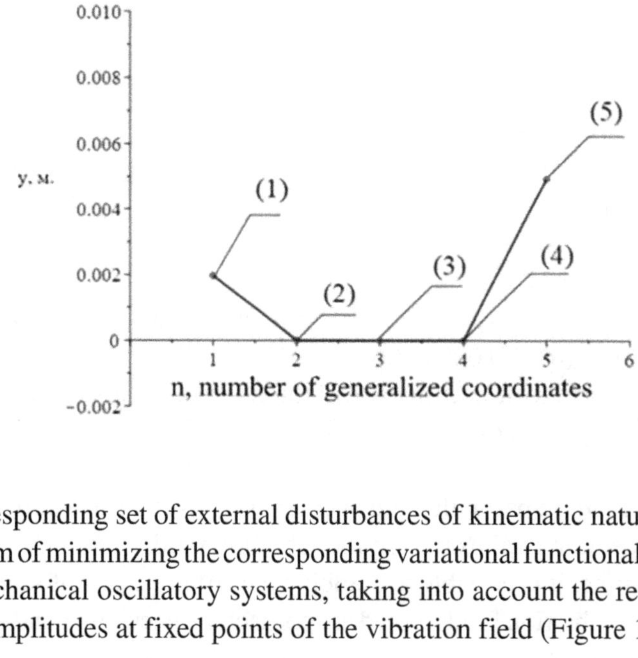

The corresponding set of external disturbances of kinematic nature is a solution to the problem of minimizing the corresponding variational functionality on dynamic states of mechanical oscillatory systems, taking into account the requirements for oscillation amplitudes at fixed points of the vibration field (Figure 17).

Figure 17. Amplitudes of external kinematic disturbances z_i, providing the necessary requirements for dynamic features of the vibration field

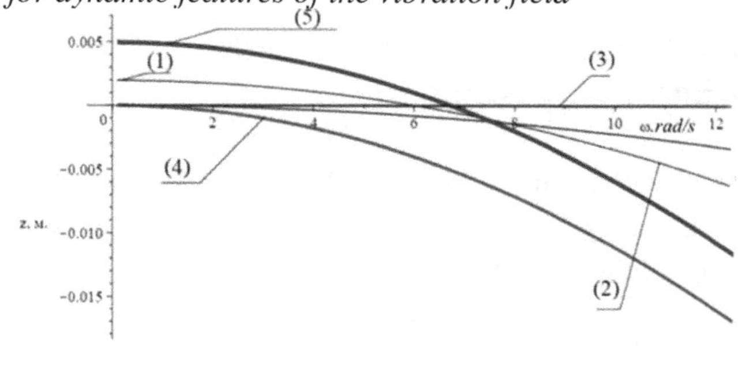

A feature of forced kinematic disturbances, which provide the required shape of the vibration field, is an increase in the amplitude of vibrations as the frequency of disturbances increases (Figure 18). External kinematic disturbances z_i provide at different frequencies a ω fixed vibration field of the technical object (Figure 18).

Figure18. Vibration field of the technical object depending on the frequency of external disturbance

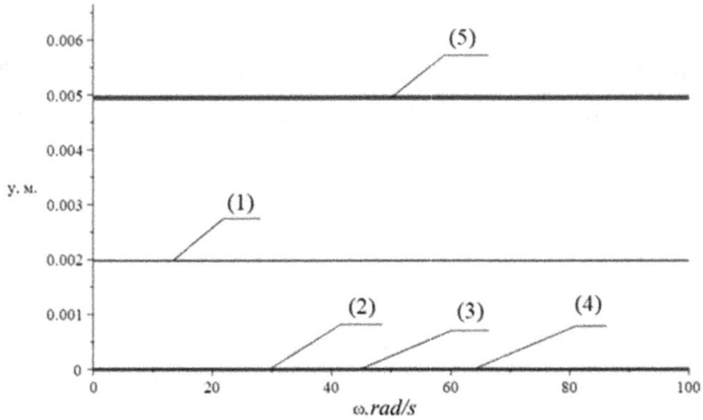

The smoothing factor given by $\alpha = 0.01$ provides an absolute deviation from the required values at $\varphi(\alpha) = 0.000053$.

Thus, it is shown that the peculiarities of the energy operator, based on the necessary values of the amplitudes of the vibration field at fixed points of the mechanical vibration system, determine the set of vibration amplitudes at the remaining points of the vibration field. In particular, the operator T_e aims to form a uniform vibration field, and the operator T_E aims to zero the vibration field.

At the same time, it is possible to combine the requirements of vibration amplitudes of points of vibration fields aimed at approaching each other and requirements aimed at zeroing the absolute amplitudes of oscillations of generalized coordinates taking into account a certain weight coefficient. This combination of requirements is actually a way to correct the vibration field.

Correction Of Vibration Field Shape With The Help Of Energy Operator

To build a family of forms of vibration fields of a technical object that meet certain conditions, energy functionality can be used, which is the sum of energy functionals, taking into account the weight coefficient:

$$\|T_\beta \bar{y}\|^2 = \|T\bar{y}\|^2 + \beta \|T_E \bar{y}\|^2 \tag{74}$$

where β is a weight coefficient reflecting the significance of the small amplitude compared to the significance of the homogeneity of the vibration field. In particular, varying the weighting factor β leads to the construction of a family of vibration fields, which are in some way "intermediate" between the constant shape and zero.

In particular, it is possible to consider a set of weight coefficients and construct for them a set of vibration fields, which are realized by specifying special forms of external disturbances. The choice of a set of weight coefficients determines the corresponding forms of vibration fields, which in turn are realized under the influence of external disturbances, which are the solution of the variation problem (Rozhenko, 2005; Vasilenko, 1983; Vasilenko & Eliseev, 1998; Bezhaev & Vasilenko, 2001).

Based on the set of β weights, vibration fields can be constructed that approximately satisfy the requirements for vibration amplitudes at fixed points (Table 3).

Table 3. Vibration field shapes for energy functional weights

N°	β	α	Not knitting φ (α), m.	Frequency ω, rad/s	$_{p1}$, m.	$_{p2}$, m.	$_{p3}$, m.	$_{p4}$, m.	$_{5}$, m.
1	1	0,01	0,0000841	10	0,0019704	0,0009850	0,0009845	0,0019686	0,0049213
2	0,5	0,01	0,0000503	10	0,0019847	0,0014467	0,0016322	0,0026337	0,0049521
3	0,25	0,01	0,0000305	10	0,0019940	0,0018969	0,0022740	0,0032196	0,0049701
4	0,125	0,001	0,0000195	10	0,0020000	0,0022503	0,0027818	0,0036611	0,0049980
5	0,0625	0,001	0,0000014	10	0,0020004	0,0024754	0,0031053	0,0039292	0,0049986

As part of the task of correcting dynamic states, the change in the weight coefficient is a method of correcting the vibration field of a technical object (Figure 19)

Figure 19. Vibration field options depending on the weight factor of the energy operator $\mathbf{T}_{\beta}$

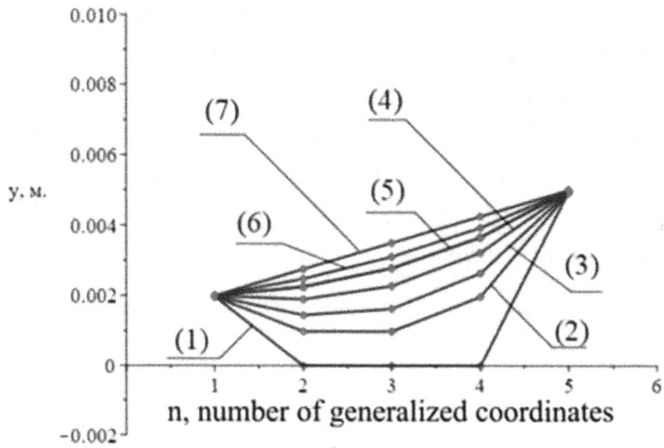

It should be noted that the amplitude-frequency characteristics of the transfer functions y_i/z_i, for which the input signal is determined on the basis of variation principles, external kinematic disturbances z_i, have one break of the 2nd kind, and the found kinematic disturbances z_i have one zero (Figure 20).

Figure 20. Amplitude-frequency characteristics β = 0.25, α = 0.01

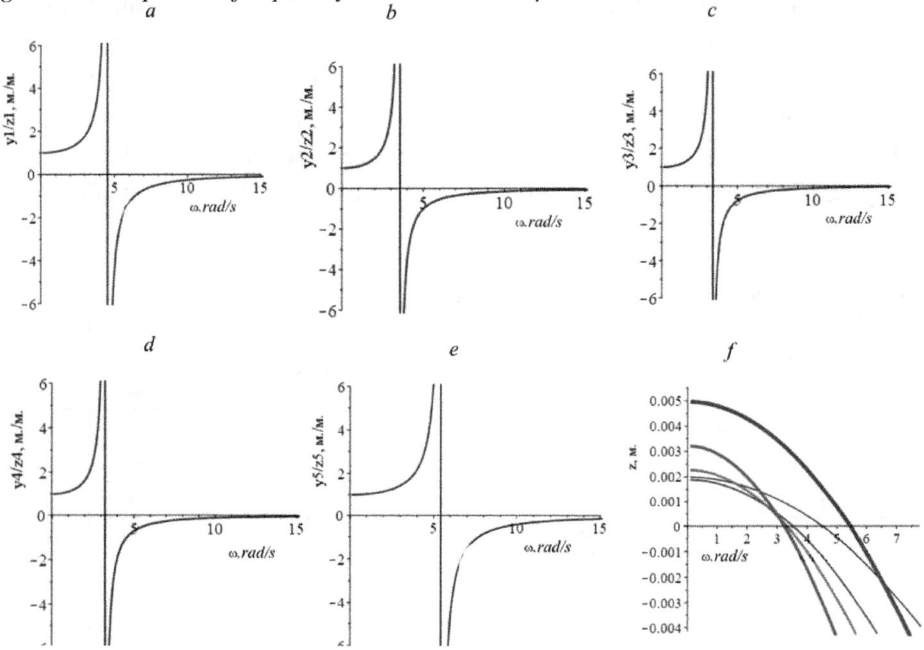

In order to provide given conditions for absolute values of vibration field amplitudes at fixed points of mechanical oscillatory system, necessary external kinematic actions can be determined on the basis of variation principles by selecting combined energy functionals taking into account weight coefficients.

In order to construct vibration fields on the basis of variation principles, the interpolating vibration field can be constructed in the form of a smoothing vibration field with a smoothing coefficient delivering sufficiently small values of deviation of the smoothing vibration field from given vibration amplitudes at fixed points of the mechanical vibration system.

Development of Ideas About The Formation Of Vibration Fields Based On The Variation Approach

In general, vibration field requirements can be a complex set of conditions, including restrictions on the amplitudes of displacements, velocities, reactions, including non-holding conditions.

A complex set of conditions may allow a single solution in the form of a vibrational field of a mechanical oscillatory system. In the case of redundancy of conditions for a vibration field, regularization methods are needed, which make it possible to build a rational solution close to the requirements in terms of delivering a minimum to a special variation functional, the structure of which takes into account the system of requirements for the desired vibration field. If the requirements for the characteristics of the vibration field are insufficient to build a single vibration field, then the choice of a single vibration field is carried out on the basis of minimizing special energy functionality. The energy functionality determines the specifics of vibration fields based on the values of the amplitudes of displacements, velocities, power disturbances or reactions that occur between the elements of the system. Energy functionals can convey ideas about the maximum potential or kinetic energy, or serve as a combination of them, taking into account the weight coefficients that act as methods for correcting vibration fields.

At the same time, conditions for the nature of external kinematic or force disturbances, which are expressed in the fact that the amplitudes of external disturbances take fixed absolute or relative values, can be added to the conditions presented to the values of amplitudes of vibration fields. The need to set requirements for input disturbances is due to the fact that in order to ensure the given shapes of vibration fields, the amplitudes of external ones can increase indefinitely as the frequency of external disturbances increases. Since one of the key factors providing the given shapes of vibration fields is associated with the ability to hold the amplitudes of external disturbances at certain boundaries.

This problem can be formulated in the form of a problem of constructing vibration fields of a given shape based on the choice of energy functionality, an operator of requirements for a vibration field, and an additional operator of requirements for external force or kinematic disturbances.

Thus, elements of the structural theory of vibration fields of mechanical oscillatory systems have been developed.

The main elements of structural theory can include methods for constructing vibration fields of mechanical oscillatory systems based on variational principles. The basic concept is the vibration field of the mechanical oscillatory system, which is the distribution of vibration amplitudes of generalized coordinates and other dynamic characteristics of the mechanical oscillatory system at the points of the object, the dynamic state of which is estimated.

It is assumed that at a number of points of the mechanical oscillatory system a system of conditions is defined, expressed in absolute and relative values of displacements, velocities and reactions. These conditions are the requirements for the vibration field characteristics of the mechanical vibration system.

Along with the requirements for the vibration field, there is a system of conditions associated with the peculiarities of external force or kinematic disturbances, which also represents a set of absolute and relative values of the amplitudes of kinematic and force disturbances.

To build an applied theory within the framework of structural representations, an analogy can be formulated between an interpolation operator and a system of conditions in a vibration field and input disturbances, the energy operator displays the shape of the vibration field associated with coordinates and their relative values. In the framework of the theory of variational spline approximation, interpolation, smoothing and mixed splines are distinguished. In the framework of ideas about vibration fields of mechanical oscillatory systems, based on structural representations, the role of a spline is played by a set of amplitudes of external force and kinematic disturbances, and the role of interpolation conditions is played by a set of requirements for the amplitudes of input or output signals.

The combination of requirements for the vibration field, requirements for external disturbances, taking into account the variation approach, makes it possible to interpret the task of determining external disturbances that form a set of dynamic states in the form of a vibration field, taking into account the system of conditions, as a spline approximation problem, covering a variety of problem statements within the framework of variation principles.

In particular, depending on the compatibility of the conditions for the input and output signals and their mutual compatibility, it is possible to set the problems of analogues to the problems of constructing a smoothing or mixed spline.

An essential feature of the problems of constructing vibration fields is the fact that the frequency of external perturbation is a parameter of the spline approximation problem, and in the case of systems with concentrated parameters, amplitude vectors serve as elements of n-dimensional Cartesian space.

Thus, the constructed analogies between the approximation tasks and the tasks of constructing vibration fields make it possible to form a new direction for the development of the methodology for assessing, forming and correcting the dynamic states of technical objects within the framework of the methodology of structural mathematical modeling.

REFERENCES

Bezhaev, A. Yu., & Vasilenko, V. A. (2001). *Variational theory of splines*. Kluwer Academic / Plenum Publishers. DOI: 10.1007/978-1-4757-3428-7

Eliseev, A. V., & Kuznetsov, N. K. (2023). Development of lever representations in the evaluation of dynamic states of mechanical oscillatory systems under connected vibrational loads. *Bulletin of Cybernetics*, *22*(2), 75–86.

Eliseev, A. V., Kuznetsov, N. K., & Mironov, A. S. (2023). Dynamic invariant maps in the evaluation of motion modes of mechanical oscillatory systems. *Proceedings of MAI*, (128).

Eliseev, A. V., & Mironov, A. S. (2022). Evaluation of the combined dynamic characteristics of mechanical oscillatory systems based on dynamic invariant maps. *Journal of Advanced Research in Technical Science*, *32*, 62–69.

Eliseev, A. V., Sitov, I. S., & Kuznetsov, N. K. (2024). Formation of dynamic effects in mechanical oscillatory systems within the framework of the generalized lever concept, considering additional kinematic constraints. Systems. Methods. *Technologies*, *1*(61), 10–20.

Rozhenko, A. I. (2005). Theory and algorithms of variational spline approximation (A. I. Rozhenko). Novosibirsk: IVMiMG SO RAN.

Vasilenko, V. A. (1983). *Spline functions: Theory, algorithms, programs (V. A. Vasilenko)*. Nauka.

Vasilenko, V. A., & Eliseev, A. V. (1998). Abstract splines with tension as functions of energy operator parameters. *Siberian Journal of Computational Mathematics*, *1*(4), 301–311.

ADDITIONAL READING:

Eliseev, A. V. (2023). Elements of structural theory of optimal vibration fields in mechanical oscillatory systems. *Fundamentals of Mechanics*, *12*, 66–72.

Eliseev, A. V. (2024). Variational approaches to the problem of forming the modes of dynamic damping of solid body oscillations under vibrational loading of power nature. Transport. *Mining and Construction Engineering: Science and Production*, *25*, 9–21.

Eliseev, A. V., & Kuznetsov, N. K. (2024). Optimal vibration fields in the modeling of dynamic states of technical objects. *Advanced Engineering Research (Rostov-on-Don)*, *24*(1), 7–22. DOI: 10.23947/2687-1653-2024-24-1-7-22

Eliseev, A. V., & Mironov, A. S. (2024). Determining the modes of dynamic damping of mechanical system oscillations as an optimization problem of the dynamic state of a technical object. *Fundamentals of Mechanics*, *13*, 14–22.

Zhao, X., Huang, Y., Dong, W., Liu, J., & Ma, G. (2024). A review of compaction mechanisms, influencing factors, and advanced methods in concrete vibration technology. *Journal of Building Engineering*, *93*, 109847. DOI: 10.1016/j.jobe.2024.109847

Conclusion

The theoretical and experimental investigations conducted within this work affirm that the development of advanced methods for assessing, monitoring, and controlling dynamic states constitutes a critical scientific imperative. This is driven by the need to ensure operational reliability, safety, and competitiveness of transport and technological machines operating under significant vibrational excitations.

Research in the field of machine dynamics requires a systematic approach based on adequate computational schemes reflecting the real behavior of technical objects. It is shown that the choice of the number of degrees of freedom, consideration of distributed parameters and methods of their reduction play a crucial role in the accuracy of modeling. An important result is the transition from traditional methods of analysis based on solving systems of differential equations to structural approaches using the apparatus of the theory of automatic control. This makes it possible to effectively solve the problems of parameter identification, frequency and waveform determination, stability analysis, and dynamic state control. The introduction of the concepts of "generalized dynamic coupling" and "generalized spring" expands the classical set of elements and makes it possible to model complex types of interactions in real structures.

A methodology for analyzing the dynamics of mechanical oscillatory systems using structural transformations and the concept of "compacts" has been developed. The introduction of compacts — dynamically equivalent blocks with frequency-dependent rigidity — makes it possible to aggregate complex systems and significantly reduce the number of degrees of freedom without losing physical adequacy. It is shown that lever connections and solid links can be effectively taken into account in models by converting kinematic and energy ratios, as well as by the method of "zeroing" coordinates in hinges. The obtained algorithms for the transformation of structural schemes systematize the approach to the analysis of systems with non-standard connections and create the basis for automating calculations.

Methods of constructing and analyzing mathematical models of mechanical systems with various types of connections have been developed and tested. It is established that additional elastic and lever connections have a significant effect on the coefficients of the equations of motion, which requires correction of the stiffness and damping matrices. It is shown that with the help of transfer functions, it is possible to effectively study the modes of dynamic damping and resonance, which is especially important when designing vibration-proof devices. The proposed algorithms for the transformation of structural schemes, illustrated by the example of a system with three degrees of freedom, demonstrate the universality of the approach and its applicability to a wide class of tasks.

The conducted research demonstrates the effectiveness of modern numerical methods, such as the finite element and boundary element method, for analyzing systems with distributed parameters. Computational algorithms have been developed for determining natural frequencies and waveforms, taking into account the influence of elastic bonds, damping and external influences. The analysis of transient and steady-state processes under various perturbations made it possible to identify patterns of behavior of systems and form criteria for their stability. The results obtained confirm that the integration of numerical methods with analytical approaches of structural modeling increases the accuracy and reliability of forecasting dynamic characteristics.

Studies of the dynamic states of systems with articulated mechanisms have made it possible to develop a methodology for simplifying structural models by formally eliminating coordinates. It is shown that such systems can be represented as simpler structures (for example, with two degrees of freedom), which simplifies the analysis. Quasi-springs and generalized levers play an important role in the formation of dynamic properties, the influence of which on transfer functions and feedback allows controlling resonant and antiresonance phenomena. The obtained results find practical application in the optimization of mechanisms with complex kinematics in the tasks of vibration protection and vibration isolation.

A methodological basis has been formed for constructing mathematical models of dyads, autonomous structural formations consisting of two inertial elements capable of elastic interaction. The study of the dynamic properties of dyads in various coordinate systems has revealed new effects related to inertial and inertial-elastic dependencies. The proposed method for determining motion parameters based on transfer and frequency-energy functions opens up prospects for creating new approaches to analysis and synthesis in the theory of mechanical vibrations, as well as in nanomechanics, where dyads can serve as basic elements of complex systems.

Methods of assessment, formation and correction of dynamic states of systems with one-way connections in which contact between elements can be disrupted have been developed. On the basis of the Dalembert principle, the dynamic component of the contact reaction is determined, and the boundaries of its change during the period of steady motion are established. It is shown that under certain motion parameters, contact between the elements of the composite body is maintained, which is critically important for ensuring the integrity and reliability of structures operating under shock and vibration loads.

Methods of controlling the dynamic states of mechanical systems based on the concepts of a generalized lever and the energetic formation of vibrational fields are investigated. It is shown that with the help of variational principles and energy criteria, vibrational fields can be formed and adjusted, providing their smoothing or interpolation. A system device for controlling dynamic states has been developed, which is applicable in the tasks of vibration protection, vibration isolation and vibration technologies (for example, transportation and processing of materials). The energy approach makes it possible to purposefully control complex oscillatory processes, increasing the efficiency and safety of technical systems.

This monograph develops the paradigm of structural mathematical modeling as a single basis for the dynamic analysis of technical objects. Combining theoretical mechanics, control theory, and circuit analysis, it offers scalable solutions for controlling vibration interactions in various fields of mechanical engineering. Future efforts should focus on non-linear extensions, experimental prototyping of systems based on motion conversion devices, and the introduction of digital twins to take advantage of predictive maintenance capabilities.

Compilation of References

Bakalov, A. N. (2006). *Theory of chains*. Nauka.

Banakh, L. Y. (2011). Vibrations of mechanical systems with self-similar structure. *Vestnik of Nizhny Novgorod Lobachevsky State University*, *4*(2), 52–53.

Banakh, L., & Kempner, M. (2010). *Vibrations of mechanical systems with regular structure*. Springer. DOI: 10.1007/978-3-642-03126-7

Baty, M. I., Djanelidze, G. Yu., & Kelzon, A. S. (1966). *Theoretical mechanics in examples and problems* (Vol. 2). Dynamics. Science.

Belokobylsky, S. V., & Eliseev, S. V. (2016). Change in the dynamic state of technical facilities based on the introduction of additional masses and leverage. Systems. Methods. *Technology*, *3*(31), 7–15.

Belokobylsky, S. V., Eliseev, S. V., & Kashuba, V. B. (2015). Impedance approaches as a form of evaluating dynamic properties of mechanical oscillatory systems in structural mathematical modeling. Systems. Methods. *Technologies*, *4*(28), 7–15.

Belokobylsky, S. V., Eliseev, S. V., & Kashuba, V. B. (2016). Elastic elements: Dynamic properties, generalized approach possibilities, and quasi-springs. Systems. Methods. *Technologies*, *2*(30), 7–17.

Belokobylsky, S. V., Eliseev, S. V., Kashuba, V. B., Nguyen, D. H., & Gypsy, V. V. (2015). On the connections between the coordinates of movement in mechanical oscillatory systems with lever devices. Systems. Methods. *Technology*, *2*(26), 7–13.

Belokobylsky, S. V., Eliseev, S. V., & Sitov, I. S. (2012). Interaction of links through a mass-inertial element in the theory of mechanical chains. Systems. Methods. *Technology*, *2*(14), 7–15.

Belokobylsky, S. V., Eliseev, S. V., & Sitov, I. S. (2013). *Dynamics of mechanical systems. Lever and inertia-elastic links*. Polytechnic.

Belokobylsky, S. V., Eliseev, S. V., & Sitov, I. S. (2013). *Dynamics of mechanical systems. Lever and inertial-elastic connections*. Polytechnika.

Belokobylsky, S. V., Eliseev, S. V., & Sitov, I. S. (2013). *Dynamics of mechanical systems: Lever and inertial-elastic connections*. Polytechnica.

Bezhaev, A. Yu., & Vasilenko, V. A. (2001). *Variational theory of splines*. Kluwer Academic / Plenum Publishers. DOI: 10.1007/978-1-4757-3428-7

Bolotin, V. V. (Ed.). (1978). *Vibrations in technology: A reference book of 6 volumes* (Vol. 1). Mechanical Engineering.

Bolshakov, R. S., & Eliseev, S. V. (2013). Methodological foundations of problems of determining bond reactions in mechanical oscillatory systems. IrGUPS. (Deposited manuscript in VINITI 29.11.2013 No. 366)

Bolshakov, R. S., & Eliseev, S. V. (2013). Methodological foundations of tasks for determining the reactions of links in mechanical oscillatory systems. Irkutsk: IrGUPS. Deposition manuscript in VINITI 29.11.2013 N°366.

Buryan, Y. A., Sorokin, B. N., & Galuza, Y. F. (2011). Active hydromechanical system for damping car vibrations. *Omsk Scientific Bulletin*, *103*, 122–126.

Clarence, W., & de Silva. (2000). Vibration. Fundamentals and Practice. Boca Raton, London, New York, Washington, D.C.: CRC Press. 957 p.

de Silva, C. W. (2000). *Vibration: Fundamentals and practice*. CRC Press.

Den-Hartog, D. P. (1960). *Mechanical vibrations*. Fizmatgiz.

Dimentberg, F. M., & Kolesnikov, K. S. (Eds.). (1980). *Vibrations in technology: A reference book in 6 volumes* (Vol. 3). Mechanical Engineering.

Druzhinsky, I. A. (1977). *Mechanical chains*. Mashinostroenie.

Druzhinsky, I. A. (1977). Mechanical chains. *Mechanical Engineering (New York, N.Y.)*.

Druzhinsky, I. A. (1977). *Mechanical circuits*. Mashinostroenie.

Dubrovsky, V. A., & Podvolotskaya, N. I. (2008). Problems of dynamics and strength of actuators and machines. *Problems of Mechanical Engineering and Automation*, *3*, 35–42.

Eliseev, A. V., Artyunin, A. I., Nguyen, D. H., & Vuong, K. C. (2016). Control of the vibration field of a technological machine based on changes in equivalent inertial properties. In Proceedings of the Conference "Vibration Technologies, Mechatronics, and Controlled Machines" (pp. 343). Kursk: Southwest State University.

Eliseev, A. V., Kopylov, Y. R., & Eliseev, S. V. (2017). Possibilities of dynamic correction of vibration fields of working elements of technological machines. In Proceedings of the International Conference on Recent Advances in Engineering, Technology and Applied Sciences (pp. 65–71). Detroit: Scientific Public Organization "Professional Science." 336 p.

Eliseev, A. V., Kopylov, Y. R., & Eliseev, S. V. (2017). Possible methods and means of assessing and controlling the vibration state of technological machines. In Proceedings of the Conference "Vibro-Wave Processes in High-Tech Parts Processing Technology" (pp. 20–22). Rostov-on-Don: DSTU. 240 p.

Eliseev, A. V., Nikolaev, A. V., & Mironov, S. V. (2018). Diads in mechanical oscillatory systems: Model of dynamic interaction formation. In Problems of Mechanics of Modern Machines: Proceedings of the VII International Scientific Conference (Vol. 1, pp. 33-40). Ulan-Ude: VSGUTU.

Eliseev, S. V. (2009). New approaches in the theory of oscillations. Tasks of controlling the dynamic state of oscillatory systems based on the introduction of additional connections. In Wiener readings: Materials of the IV All-Russian Scientific-Practical Conference (pp. 46-60). IrSTU.

Eliseev, S. V. (2018). Applied system analysis and structural mathematical modeling (dynamics of transport and technological machines: connectivity of movements, vibration interactions, leverage) (A. I. Artyunin, Ed.). IrGUPS.

Eliseev, S. V. (2018). Applied system analysis and structural mathematical modeling (dynamics of transport and technological machines: movement connectivity, vibration interactions, lever connections). Irkutsk: IrGUPS.

Eliseev, S. V., & Kargapoltsev, S. K. (2014). Representation of mechanical element connection features in the theory of mechanical chains. In L. V. Massel (Ed.), Information and mathematical technologies in science and management: Proceedings of the XIX Baikal All-Russian Conference (pp. 100–108).

Eliseev, S. V., Artyunin, A. I., Ermoshenko, Yu. V., et al. (2013). Methodological approaches in system analysis and mathematical modeling of mechanical oscillatory systems. Irkutsk: IrGUPS. (Deposited in VINITI 07.02.2013 No. 37).

Eliseev, S. V., Artyunin, A. I., Logunov, A. S., et al. (2013). Dynamic interactions of machine elements: design schemes and mathematical models of vibration states. Irkutsk: Deposited in VINITI 08.11.13 No. 313.

Eliseev, S. V., Eliseev, A. V., & Mironov, A. S. (2018). Approaches to evaluating the features of vibration fields during the interaction of a working medium with the working body of technological machines. In Proceedings of the IV International Scientific and Practical Conference "Open Innovation" (pp. 318). Penza: MCSNS "Science and Education."

Eliseev, S. V., Eliseev, A. V., & Mironov, A. S. (2018). Possibility of changing the vibratory field of working bodies of technological machines. Journal of Advanced Research in Technical Science, Issue 9–1, 50–56.

Eliseev, S. V., Kuznetsov, N. K., Bolshakov, R. S., et al. (2015). Justification and technologies of equivalent transformations of structural mathematical models of technical objects. Irkutsk: Dep. VINITI 20.10.2015 No. 179.

Eliseev, S. V., Loncikh, P. A., Kashuba, V. B., Parshuta, E. A., Kaimov, E. V., & Nguyen, D. H. (2015). New approaches in the dynamics of mechanical elastic-inertial systems with lever connections. In Science and education in the life of modern society: Collection of scientific papers from the International Scientific and Practical Conference (Vol. 12, pp. 75–85).

Eliseev, S. V., Moskovskikh, A. O., Bolshakov, R. S., & Savchenko, A. A. (2012). Possibilities of integration of methods of circuit theory and theory of automatic control in machine dynamics problems. *Science and Education: Electronic Scientific and Technical Publication*, 6. http://technomag.edu.ru/doc/378699.html

Eliseev, S. V., Trofimov, A. N., Bolshakov, R. S., & Savchenko, A. A. (2012). The concept of feedback in the dynamics of mechanical systems and dynamic damping of vibrations. *Science and Education: Electronic Scientific and Technical Publication*, (5). http://technomag.edu.ru/doc/378353.html

Eliseev, S. V., Zasyadko, A. A., Karpukhin, E. L., Kukharenko, V. P., & Metzeger, P. K. (1988). *Software-algorithmic complex of optimization design of complex vibration protection systems. In Sixth All-Union Conference on Control in Mechanical Systems* (p. 61). Institute of Applied Mechanics of the Academy of Sciences of the Ukrainian SSR.

Eliseev, S.V., Trofimov, A.N., & Kaimov, E.V. (2014). Forms of partial connectivity in mechanical system oscillations. International Journal of Applied and Fundamental Research, 5-1, 15-25.

Eliseev, A. V. (2017). Diads in mechanical systems: Features of dynamic properties. Part I. *Vestnik Irkutskogo Gosudarstvennogo Tehniceskogo Universiteta, 21*(7), 26–38. DOI: 10.21285/1814-3520-2017-7-26-38

Eliseev, A. V. (2017). Diads in mechanical systems: Features of dynamic properties. Part II. *Vestnik Irkutskogo Gosudarstvennogo Tehniceskogo Universiteta, 21*(8), 22–37. DOI: 10.21285/1814-3520-2017-8-22-37

Eliseev, A. V. (2017). Dynamic properties of a dyad with heterogeneous partial systems. *Vestnik Irkutskogo Gosudarstvennogo Tehniceskogo Universiteta, 21*(5), 32–53. DOI: 10.21285/1814-3520-2017-5-32-53

Eliseev, A. V., & Eliseev, S. V. (2017). Diad as a basis for forming mechanical oscillatory systems. Systems. Methods. *Technologies*, *4*(36), 25–38.

Eliseev, A. V., & Eliseev, S. V. (2017). The dyad as the basis for the formation of mechanical oscillatory systems. Systems. Methods. *Technologies*, *36*(4), 25–38.

Eliseev, A. V., & Kuznetsov, N. K. (2023). Development of lever representations in the evaluation of dynamic states of mechanical oscillatory systems under connected vibrational loads. *Bulletin of Cybernetics*, *22*(2), 75–86.

Eliseev, A. V., Kuznetsov, N. K., & Mironov, A. S. (2023). Dynamic invariant maps in the evaluation of motion modes of mechanical oscillatory systems. *Proceedings of MAI*, (128).

Eliseev, A. V., & Mironov, A. S. (2022). Evaluation of the combined dynamic characteristics of mechanical oscillatory systems based on dynamic invariant maps. *Journal of Advanced Research in Technical Science*, *32*, 62–69.

Eliseev, A. V., Selivinsky, V. V., & Eliseev, S. V. (2015). *Dynamics of vibrational interactions of elements of technological systems with consideration of non-retaining connections: Monograph*. Nauka.

Eliseev, A. V., Selvinsky, V. V., & Eliseev, S. V. (2015). *Dynamic interactions of elements in technological systems considering non-retentive connections*. Nauka.

Eliseev, A. V., Selvinsky, V. V., & Eliseev, S. V. (2015). *Dynamics of vibration interactions of elements of technological systems, taking into account non-holding bonds* [Monograph]. Nauka.

Eliseev, A. V., Selvinsky, V. V., & Eliseev, S. V. (2015). *Dynamics of vibration interactions of elements of technological systems, taking into account non-holding ties*. Nauka.

Eliseev, A. V., Sitov, I. S., & Kuznetsov, N. K. (2024). Formation of dynamic effects in mechanical oscillatory systems within the framework of the generalized lever concept, considering additional kinematic constraints. Systems. Methods. *Technologies*, *1*(61), 10–20.

Eliseev, A. V., & Vuong, K. C. (2016). Some possibilities for controlling one-dimensional vibration fields of technological machines. Modern Technologies. Systems Analysis. *Modeling*, *49*(1), 33–41.

Eliseev, S. V. (1978). *Structural theory of vibration protection systems*. Nauka.

Eliseev, S. V. (2018). *Applied systemic analysis and structural mathematical modeling (dynamics of transport and technological machines: movement connectivity, vibrational interactions, lever connections) (Monograph)*. Irkutsk State University of Railway Transport.

Eliseev, S. V., Artunin, A. I., & Bolshakov, R. S. (2015). Mechanical chains in structural diagrams of vibration protection systems: Methodology for determining dynamic reactions. *Bulletin of the All-Russian Scientific Research and Design Institute of Electric Locomotive Engineering*, *1*(69), 93–111.

Eliseev, S. V., & Artyunin, A. I. (2016). *Applied theory of oscillations in problems of dynamics of linear mechanical systems*. Nauka.

Eliseev, S. V., & Artyunin, A. I. (2016). *Applied theory of oscillations in the dynamics of linear mechanical systems*. Nauka.

Eliseev, S. V., & Artyunin, A. I. (2016). *Applied theory of vibrations in the dynamics of linear mechanical systems*. Nauka.

Eliseev, S. V., Belokobylsky, S. V., & Upyr, R. Y. (2009). Generalized spring in the tasks of machinery and equipment. Zbirnik Naukovy Pratsy (Galuzev Machine-Working, Budivnitsvo). *Poltava National Technical University Imeni Yuri Kondratyuk*, *1*(3(25)), 79–89.

Eliseev, S. V., Bolshakov, R. S., Nguyen, D. H., & Vyong, K. Ch. (2016). Determination of natural vibration frequencies of mechanical oscillatory systems: Features of using the frequency energy function. *Bulletin of Irkutsk State Technical University*, *6*(113, PART I), 26–33.

Eliseev, S. V., Kashuba, V. B., & Sitov, I. S. (2013). Reduced stiffness of the feedback circuit. Determination of dynamic reactions in a mechanical oscillatory system. Systems. Methods. *Technology*, *2*(18), 15–22.

Eliseev, S. V., & Khomenko, A. P. (2014). *Dynamic damping of oscillations feedback concept and structural methods of mathematical modeling*.

Eliseev, S. V., & Khomenko, A. P. (2014). *Dynamic damping of vibrations: Feedback concept and structural methods of mathematical modeling*. Nauka.

Eliseev, S. V., Khomenko, A. P., & Kaimov, E. V. (2016). Forms of manifestation of leverage in dynamic interactions of elements of mechanical oscillatory systems. Modern technologies. Systems analysis. *Modeling*, *3*(51), 8–18.

Eliseev, S. V., Khomenko, A. P., & Upyr, R. Y. (2009). Mechatronics of vibration protection systems with lever links. Modern Technologies. Systems Analysis. *Modeling*, (3), 104–119.

Eliseev, S. V., Khomenko, A. P., & Upyr, R. Yu. (2009). Mechatronics of vibration protection systems with lever connections. Contemporary Technologies. System Analysis. *Modeling*, *3*, 104–119.

Eliseev, S. V., Kovyrshin, S. V., & Bolshakov, R. S. (2012). Features of constructing compacts of elastic elements in mechanical oscillatory systems. Interactions with system elements and connection forms. Modern Technologies. *Systems Analysis. Modeling*, (4(36)), 61–70.

Eliseev, S. V., Kuznetsov, N. K., & Kaimov, E. V. (2015). On the theory of lever connections in the dynamics of mechanical oscillatory systems. *Bulletin of the Irkutsk State Technical University*, *12*(107), 30–40.

Eliseev, S. V., & Lonchikh, P. A. (2012). Evaluation of interaction forms between partial systems in mechanical chains. Possible simplifications. *Vestnik Irkutsk State Technical University*, *6*(65), 17–21.

Eliseev, S. V., Lukyanov, A. V., Reznik, Yu. N., & Khomenko, A. P. (2006). *Dynamics of mechanical systems with additional ties*. ISU.

Eliseev, S. V., Orlenko, A. I., & Eliseev, A. V. (2017). Structural formations in mechanical oscillatory systems: Dyads, their properties, and the possibilities for changing dynamic states. *Transport of the Urals*, *3*(54), 56–63. DOI: 10.20291/1815-9400-2017-3-56-63

Eliseev, S. V., Orlenko, A. I., & Nguyen, D. H. (2017). Devices for motion transformation in the structure of a diad in a mechanical oscillatory system. *Vestnik of Don State Technical University*, *17*(3), 46–59. DOI: 10.23947/1992-5980-2017-17-3-46-59

Eliseev, S. V., Parshuta, E. A., & Bolshakov, R. S. (2013). Transfer functions of mechanical oscillatory systems. Ability to assess reduced stiffness. *International Journal of Applied and Basic Research*, (1C), 11–18.

Eliseev, S. V., Parshuta, E. A., Kaimov, E. V., & Kinash, N. Z. (2014). Mechanisms in the structure of vibration protection systems: Mathematical models, dynamic property assessment (Part I). *Bulletin of VSGUTU*, *6*(51), 37–44.

Eliseev, S. V., Parshuta, E. A., Kaimov, E. V., & Kinash, N. Z. (2015). Mechanisms in the structure of vibration protection systems: Mathematical models, dynamic property assessment (Part II). *Bulletin of VSGUTU*, *1*, 52–60.

Eliseev, S. V., Reznik, Y. I., Khomenko, A. P., & Zasyadko, A. A. (2008). *Dynamic synthesis in generalized problems of vibration protection and vibration isolation of technical objects*. ISU.

Eliseev, S. V., Reznik, Y. N., & Khomenko, A. P. (2011). *Mechatronic approaches in the dynamics of mechanical oscillatory systems* [Monograph]. Nauka.

Eliseev, S. V., Reznik, Y. N., Khomenko, A. P., & Zasyadko, A. A. (2008). *Dynamic synthesis in generalized vibration protection and vibration isolation problems for technical objects*. ISU.

Eliseev, S. V., Reznik, Y. N., Khomenko, A. P., & Zasyadko, A. A. (2008). *Dynamic synthesis in generalized vibration protection and vibration isolation problems of technical objects*. Irkutsk State University Publishing.

Eliseev, S. V., Volkov, L. N., & Kukharenko, V. P. (1990). Dynamics of mechanical systems with additional connections. *Science*.

Elyseev, S. V., & Elyseev, A. V. (2013). Methodological foundations for determining contact static and dynamic reactions considering non-holding connections. Irkutsk: Irkutsk State University of Railway Transport. 138 pages. Illustrated. Dep. in VINITI 16.12.13 No. 369.

Elyseev, S. V., & Elyseev, A. V. (2014). Conditions for contact vibrations of composite solid bodies under dynamic loads considering non-holding connections. Irkutsk: Irkutsk State University of Railway Transport. 42 pages. Illustrated. Dep. in VINITI 27.01.2014 No. 31.

Elyseev, A. V., & Elyseev, S. V. (2014). *Determining contact reactions in composite solid bodies under dynamic loads considering non-holding connections. Modern Technologies. System Analysis. Modeling, 1, 45–54*. Irkutsk State University of Railway Transport.

Elyseev, S. V., & Elyseev, A. V. (2013). *Generalized approaches in the determination of contact reactions in solid bodies under static loads considering non-holding connections. Modern Technologies. System Analysis. Modeling, 4, 51–60.* Irkutsk State University of Railway Transport.

Elyseev, S. V., Reznik, Yu. N., & Khomenko, A. P. (2011). *Mechatronic approaches in the dynamics of mechanical systems*. Nauka.

Emelyanov, S. V., & Korovin, S. K. (1997). *New types of feedback: Control under uncertainty*. Nauka, Fizmatlit.

Frolov, K. V., & Furman, F. A. (1980). *Applied theory of vibration protection systems*. Mashinostroenie.

Galiev, I. I., Nekhaev, V. A., & Nikolaev, V. A. (2010). *Methods and means of vibration protection of railway crews*. SEI "Educational and Methodological Center for Education in Railway Transport."

Galperin, I. I. (1964). *Automation as one-sided mechanics*. Energiya.

Ganiev, R. F., & Kononenko, V. O. (1976). *Vibrations of solid bodies*. Nauka.

Gardner, F., & Bairns, J. L. (1949). *Transients in linear systems with concentrated constants*. GITTL.

Gardner, F., & Barnes, J. L. (1949). *Transient processes in linear systems with concentrated constants*. GITTL.

Genkin, M. D. (1981). Vibrations in technology. (In 6 volumes). Moscow: Mechanical Engineering. Vol. 5: Measurements and tests.

Genkin, M. D., & Ryaboy, V. M. (1988). *Elastic-inertial vibration isolation systems*. Ultimate capabilities, optimal structures. Nauka.

Genkin, M. D., & Ryaboy, V. M. (1988). *Elastic-inertial vibration isolation systems: Ultimate capabilities, optimal structures*. Nauka.

Guskov, A. M., Panovko, G. Ya., & Shokhin, A. E. (2012). Calculation of a rod spatial system of vibration isolation for solid bodies under transport vibration. *Problems of Machine Engineering and Reliability of Machines*, *2*, 17–24.

Halperin, I. N. (1968). *Automation as one-way mechanics*. Mechanical Engineering.

Harris, C. M., & Allan, G. (2002). *Shock and vibration handbook*. McGraw-Hill.

Ivanov, B. G. (2007). Development of methods for calculating the dynamics and strength of transport equipment units with lever-hinge links [Doctoral dissertation, Samara State Transport University]. Samara.

Jie, F. S., Morse, I. E., & Hinkle, R. T. (1966). *Mechanical oscillations* (Obraztsov, I. F., Ed.). Mashinostroenie.

Karamyshkin, V. V. (1988). Dynamic vibration dampers. *Mechanical Engineering (New York, N.Y.)*.

Karnovsky, I. A., & Lebed, E. (2016). *Theory of vibration protection*. Springer International Publishing. DOI: 10.1007/978-3-319-28020-2

Kashuba, V. B., Eliseev, S. V., & Bolshakov, R. S. (2017). *Dynamic reactions in the connections of elements of mechanical oscillatory systems*. Nauka.

Khokhlov, A. A. (2002). *Dynamics of complex mechanical systems*. MIIT.

Khomenko, A. P., Eliseev, S. V., & Yermoshenko, Y. V. (2012). System analysis and mathematical modeling in mechatronics of vibration protection systems. Irkutsk: IrGUPS. 288 p.

Khomenko, A. P., Eliseev, S. V., Artyunin, A. I., Bolshakov, R. S., Kaimov, E. V., Kinash, N. Z., & Nguyen, D. H. (2015). Structural mathematical models in problems of dynamics of mechanical oscillatory systems. Irkutsk: Dep. VINITI RAN 30.03.2015, No. 62.

Khomenko, A. P., Eliseev, S. V., Artyunin, A. I., Kaimov, E. V., & Eliseev, A. V. (2015). Device for damping oscillations (Patent No. 150331 RUS, IPC F16F 15/04). Russian Federation. Application No. 2014138832/11, filed September 25, 2014, and published February 10, 2015. Bulletin No. 4.

Khomenko, A. P. (2000). *Dynamics and control in the tasks of vibration protection and vibration isolation of mobile objects*. ISU.

Khomenko, A. P., & Eliseev, S. V. (2015). On some properties of dynamic damping of vibrations in mechanical systems. Modern Technologies. Systems Analysis. *Modeling*, *2*(46), 8–19.

Khomenko, A. P., & Eliseev, S. V. (2016). Characteristic frequency equation: Structure, dynamic stiffness, and features of system element interaction. Modern Technologies. Systems Analysis. *Modeling*, *2*(50), 12–24.

Khomenko, A. P., & Eliseev, S. V. (2016). Development of the energy method for determining the frequencies of free oscillations of mechanical systems. Modern Technologies. System Analysis. *Modeling*, *1*(49), 8–19.

Khomenko, A. P., & Eliseev, S. V. (2016). Lever connections in mechanical chains: Dynamic effects. Modern Technologies. Systems Analysis. *Modeling*, *4*(52), 8–16.

Khomenko, A. P., & Eliseev, S. V. (2017). Structural formations in mechanical oscillatory systems: Diad as a form of interaction of elements. Modern Technologies. Systems Analysis. *Modeling*, *2*(54), 8–14.

Khomenko, A. P., Eliseev, S. V., & Bolshakov, R. S. (2014). Features of the interaction of partial systems in a vibration protection circuit with two degrees of freedom: Leverage in dynamic damping of vibrations. Modern Technologies. Systems Analysis. *Modeling*, *3*(43), 8–19.

Kim, P. D. (2003). Theory of automatic control. Vol. 1. Linear systems. Fizmatgiz.

Kiryukhin, A. V., Tikhonov, V. A., Chistyakov, A. G., & Yablonsky, V. V. (2012). Active vibration protection - purpose, principles, condition. Active vibration protection and noise insulation of pipelines and experimental research. *Problems of Mechanical Engineering and Automation*, (4), 102–110.

Kolovsky, M. Z. (1976). *Automatic control of vibration protection systems*. Nauka.

Kopylov, Y. R. (2011). Dynamics of vibration shock hardening processes [Monograph]. CPI "Scientific Book."

Kopylov, Y. R. (2011). *Dynamics of vibro-impact hardening processes*. IPC Scientific Book.

Kopylov, Y. R., & Eliseev, A. V. (2014). Features of the technical implementation of vibration strengthening technology. *Modern High Technologies*, *12*(2), 152–155.

Kron, G. (1972). *Study of complex systems in parts - diakoptics (Trans. from English)*. Nauka.

Lavrus, V. V. (2006). Improvement of pneumatic lever-hinge systems of railway transport [Doctoral dissertation, Orel State Technical University]. Orel.

Lenck, A. (1978). Electromechanical systems. Systems with concentrated parameters. Moscow: Mir. 288 p.

Lurie, A. I. (1959). *Operational calculus and its applications to problems of mechanics*. Nauka.

Lurie, A. I. (1968). *Analytical mechanics*. Nauka.

Naidenko, O. K., & Petrov, P. P. (1962). *Damping of ship engines and mechanisms*. Sudpromgiz.

Orlenko, A. I., Eliseev, A. V., & Eliseev, S. V. (2018). Additional connections: Possibilities for changing dynamic states of mechanical oscillatory systems. In Proceedings of the 1st International Scientific and Practical Conference "Machines, Units, and Processes: Design, Creation, and Modernization" (pp. 31–37). Saint Petersburg: SPbSF NII MS.

Panovko, G. Y. (2006). *Dynamics of vibrational technological processes (Text).* NITS Regular and Chaotic Dynamics, Institute of Computer Technologies.

Panovko, Y. G. (1991). *Introduction to the theory of mechanical vibrations.* Nauka.

Rocard, Y. (1949). *General dynamics of vibrations.* Masson.

Rozhenko, A. I. (2005). Theory and algorithms of variational spline approximation (A. I. Rozhenko). Novosibirsk: IVMiMG SO RAN.

Ryabov, I. M., Chernyshov, K. V., Eliseev, S. V., & Upyr, R. Y. (2010). Experimental studies of inertia-friction dampers in vehicle suspension systems (part 1). *Bulletin of Irkutsk State Technical University, 2*(42), 146–154.

Samarsky, A. A., & Mikhailov, A. P. (2005). *Mathematical modeling: Ideas.* Methods. Examples. Fizmatlit.

Shatalov, A. S. (1962). *Structural methods in control theory and electroautomatics.* Gosenergoizdat.

Sommerfeld, A. (2001). Mechanics. Nizhniy Novgorod: NITS RKhD.

Tarasik, V. P. (2004). Mathematical modeling of technical systems (2nd ed., revised and expanded). Minsk: Design PRO. 640 p.

Tarasik, V. P. (2004). Mathematical modeling of technical systems: A textbook for technical universities (2nd ed., revised and expanded). Minsk: Design PRO.

Tse, F. S., Morse, I. E., & Hinkle, R. T. (1966). Mechanical vibrations (Edited by corresponding member of the USSR Academy of Sciences I.F. Obraztsov). Moscow: Mashinostroenie. 508 p.

Vargunin, V. N., Gusarov, V. N., Ivanov, B. G., Levchenko, A. S., & Mulyukin, O. P. (Eds.). (2006). *Design and calculation of lever-articulated means and units.* SamGAPS.

Vasilenko, V. A. (1983). *Spline functions: Theory, algorithms, programs (V. A. Vasilenko).* Nauka.

Vasilenko, V. A., & Eliseev, A. V. (1998). Abstract splines with tension as functions of energy operator parameters. *Siberian Journal of Computational Mathematics*, *1*(4), 301–311.

Veitz, V. L., Kachura, A. E., & Martynenko, A. M. (1971). Dynamic calculations of machine drives. *Mechanical Engineering (New York, N.Y.)*.

Vershinsky, S. V., Danilov, V. I., & Khusidov, V. D. (1991). Wagon dynamics. *Transport*.

About the Authors

Andrey V. Eliseev, Cand. Sc. (Eng.), is Associate Professor at two institutions: Dept. of Mathematics, Irkutsk State Transport University, and Dept. of Engineering & Standardization, Irkutsk National Research Technical University. He holds an MSc in Mechanics & Mathematics from Novosibirsk State University and additional qualifications in economic theory. He defended his PhD (2015) at Southwestern State University (Kursk) on vibrational dynamics in technological systems with unilateral constraints. From 2011-2021, he worked under Prof. Sergey V. Eliseev, a leading expert in vibration protection. Eliseev's primary research focus is developing structural mathematical models for intensely loaded technical systems, uniquely applying automatic control theory to mechanical oscillatory dynamics. He is a prolific co-author of over 300 scientific papers, 30+ inventions, and co-authored monographs on vibration isolation and dynamic condition assessment for machinery.

Nikolay K. Kuznetsov, Dr. Sc. (Eng.), Professor, heads the Department of Engineering and Standardization at Irkutsk National Research Technical University (IRNTU). An alumnus of Irkutsk Polytechnic Institute (1973, Aviation Engineering), his career began at the Irkutsk branch of NIAT (1973-1976). He has been associated with Irkutsk Polytechnic University (now IRNTU) since 1976, progressing from graduate student to Associate Professor in the Department of Theoretical Mechanics (1976-1984). Following significant leadership of the mechatronics laboratory at the Irkutsk Scientific Center, SB RAS (1984-1996), he returned to IRNTU, rising to Professor and subsequently Head of his current department (since 2008). A prominent follower of S.V. Eliseev's school in active vibration protection, Kuznetsov has made substantial contributions to developing methods for the dynamic synthesis of controlled machines, specifically considering elasticity of the links. His prolific output includes co-authoring over 230 scientific works: 7 monographs, 8 textbooks, 15 patents, and 30 articles indexed in WoS/Scopus. He has supervised 5 successful PhD candidates and chairs the IRNTU Dissertation Council.

Index

A

B

C

D

F

G

I

K

L

M

N

O

S

T

V

www.ingramcontent.com/pod-product-compliance
Lightning Source LLC
LaVergne TN
LVHW080906300926
849162LV00003B/580
* 9 7 9 8 3 3 7 3 0 4 4 7 2 *